Lithium Across the Universe

Online at: https://doi.org/10.1088/2514-3433/acd8bb

AMERICAN ASTRONOMICAL SOCIETY | IOP | ebooks™

AAS Editor in Chief

Ethan Vishniac, Johns Hopkins University, Maryland, USA

About the program:

AAS-IOP Astronomy ebooks is the official book program of the American Astronomical Society (AAS) and aims to share in depth the most fascinating areas of astronomy, astrophysics, solar physics and planetary science. The program includes publications in the following topics:

- GALAXIES AND COSMOLOGY
- INTERSTELLAR MATTER AND THE LOCAL UNIVERSE
- STARS AND STELLAR PHYSICS
- EDUCATION, OUTREACH, AND HERITAGE
- HIGH-ENERGY PHENOMENA AND FUNDAMENTAL PHYSICS
- THE SUN AND THE HELIOSPHERE
- THE SOLAR SYSTEM, EXOPLANETS, AND ASTROBIOLOGY
- LABORATORY ASTROPHYSICS, INSTRUMENTATION, SOFTWARE, AND DATA

Books in the program range in level from short introductory texts on fast-moving areas, graduate and upper-level undergraduate textbooks, research monographs, and practical handbooks.

For a complete list of published and forthcoming titles, please visit iopscience.org/books/aas.

About the American Astronomical Society

The American Astronomical Society (aas.org), established 1899, is the major organization of professional astronomers in North America. The membership (~7,000) also includes physicists, mathematicians, geologists, engineers, and others whose research interests lie within the broad spectrum of subjects now comprising the contemporary astronomical sciences. The mission of the Society is to enhance and share humanity's scientific understanding of the universe.

Lithium Across the Universe

Eduardo Martín
Instituto de Astrofísica de Canarias (IAC) (Spain), Canary Islands, Spain

IOP Astronomy, Bristol, UK

ISBN 978-0-7503-3623-9 (ebook)
ISBN 978-0-7503-3621-5 (print)
ISBN 978-0-7503-3624-6 (myPrint)
ISBN 978-0-7503-3622-2 (mobi)

DOI 10.1088/2514-3433/acd8bb

Version: 20230901

AAS–IOP Astronomy
ISSN 2514-3433 (online)
ISSN 2515-141X (print)

British Library Cataloguing-in-Publication Data: A catalogue record for this book is available from the British Library.

Published by IOP Publishing, wholly owned by The Institute of Physics, London

IOP Publishing, No.2 The Distillery, Glassfields, Avon Street, Bristol, BS2 0GR, UK

US Office: IOP Publishing, Inc., 190 North Independence Mall West, Suite 601, Philadelphia, PA 19106, USA

The author wishes to dedicate this book to the memory of France Allard, George Herbig, Francesco Palla, François Spite, and John Stauffer, dear colleagues who were sources of inspiration, and with whom he enjoyed enriching conversations and correspondence.

They made fundamental contributions to advancing the frontier of knowledge in many aspects of astrophysics, including the physics of lithium in young stars, very low-mass stars and brown dwarfs, and the oldest stars in the Galaxy.

They will never be forgotten.

Contents

Preface

Preface (by John E. Beckman)

After the discovery of the cosmic microwave background radiation by Penzias and Wilson in 1965, the plausibility of a universe which initiated in an ultrahigh-density, ultrahigh-temperature primordial fireball (the "Big Bang") increased to the point where even those sceptical of the hypothesis felt obliged to take note. A prime example was the predictive paper by Fowler, Wagoner, and Hoyle, of the chemical elements which should have been produced in the fireball. They predicted that significant quantities of only D, ^{3}He, ^{4}He and ^{7}Li could have been synthesized, and stated that the abundances of these nuclides are in "reasonable agreement" with solar system abundances. In order to make these predictions they needed to make specific physical and cosmological assumptions. The key physical assumptions were the value of the half-life for decay of free neutrons, which would limit the timescale of their participation in reactions during the rapid cool-down of the fireball, and the number of independent neutrino families governing the overall reaction chains. The cosmological assumptions were observational: the ratio of the number of photons to the number of baryons in the universe, the Hubble parameter, which measures its present expansion rate, and the acceleration parameter, which measures how this rate changes on cosmological timescales.

Following these predictions observational astronomers initiated programs to determine the abundances of primordially produced D, ^{3}He, ^{4}He, and ^{7}Li with the aim of testing what was becoming the "standard model" of cosmology. The method was to observe these elements in a range of objects, stars and the interstellar medium, of varying metallicity, and extrapolate to zero metallicity, on the assumption that all the remaining elements, ("metals") have been synthesized in stellar and interstellar processes subsequent to the Big Bang. The results were encouraging, especially so for ^{4}He, which makes up around a quarter of the baryonic mass of the universe, far too much to have been produced in stars, and for D and ^{3}He, with much smaller, but still measurable, abundances. By 1980 the subject of primordial nucleosynthesis had achieved a fair observational basis. Indeed, the measured values of the primordial abundances had obliged laboratory physics to remeasure the value of the half-life of free neutrons, the first contribution of observational cosmology to basic physics.

In 1982 the observational extrapolation to low metallicity for ^{7}Li appeared to reach a conclusion for primordial ^{7}Li. François and Monique Spite measured this nuclide in stars of the Galactic halo with low metallicity and found an abundance that did not appear to decrease as the stars' metallicities fall below around one hundredth of the solar value. This ^{7}Li abundance could be attributed to primordial lithium, and it gave rise to what seemed to be a concordance between the predictions of the fireball models and the observed abundances of the four nuclides produced. This concordance included the inference, previously obtained from ^{4}He alone, that the mean baryon density of the universe is only some 5% of that required to close it. Even so, ^{7}Li fitted the concordance model the least well of the primordial nuclides.

When I arrived at the Instituto de Astrofísica de Canarias (IAC) in 1984, the observational facilites at the Instititute's observatory on La Palma presented an excellent opportunity to pin down the primordial lithium abundance. Encouraged by my young colleague Paolo Molaro of the Trieste Observatory, I suggested to Rafael Rebolo, then my graduate student at the IAC, that he should pick up the primordial lithium problem. At the same time, with Paolo, we observed beryllium (^{9}Be) in low-metallicity stars, using the outstanding near-UV transparency at the La Palma observatory to make the spectroscopic observations. The aim was to determine whether the ^{7}Li in the "Spite plateau" constituted "the primordial lithium, all of the primordial lithium, and nothing but the primordial lithium" to paraphrase the British legal system. If there was no ^{9}Be in the low-metallicity stars, this should mean that those galactic processes (above all spallation of C and N in the interstallar medium by high-energy cosmic rays) that produce ^{9}Be, but also ^{7}Li, had not contributed significant ^{7}Li to the stars in the plateau. This work, with Paolo and Rafael, not only confirmed the absence of ^{9}Be at the lowest metallicities, but produced the first plot of ^{9}Be against metallicity over a wide metallicity range. At the same time Rafael made multiple observations of the ^{7}Li abundance in stars over a range of metallicity, both in the general field, and in clusters. There was a period in the late 1980s when we thought that the primordial lithium problem was resolved, to the point where we could even set an upper limit of 3 to the number of distinct neutrino families, making what seemed to be another astrophysical contribution to basic physics.

This book starts where this simple picture began to break down. It covers over three decades of astrophysical lithium studies showing that this element is an intriguing component of many processes in, especially, cool stars, brown dwarfs, and exoplanets. A reason why lithium is fascinating is that it is fragile, and can be quite easily destroyed in the nuclear core zones of stars. So its measured abundance in any given object is the result of an interplay between creative nucleosynthesis, and destructive processes within stars. Lithium can be created not only in the interstellar medium, but also in novae, in supernovae, and in the accretion disks of the neutron stars or black holes in X-ray binaries. Its rate of destruction in a star, or a substellar object depends on the rate of convective mass interchange between the nuclear zone and the surface, which is in turn a function of stellar mass and temperature. It is, in general, higher for cooler, less massive objects which have deeper convection zones. The observed abundance of lithium in an atmosphere may depend not only on these processes but also on effects due to planets which they harbor. Swallowing a planet with a high lithium content will, naturally, tend to augment the lithium abundance in a stellar atmosphere, but tidal interactions with planets may stir up convection in their host stars, increasing the lithium destruction rate.

Eduardo Martín was Rafael Rebolo's first graduate student; together with María Rosa Zapatero Osorio, another of Rafael's students, the three discovered the first brown dwarf, Teide 1, at the IAC's Teide observatory in Tenerife, published in 1995. Their experience with lithium had previously led Rafael, Eduardo, and Antonio Magazzù toward the "lithium test" for brown dwarfs. Even though a brown dwarf is cool, and has a very deep convection zone, its core does not reach the temperature needed to sustain nuclear fusion, so any lithium in its atmosphere when it formed will not be convectively destroyed. Even though its surface temperature may not be lower than those of the coolest stars, the presence of lithium in its spectrum is the “giveaway” whose presence affirms that the object is in fact a brown dwarf. During the subsequent three decades, Eduardo has committed his research time to studying how lithium may be produced or destroyed in many types of objects and conditions, often collaborating with Rafael, who has widened his scope to take in the field of exoplanet hunting and diagnostics. They have seen the fields of low-mass stars and substellar objects grow rapidly during this time and have successfully maintained leading roles in their work, (as indeed has Paolo). Eduardo is in an unequalled position to produce the wide but detailed survey on lithium which is the subject of this book.

One of the leading developments in Big Bang cosmology during the past 30 years has been the exploration in ever increasing detail of the cosmic microwave background, above all through satellite observations, but also from the ground. The results on its angular distribution have led to a new, and effective approach to the measurement of key cosmological parameters, such as the mean baryon density, the mean overall density as a fraction of the value needed to produce a closed universe, and the Hubble parameter. It is now possible to predict the primordial light element abundances using these parameters. The results show very good agreement with the observed values for the abundances of ^{4}He, D, and ^{3}He, but predict a value for ^{7}Li which is nearly half an order of magnitude bigger than that measured in the low-metallicity “plateau”, in which a downward slope to the lowest metallicities has since been observed. The efforts of the observationalists to wrestle with this difficulty are well described in this book, as part of more general descriptions of how lithium presents a unique interpretational challenge to astrophysics. These are detective stories, but with an advantage: the culprit has in many cases not yet been identified. I am sure that any young astrophysicist who reads this will emerge with an appetite whetted for research.

Acknowledgements

The author is grateful for the help and support of his wife, Ana García Bouza.

IAC colleagues John Beckman, Nicolas Lodieu, Rafael Rebolo, Tarik Shahbaz, and Jun-Yan Zhang provided encouraging and insightful comments.

John also wrote the preface.

The IAC's scientific editor Terry Mahoney made numerous corrections. IAC graphics designer Gaby Pérez helped to make the original figures included here.

CAB colleagues David Barrado and Maria Rosa Zapatero Osorio provided numerous comments and corrections.

Long-time collaborators and friends Antonio Magazzù and Yakiv Pavlenko provided corrections and comments.

Paolo Molaro (Osservatorio de Trieste, Italy) contributed very useful remarks and corrections.

Author biography

Eduardo Martín

IAC Professor of Research Eduardo Martín is an expert on optical and infrared observational spectroscopy of cool stars and ultracool dwarfs.

He has led several key studies of the properties of substellar-mass objects, including pioneering the well-known lithium test for brown dwarfs. Over three decades, his research activity has included topics in astrobiology, astronomical instrumentation and surveys, brown dwarfs, exoplanets, planet and star formation, and ultracool atmospheres.

He started his research career at the Institut d'Astrophysique de Paris (IAP) in 1988 with a grant from the Institut des Sciences de l'Univers (INSU).

At IAP, he worked with and learned from Jean Audouze, Gibor Basri, Claude Bertout, Jerôme Bouvier, Michel Cassé, Roger Ferlet, Niko Prantzos, Eduardo Simoneu, François and Monique Spite, and Elizabeth Vangioni-Flam.

From 1990 to 1994, he carried out his doctoral studies and research work at the IAC under the guidance of Professor Rafael Rebolo. During those years, he worked with and learned from John Beckman, Ramón García López, Antonio Magazzù, and Yakiv Pavlenko.

He received his PhD degree in Astrophysics in 1994 from the University of La Laguna (Tenerife, Spain). He has been a support astronomer in the observatories of the Canary Islands, and has enjoyed postdoctoral fellowships at the Anton Pannekoek institute of the University of Amsterdam where he was sponsored by Henk Spruit and Jan van Paradijs, the Astronomy Department at the University of California in Berkeley under the supervision of Professor Gibor Basri, and the division of Planetary Sciences in Caltech with Professor Shri Kulkarni.

In 2000, he became a faculty member at the Institute of Astronomy of the University of Hawaii, where he stayed until moving back to his native island of Tenerife to take a staff research position at the IAC in 2004. In 2009 he obtained a full professor position of the Spanish Research Council (CSIC) at the INTA-CSIC Centro de Astrobiología.

Currently, he is a full professor at the IAC, and the head of the Independent Legacy Group on ultracool objects selected in 2012 by the European Space Agency (ESA) to be part of the Euclid Science Team (EST). In 2022 he was awarded a European Research Council (ERC) Advanced Grant to support his investigations into the nature of substellar objects.

Contact information: ege@iac.esege1008@gmail.com
Researcher unique identifier(s): https://orcid.org/0000-0002-1208-4833
URLs: http://research.iac.es/galeria/ege; http://substellar.net

Acronyms

AAT	3.9 m Anglo-Australian telescope
CAB	Centro de Astrobiología
CFHT	3.6 m Canada France Hawai Telescope
CMD	Color–magnitude diagram
CoG	Curve of growth
CTIO	Cerro Tololo Inter-American Observatory (Chile)
EMP	Extremely metal-poor
EW	Equivalent width
pEW	Pseudo-equivalent width
ESO	European Southern Observatory
GTC	10.4 m Gran Telescopio Canarias
HET	10 m Hobby–Eberly Telescope
HRD	Hertzsprung–Russell diagram
HST	2.4 m Hubble Space Telescope
IAC	Instituto de Astrofísica de Canarias
INT	2.5 m Isaac Newton Telescope
ISM	Interstellar medium
JWST	James Webb Space Telescope
KPNO	Kitt Peak National Observatory (Arizona-Sonoran desert, Tohono-O'odham nation, USA)
MKO	Maunakea Observatory (Big island of Hawai, USA)
LDB	Lithium depletion boundary
LBT	11.8 m Large Binocular Telescope
ORM	Observatorio del Roque de los Muchachos (La Palma, Canary islands, Spain)
PMS	Pre-main-sequence
QSO	Quasi-stellar object
RGB	Red giant branch
S/N	Signal-to-noise ratio
TNG	3.5 m Telescopio Nazionale Galileo
VLT	8.2 m Very Large Telescope
WHT	4.2 m William Herschel Telescope

Conventions

A(Li) lithium abundance

Log N(Li) = $\log_{10} N_{Li}$ = lithium abundance

Log N(H) = $\log_{10} N_H$ = hydrogen abundance

Lithium abundance scale = A(Li) = log N(Li) = $\log_{10} (N_{Li}/N_H) + 12$ where N_{Li} and N_H are the number densities of Li and H atoms, respectively, in a scale where $\log_{10} N(H) = 12$.

Iron abundance = [Fe/H] = $\log_{10} (N_{Fe}/N_H) - \log_{10} (N_{Fe}/N_H)_{Sun}$ where N_{Fe} and N_H are the number densities of iron and hydrogen atoms, respectively. This convention for the iron abundance is relative to the Sun so that a value of [Fe/H] = 0 means the solar abundance. It is frequently used as proxy for the overall stellar metallicity

Surface gravity = $\log g = 4 \log T_{eff} + \log (M/M_\odot) - 10.61 + 0.4 (BC_V + m_V - 5 \log (d) + 5 - 3.1 E(B–V) - M_{bol,\odot})$ where $M_{bol,\odot} = 4.75$ is the solar bolometric magnitude, the constant 10.61 is determined from the solar effective temperature $T_{eff,\odot} = 5777$ K and the solar surface gravity $\log g_\odot = 4.438$, BC_V is the bolometric correction of the star in the visual band, m_V is the visual apparent magnitude of the star, d is the distance, and E(B–V) is the excess in B–V color due to interstellar reddening.

Symbols

α	Alpha particle (^{4}He nucleus)
Å	Ångstrom = 0.1 nanometers (nm)
μm	Micron = 1000 nm
e^{-}	Electron
e^{+}	Positron
K	Kelvin (absolute temperature unit)
γ	Gamma ray
λ	Wavelength (μm, nm or Å)
ν	Neutrino
p	Proton
ρ	Density (kg m^{-3})
T_{eff}	Effective temperature (K)
$V \sin i$	Projected rotational velocity (km s^{-1})
σ	Standard deviation
H	Hydrogen
D	Deuterium
Li	Lithium
^{6}Li	Lithium 6, less abundant lithium isotope
^{7}Li	Lithium 7, most abundant lithium isotope
R_{I}.	Li isotope ratio = ^{6}Li/^{7}Li
Li I	Neutral lithium atom
Li II	Ionized lithium atom
^{11}B	Boron 11, most abundant boron isotope
^{7}Be	Beryllium 7, unstable beryllium isotope that decays into ^{7}Li
^{9}Be	Beryllium 9, most abundant beryllium isotope
^{3}He	Helium 3, less abundant helium isotope
^{4}He	Helium 4, more abundant helium isotope
η	Ratio of baryons to photons in the universe
R	Resolving power = λ/delta λ
M_{E}	Earth mass
R_{E}	Earth radius
M_{J}	Jupiter mass
R_{J}	Jupiter radius
$M_{\odot}$	Solar mass
$R_{\odot}$	Solar radius
Q	Quettagram = 1 quadrillion T (metric tons) = 10^{30} grams = 0.5 M_{J}

AAS | IOP Astronomy

Lithium Across the Universe

Eduardo Martín

Chapter 1

Background

This chapter provides the reader with basic information about the light chemical element called lithium (Li). It is recommended to read this chapter first for those who do not have previous knowledge of this field. Advanced students and researchers may skip it and proceed to the rest of the book.

In the following pages, a historical perspective is presented on developments in the understanding of lithium across the universe, and the applications that its study has in many branches of astrophysics.

With an atomic number $Z = 3$, lithium is the third element in the periodic table and the lightest of the six alkali metals.

Alkalis are very soluble in water and form strong bases that react with acids to form neutral salts. In contrast with the heavier alkali metals, potassium (K) and sodium (Na), lithium is a rare element that represents only 0.007% of the Earth's crust.

Even rarer alkali metals are cesium (Cs), francium (Fr), and rubidium (Rb). The most abundant alkalis are potassium (K) and sodium (Na). Relative comparisons between these elements provide a great deal of information about them.

Lithium on Earth is found as a solid metallic compound in certain minerals. In brown dwarfs, giant planets, interstellar clouds, and stellar atmospheres, however, it is detected in its gaseous form.

The many roles played by lithium throughout the study of the universe that are reviewed in this book include the following:

In cosmology, lithium is one of the few chemical elements produced by Big Bang nucleosynthesis. The other primordial elements are D and He, but Li is the only one that clearly reveals disagreement between the predictions from Big Bang nucleosynthesis and the observational abundances.

In the Milky Way, lithium poses a challenge to models of chemical evolution. There is consensus that the abundance of lithium has increased over the evolution of

doi:10.1088/2514-3433/acd8bbch1

the Galaxy. Some authors have proposed that the dominant source of lithium enrichment is nova explosions, but the jury is still out on this question.

In open clusters and stellar associations, lithium provides an alternative clock that complements other dating methods. The wide spread in lithium abundance among stars of similar age and mass is not yet well understood.

In stellar evolution, lithium is a tool for probing the mixing processes between the interior and the surface of the stars, including the Sun.

In the interstellar medium, lithium provides information about cosmic rays and spallation nuclear reactions.

In brown dwarfs, lithium probes the maximum central temperature that sub-stellar-mass bodies attain during their lifetime.

In stars hosting exoplanets, lithium could be a tracer of stellar–planet interactions and serves as a witness of catastrophic collisions that may result in planetary engulfment by the star.

In stellar remnants (black holes, neutron stars, and white dwarfs), lithium provides information about explosions, mergers and other violent events.

On Earth, lithium is recognized as white gold, a rare and precious chemical element in the making of batteries and as brain medication.

Most of the information that we have on the presence of lithium across the universe reaches to us in the form of light absorption from the resonance doublet of the neutral Li I atom at 670.8 nm. This line has been detected in many astrophysical environments, from giant stars located in the far reaches of the Milky Way, or even in other galaxies, to stars orbiting black holes, brown dwarfs in the solar vicinity, the atmospheres of exoplanets and solar system objects such as comets.

There is a direct connection between the strength of Li I lines and the number of lithium atoms in the line of sight. Chemical abundances can be obtained from spectroscopic observations of Li I lines using models that incorporate a variety of chemical and physical processes as well as the fundamental atomic data.

The most sophisticated models include three-dimensional (3D) non-local thermodynamic equilibrium (NLTE) effects. Even with formidable analysis tools and high-quality data, firm detection of the rarer isotope (^{6}Li) has proved elusive in stellar atmospheres. It is still a formidable challenge. So far the only conclusive evidence for it outside the solar system has been obtained in the interstellar medium (ISM).

While the information that we have on lithium comes mainly from its neutral atomic state, it is well known that this is usually not the dominant state in which Li atoms are found.

For example, in the atmospheres of the Sun and solar-type stars, most of the Li is ionized (Li II). In the ISM, lines of LiH and Li I are weak.

In the atmospheres of brown dwarfs and planets most of the Li is in molecules or locked in dust grains. The ionized, molecular-bound and dust-locked states of lithium are very difficult to observe directly, and hence the lines of the neutral state continue to be the most widely used means of obtaining information on the abundance of this element across the cosmos.

1.1 A Brief History of Lithium

The universe, as it is currently understood in the framework of contemporary cosmology, is made up mostly of dark energy and dark matter, which are perceived by their effects on ordinary matter and light, but their true nature has not been revealed yet.

Baryonic matter (neutrons and protons), the stuff that we are all made of, comes from a very tiny asymmetry between matter and antimatter that originated within the first second of the Big Bang.

After most of the matter annihilated with antimatter, a small fraction of order of 10^{-9} survived and went through a set of nuclear reactions in the first few minutes that produced the first chemical elements.

Most of the hydrogen (H) and helium (He) atoms that are still with us today are thought to come from Big Bang baryogenesis and nucleosynthesis. For alternative cosmological scenarios the reader is invited to check Martín López Corredoira's e-book in this series.

All other chemical elements in the periodic table, known as metals by the astrophysicists, were made later in the furnaces provided by the birth and evolution of massive astrophysical objects.

All but one, in fact, the single exception being the subject of this book.

Lithium (Li), the lightest of metals, was created in the Big Bang, together with H and He, but not only there. According to standard Big Bang nucleosynthesis the universe was seeded with a homogeneous lithium abundance log N(Li) = 2.7. This abundance is dubbed the primordial lithium abundance and it is assumed that the first stars formed with this Li content.

Since the Big Bang, the Li abundance in the Galaxy has been on the rise (Figure 1.1). The present lithium abundance in the Galaxy is dubbed the cosmic Li abundance and its value is log N(Li) = 3.3. The solar system formed with this lithium abundance.

Most of the Li present today in the Milky Way does not come from the Big Bang. One of the leading contenders to explain the lithium enrichment in the Galaxy is nova explosions, but there are others.

Lithium has a rich and diverse biography because it is one of the few primordial chemical elements (coming from the Big Bang), and is created and destroyed in many astrophysical environments with a net positive increment over the lifetime of the Galaxy.

The initial lithium abundance created by Big Bang nucleosynthesis early in the Cosmic Dark Ages is called the "***primordial lithium abundance***". The present lithium abundance observed in the solar system (meteorites), the ISM, and the youngest stars in the Galaxy is called the "***cosmic lithium abundance***". The abundance of lithium has grown owing to net production in energetic sources such as black holes, cosmic rays, thermonuclear flashes in red giants, stellar flares, and nova and supernova explosions. Nova explosions are one of the leading contenders as the dominant source of Li enrichment in the Galaxy, but others remain viable as well.

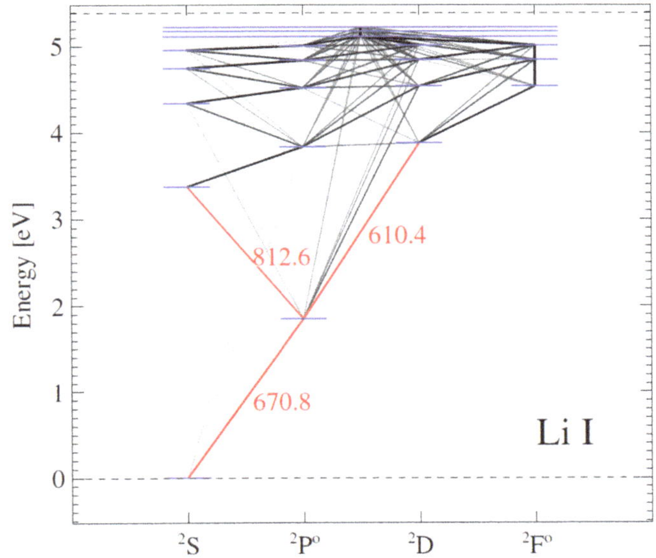

Figure 1.1. Simplified Grotrian diagram of the neutral lithium atom (Li I). The model contains over 20 energy levels, but the upper ones are collapsed for simplicity. The allowed energy levels for the outer electron are measured in units of electronvolts (eV). The wavelengths of the transitions between those levels are labeled in units of nanometers as recommended by the International System of Units. The most observed transition is the Li I resonance doublet at 670.8 nm, which arises between the first excited level and the ground level. The second most observed Li I transitions are the subordinate lines at 610.4 nm and 812.6 nm. In the most lithium-rich star, Li I lines corresponding to up to seven different transitions have been observed (Kowkabany et al. 2022). Figure credit: Wang, E. X. et al. 2021 "3D NLTE spectral line formation of lithium in late-type stars" MNRAS 500 2159.

Fresh production of lithium has been observed directly via gamma-ray spectroscopy in solar flares and x-ray outbursts of binaries with black holes and neutron stars.

Observing lithium, and estimating its abundance, across the universe is quite straightforward because the neutral lithium atom (Li I) has a very clear fingerprint in the optical spectra of many astrophysical sources (Chapter 1).

Modeling lithium destruction and production in a variety of environments is also quite feasible because the nuclear reactions involved are well known (Chapter 2).

The transport of Li atoms from places where they are created or destroyed to places where they can be observed is a fascinating and complex topic, particularly when these processes take place inside stars in different evolutionary phases (Chapter 3).

As a result of the incredibly rich biography of lithium, it has become a versatile element with many applications in astrophysics. It is observed in the ISM and the youngest stars where it defines the starting chemical abundance content of new generations of stars (Chapter 4).

Lithium is useful as a chronometer to measure the ages of young stellar associations, moving groups and open clusters (Chapters 4, 5, and 7).

It is also used as a probe to investigate the lives of stars such as the Sun (Chapter 5).

Lithium is a relic of the primordial chemical enrichment of the universe (e.g. Big Bang nucleosynthesis), as it can be found in the oldest stars (Chapter 6).

In the substellar domain, lithium becomes a tool for investigating the properties of brown dwarfs (Chapter 7), and the formation and evolution of exoplanets (Chapter 8).

1.2 Atomic Lithium in the Gas Phase

Neutral monatomic lithium (Li I) in the gas phase is the most abundant form of lithium under the conditions prevalent in the most numerous stars, (M-type) red dwarfs and in sunspots.

In the quiet solar photosphere and in solar-type stars, lithium is mostly ionized (Li II). In cooler objects, such as brown dwarfs and planets, lithium begins to combine into molecules and dust grains.

Where it is in the form of gas, the abundance of lithium can be determined from the strength of spectral lines measured by absorption spectroscopy. Most astrophysical lithium abundance determinations across the universe derive from measurements of the Li I resonance doublet.

1.2.1 The Li I Resonance Doublet and Other Lithium Lines

Alkali metals are hydrogen-like atoms because in the neutral state they consist of a Bohr-like shell with a single outer electron. A simplified representation (Grotrian diagram) of the electronic transitions allowed between the energy levels of Li I is shown in Figure 1.1.

The wavelength of the Li I resonance feature occurs at 670.8 nm, which is a region of high transparency in the Earth's atmosphere. Furthermore, astronomical telescopes are usually equipped with optical instruments that use charge coupled devices (CCDs) as detectors. CCDs have a very high sensitivity around 600 nm. As a result, the Li I resonance feature is the most commonly used transition to study the presence of lithium in the atmospheres of a wide variety of stars, brown dwarfs, and giant planets.

A recent analysis of 314 high-resolution spectra of very young stars with strong lithium lines has determined the mean air wavelength of the Li I resonance doublet to be located at 6707.856 ± 0.026 Å (Campbell-White et al. 2023).

The subordinate transitions of Li I at 610.4 nm and 812.6 nm have also been observed, albeit they are usually weak and not as frequently studied as the resonance doublet. Recently, seven lines from Li I have been detected in the remarkably metal-poor super-lithium-rich giant star 2MASS J05241392-0336543 (hereafter 2MASS 0524-03; Kowkabany et al. 2022).

Five of these lines (460.3, 497.2, 610.4, 670.8, and 812.6 nm) were sufficiently strong that these authors could isolate them and obtain independent measurements of the lithium abundance in the stellar atmosphere (see Figure 1.2). This star is one of the objects with the highest lithium content in the universe. It will be interesting to check whether or not additional Li I lines can be detected, in particular at longer

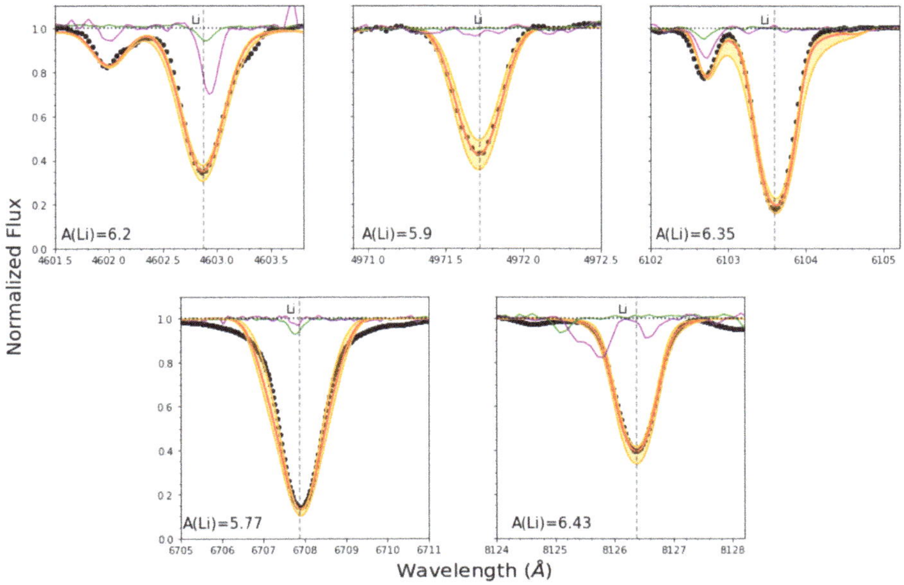

Figure 1.2. Five different lines of the neutral lithium atom (Li I) observed in high-resolution (R = 35,000) spectra of the giant star 2MASS 0524-03 obtained with the Magellan Inamori Kyocera Echelle (MIKE) spectrograph on the 6.5 m Magellan–Clay telescope at Las Campanas observatory (Chile). The MIKE data are denoted by black thick points. The best-fit synthetic spectra are shown with red lines, and their uncertainty is shaded in orange. Purple and green lines are the MIKE spectra of two giants with similar metallicity to 2MASS 0524-03 but lower lithium abundance of A(Li) = 2.34 and 0.78, respectively. Note that the strongest Li I line in 2MASS 0524-03, the resonance feature at 670.8 nm, yields the lowest lithium abundance. Figure credit: Kowkabany, J. et al. 2023 "Discovery of an ultra lithium-rich metal-poor giant star" ApJ in press.

wavelengths where there is less blending with atomic and molecular lines from other chemical elements.

A full digitized list of all the allowed lithium atom transitions can be found in the Vienna Atomic Line Database (VALD), which includes relevant information about the lines from the two stable lithium isotopes, ^{6}Li and ^{7}Li. A compilation of the air wavelengths of the Li I resonance lines from both isotopes can be found in Morton (2003).

Updated information about the lines of ^{6}Li and ^{7}Li, respectively, including wavelengths and oscillator strengths are available at the following websites:

http://kurucz.harvard.edu/atoms/0300/gf03006.lines
http://kurucz.harvard.edu/atoms/0300/gf03007.lines

The layout of those data files is available at:
http://kurucz.harvard.edu/linelists.html

The resonance lines of Li I and other alkali elements are multiplets, owing to the spin–orbit splitting of the first excited state of the unpaired outer electron. The gyromagnetic ratios cause the Li I doublet to be in a 2:1 resonance.

The strongest alkali element resonance multiplet in the solar spectrum is that of sodium (Na) because its abundance is the highest of the alkali elements in the atmosphere of the Sun. Observation of the Na I resonance multiplet in the Sun led to the discovery of the Zeeman effect, a splitting of the lines caused by a magnetic field. The complex details of the atomic fine and hyperfine structure have been analyzed many times as they are a good testbed for comparing observations and theory.

The wavelengths and the oscillator strengths of the lithium spectral lines have been gradually refined over time in the literature.

At the turn of the twentieth century, the Li I resonance doublet was first identified in photographic plates of the Sun obtained by Ellerman and Nicholson using the spectrograph mounted on the 22,86 m high solar tower at Mount Wilson Observatory in southern California.

The Mount Wilson solar spectra were compared with laboratory spectra by Arthur S. King (1916) and the presence of a strong Li I feature at 670.8 nm was noted in sunspots, but not in the photosphere.

The strong temperature dependence of the strength of the Li I resonance doublet expected from its low ionization potential (5.4 eV) is confirmed by the observations of laboratory spectra at different temperatures and from the solar data, because sunspots are much cooler than the solar photosphere.

The Sun is characterized by an effective temperature value of 5772 K currently recommended by the International Astronomical Union (IAU). Surface magnetic fields on the solar surface concentrate on spots that are darker than the surrounding areas because they are cooler by about 1500 K. An overview of magnetic fields in the Sun and solar-like stars can be found in the e-book of this series authored by Gibor Basri (2021).

1.2.2 Lithium Abundances

Estimates of the lithium abundance and lithium isotope ratio in the Sun and the brightest stars was stimulated by the theoretical suggestion made by Gamow & Landau (1933) that the presence of this light element in the stellar surface could be an indicator of the internal temperature of the star. They calculated the rate of the lithium destruction by proton bombardment (^{7}Li + p => 2 ^{4}He) in stellar interiors and concluded that it would be efficiently destroyed at temperatures of a few 10^6 K.

The question of whether or not the lithium destruction in stellar interiors would be detectable at the stellar surface was addressed by Eddington (1933), who showed that diffusion would take much longer than the lifetime of most stars. Although Sir Arthur Stanley Eddington was not aware of the role of convective transport, his calculations indicated that in the radiative regions of stellar interiors lithium-depleted material would have a hard time in getting to the photosphere.

The abundance of lithium in meteorites and on Earth was given as 1.7×10^{-4} times lower than that of calcium in the compilation of Brown (1949), but the lithium abundance of the Sun was not available at that time.

Greenstein & Richardson (1951) demonstrated that the surface abundance of lithium can be estimated from spectroscopic measurement of the equivalent width (EW) of Li I resonance lines present in the solar spectrum, and they estimated that in the solar photosphere the abundance of lithium is two orders of magnitude lower than in the meteorites.

Since the whole solar system is thought to have formed from the same fragment of a molecular cloud, the lithium abundance should initially be the same for the Sun, the planets and the minor bodies such as asteroids, comets, and dwarf planets.

The low lithium abundance in the Sun supports Gamow and Landau's idea that the solar interior is hotter than the temperature required for thermonuclear lithium burning and that lithium-depleted material somehow gets mixed all the way to the surface. Exactly how this happens remains unclear, as we shall see later in this book.

Greenstein & Richardson (1951) calculated the lithium ionization equilibrium to be (Li II + Li I)/Li I = 616 meaning that most of the lithium atoms in the solar surface are ionized. Therefore, ionization corrections, which are highly dependent on the effective temperature ($T_{\rm eff}$), are of utmost importance for deriving lithium abundances from the equivalent width (EW) of neutral lithium lines.

The EW is frequently used to measure the strength of a spectral line in units of wavelength. In the literature it is most frequently found in units of ångström (1 Å = 10^{-8} cm), although it can also be found in milliångströms (1 mÅ = 1 Å/1000) for small values of the EW, and in nanometers (1 nm = 1 Å × 10), which is the unit recommended by the IAU.

Sources of uncertainties in the determination of the EWs of any spectral line include the following (Houtgast 1966):

1. Instrumental (background noise, detector noise, spectral resolution, scattered light)
2. Astrophysical (line blends, continuum shape)
3. Interpretation (far wings, profile irregularities)

The minimum spectral resolution and signal-to-noise ratio (S/N) required to measure lithium abundance depend strongly on the type of object under study. As a simple rule of thumb, Cayrel (1988) proposed that the uncertainty in measuring the EW of a spectral line with Gaussian shape depends on the spectral dispersion of the spectrograph, the spectral resolution of the instrumental + telescope configuration and the S/N according to the following simple formula:

$$(\mathrm{EW}) = 1.6\ \sqrt{\mathrm{FWHM} \times\ D}\ \div \mathrm{S/N} \tag{1.1}$$

In Equation (1.1), σ (EW) is the standard deviation of an equivalent width measured in Å; FWHM is the full width half maximum of a single spectral line (in units of Å), assuming that its profile has an approximately Gaussian shape (which usually is the case); D is the dispersion provided by the spectrometer (in units of Å per pixel); and S/N is the signal-to-noise ratio in the continuum.

Equation (1.1) is useful for obtaining a rough estimate of what spectroscopic resolution and S/N are required to detect a feature with a given EW. As an example, Cayrel (1988) stated that for the typical spectra obtained with the ESO CAT spectrograph—FWHM = 0.1 Å, D = 0.035 Å, S/N = 200—the predicted 1σ

accuracy on the EW is 0.4 mÅ. Repeated observations of the same star showed that this estimate was too optimistic by at least a factor of two in the case of the spectra analyzed by Roger Cayrel.

Two simplified versions of Equation (1.1) are the following:

$$(\mathrm{EW}) = 2.26 \times D \div \mathrm{S/N} \tag{1.2}$$

Equation (1.2) is a particular case of Equation (1.1). It relates the precision (1) that can be achieved on the measurement of an EW to the dispersion of the spectrograph and the S/N achieved. It applies to the particular case of a well-designed spectrograph that has a slit width in the telescope's focal plane that is equal to the typical seeing of the particular observing site, and this slit width projects onto two pixels on the detector (FWHM = 2 D).

$$(EW) = 7590 \div (R \times \mathrm{S/N}) \tag{1.3}$$

Equation (1.3) is the same particular case of Equation (1.1) (when FWHM = 2 D), but rewritten as a function of the resolving power where $R = \lambda$/delta λ, with $\lambda = 6708$ Å (the central wavelength of the Li I resonance doublet).

Another expression that has been used to separate detection of the Li I resonance doublet at 670.8 nm from upper limit (3 confidence) was given by Deliyannis et al. (1993), and it is written as follows:

$$3(\mathrm{EW}) = 4.51\sqrt{\mathrm{FWHM} \times D} \div \mathrm{S/N} \tag{1.4}$$

The following simplified version of Equation (1.1) was published by Kenyon et al. (2005) and was used to estimate the uncertainties (1 confidence) of their EW measurements of the Li I feature at 670.8 nm and the Na I doublet at 818.3/819.5 nm:

$$(\mathrm{EW}) = \sqrt{\mathrm{FWHM} \times D} \div \mathrm{S/N} \tag{1.5}$$

Equation (1.5) was adopted by Hernández et al. (2014) to estimate the errors in EW of the Li I doublet at 670.8 nm measured in high-resolution ($R = 34{,}000$) spectra of Ori cluster candidate members observed with Hectochelle multiobject spectrograph at the 6.5 m Multiple Mirror Telescope Observatory at the Fred Lawrence Whipple Observatory (Mount Hopkins, Arizona, USA).

An empirical method of estimating the errors of the EWs of atomic lines in the spectral region around the Li I resonance doublet was presented by Soderblom et al. (1993). They measured the EW of the Ca I line at 671.7 nm and the Fe I line at 670.3 nm using Gaussian fitting to the line profiles.

They derived a relationship between the scatter around the mean EW versus stellar color. Assuming the Ca and Fe abundances of the stars in their sample of Pleiades cluster members to be constant, they found a 1σ (EW) value of 15 mÅ for their Hamilton Echelle spectra obtained at the Shane 3 m telescope in Lick Observatory with R = 50,000. Equation (1.3) gives an error of 3 mÅ for R = 50,000 and S/N = 50. In this case, Equation (1.3) provides a conservative error estimate. The significance value (3 confidence) from Equation (1.4) would be closer, but still smaller, than the error bar given in Soderblom et al.

Soderblom et al. (1993) provided a mean correction for the Fe I line at 670.7441 nm, which is blended with the Li I resonance feature.

Some authors choose not to measure EW, but instead they rely on comparisons of observational and synthetic spectra to derive the lithium abundance using best-fit techniques, such as chi-squared minimization. This method is quite efficient in the case of large spectroscopic surveys, for example, the LAMOST dataset of 169,479 FGK stars analyzed by Gao et al. (2021). It works quite well for stars where the determination of the continuum can be made in an automated way, and where there is a very good agreement between empirical and theoretical spectra.

As the lithium abundance increases, the absorption strength and the EW of a given Li I line also increases. This correspondence between line strength and chemical abundance is called the curve of growth (CoG) of the spectral line. In the case of the Sun, the CoGs of two Li I spectral lines are shown in Figure 1.3.

CoG computations done under the simplifying assumption of Local Thermodynamic Equilibrium (LTE) are generally adequate for weak absorption lines.

However, for strong absorption lines such as those observed in very young stars and red giants, departures from LTE can be significant. Non-LTE calculations of the CoG become necessary for estimating realistic lithium abundances.

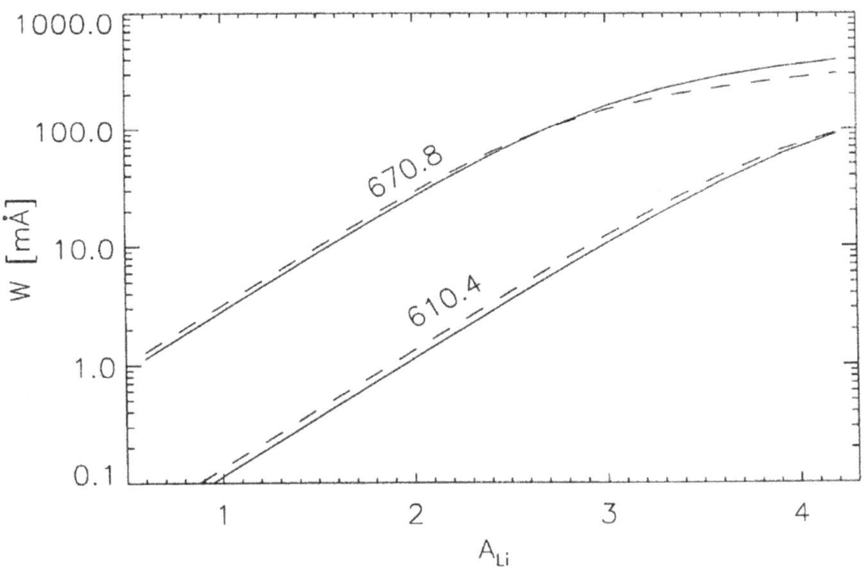

Figure 1.3. CoG for two Li I spectral lines under solar conditions. In the abscissa the lithium abundance scale is $A(\text{Li}) = \log N(\text{Li}) = \log (N_{\text{Li}}/N_{\text{H}}) + 12$ where N_{Li} and N_{H} are the number densities of Li and H atoms, respectively, and the hydrogen abundance is set to $A(\text{H}) = \log N(\text{H}) = 12$. For example, an Li abundance a factor of 10^9 lower than that of hydrogen is denoted as $A(\text{Li}) = \log N(\text{Li}) = 3.0$. In the ordinate the computed EW from theoretical spectra is given in units of mÅ. Solid lines denote computations made under local thermodynamic equilibrium (LTE) and dashed lines denote computations under non-LTE. Figure credit: Carlsson, M. et al. 1994 "The non-LTE formation of Li I lines in cool stars" A&A 288 860, reproduced with permission © ESO.

Tabulated LTE and non-LTE CoGs computed using the Kurucz (1993) atmospheric models can be found in the paper by Pavlenko & Magazzù (1996) for effective temperatures ($T_{\rm eff}$) between 6000 K and 3500 K, surface gravities between log g = 3.0 (giant stars) and log g = 4.5 (main-sequence stars), and lithium abundances between log N(Li) = 4.0 (about an order of magnitude higher than Cosmic) and log N(Li) = −1.5 (more than four orders of magnitude lower than Cosmic). They provide calculations for the Li I resonance doublet at 670.8 nm and the subordinate lines at 610.4 nm and 812.6 nm.

Custom-made COGs have been made for the lithium abundance analysis of thousands of stars observed with the GAIA-ESO survey (Franciosini et al. 2022).

Stars where the Li I lines are expected to be weak because of high temperature (most of Li is ionized), or because of low lithium abundance due to Li depletion require high spectral dispersion and resolution, and high S/N.

In practice the expectations of what can be achieved can be guided by a variety of examples from the literature.

Figure 1.4 illustrates examples of solar-type stars with high and low lithium abundance.

The capability of a spectrograph to resolve spectral lines is usually represented by the dimensionless quantity called resolving power (R), which is defined as follows:

Resolving power = $R = \lambda$/delta λ; where λ is the central wavelength of the spectral region under study and delta λ is equal to the FWHM defined in Equation (1.1). The concept of R was already introduced in Equation (1.3).

As an example, for solar-type stars, the combination of $R = 10^5$ with S/N = 10^2 has allowed the determination of lithium abundance down to values of log N (Li) = 1.5 in G-type stars located at distances of up to 130 pc with HARPS on the 3.6 m ESO telescope at La Silla (see Figure 1.4 from Pavlenko et al. 2018).

A recent review of the stellar abundances of Li based on high-resolution spectroscopy, together with those of B and Be, is available in Boesgaard (2023).

For stars significantly cooler than the Sun, namely the K-type (also called orange dwarfs) and M-type (known as red dwarfs) stars, the Li I lines are expected to get stronger (see CoGs as a function of stellar effective temperature in Figure 1.5) and thus the requirements on spectral resolution and S/N can be relaxed.

For example, the combination of R = 9000 with S/N in the range 20–100 has allowed the determination of lithium abundance up to values of log N (Li) = 1.0 in young K-type and M-type stars located at distances of up to 150 pc with ISIS at the WHT (Martín et al. 1994). A sample of representative spectra is shown in Figure 1.6.

Another example of a spectroscopic lithium observation taken with a resolving power of R = 9300 are the data obtained with the FLECHAS spectrograph (Mugrau et al. 2014). This is an Echelle spectrograph coupled with a fiber-link to the 0.90 m telescope of the University Observatory in Jena. Bischoff et al. (2020) used FLECHAS to search for the Li I resonance doublet in runaway stars.

Figure 1.7 shows an example of the FLECHAS spectra. It features the detection of a strong Li I resonance doublet in HIP 46843. This a G8V runaway star where the authors measured a Li I EW = 200 ± 12 mÅ using the option for direct integration of the line profile available in the IRAF task splot.

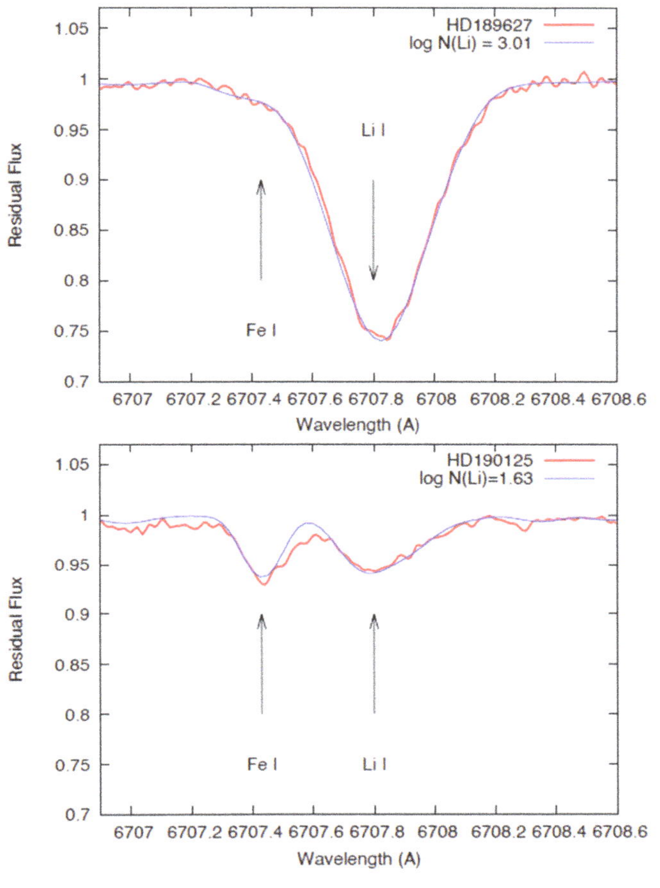

Figure 1.4. Li I resonance lines of two solar-type stars observed with HARPS. The spectrum of HD189627 is fitted with a non-LTE synthetic spectrum with the following parameters: log N(Li) = 3.01, $T_{\rm eff}$ = 6210 K, surface gravity = log g = 4.40, microturbulent velocity = Vm = 1.4 km s^{-1}, rotational broadening = V sin i = 7.8 km s^{-1}, iron abundance = [Fe/H] = 0.07. The spectrum of HD190125 is fitted with a non-LTE synthetic spectrum with the following parameters: log N(Li) = 1.63, $T_{\rm eff}$ = 5644 K, log g = 4.53, Vm = 1.0 km s^{-1}, V^{-1} sin i = 3.3 km s^{-1}, [Fe/H] = 0.04. Note that the high spectral resolution is clearly needed to resolve the Li I resonance doublet from the nearby Fe I line at 670.7473 nm, which is particularly critical at low lithium abundance. Figure credit: Pavlenko Ya et al. 2018 "A detailed study of lithium in 107 CHEPS dwarf stars" A&A 611 A27, reproduced with permission © ESO.

To convert from the Li I EW values to photospheric lithium abundances, these authors used the CoG values published by Soderblom et al. (1993).

The coolest objects for which lithium abundance has been determined are L-type and early-T type brown dwarfs with temperatures in the range from 1800 K to 1200 K (Lodieu et al. 2015, 2018; Martín et al. 2018, 2022).

In these ultracool objects the Li I resonance doublet is very strong and can be detected with low spectral resolution (R = 500–600) and modest S/N (> 10). This fact is illustrated in Figure 1.8 for L-type brown dwarfs in the Hyades open cluster.

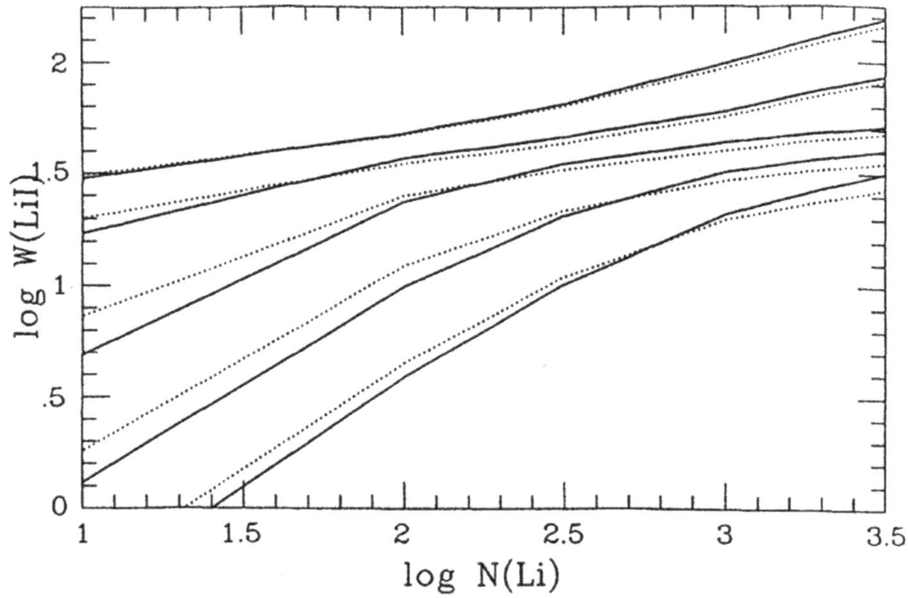

Figure 1.5. CoGs of the Li I resonance doublet for different effective temperatures. As in Figure 1.3, solid lines denote calculations made in LTE conditions and dotted lines in non-LTE conditions. Figure credit: Martín, E. L. et al. 1994 "Pre-main sequence lithium burning. I. Weak T Tauri stars" A&A 282 503, reproduced with permission © ESO.

Some of the largest lithium abundances that have been observed in the universe are located in the atmospheres of red giant stars.

Figure 1.9 shows the photographic plates where a very strong Li I resonance doublet absorption at 670.8 nm was noted by Sanford (1950) in the C-type star WX Cygni using the high-resolution spectrograph mounted at the coudé focus of the Hooker 100 inch telescope at Mount Wilson Observatory in California. With an aperture of 2.54 m, this telescope at Mount Wilson was the largest in the world from 1917 to 1948.

The Li abundance of WX Cygni and other lithium super-rich carbon stars were shown to be 10–100 times larger than the cosmic abundance (Abia et al. 1991).

The Large Sky Area Multi-Object Fiber Spectroscopy Telescope (LAMOST) has found some of the most extreme examples of Li super-rich red giant stars known to date (Yan et al. 2018; Kowkabany et al. 2022).

Figure 1.10 shows the Li I resonance feature detected in TYC429-2097-1 using LAMOST and the follow-up higher-resolution spectra (R = 80,000) obtained with the 2.4 m Automated Planet Finder (APF) telescope at Lick Observatory.

An average Li abundance of $N(\text{Li}) = 4.51 \pm 0.09$ for this star was obtained from NLTE calculations using three Li I lines (610.36, 670.78, and 812.63 nm); that is 15 times higher than the Li abundance in meteorites and T Tauri stars, strongly indicating that a lithium enhancement process has taken place in this red giant.

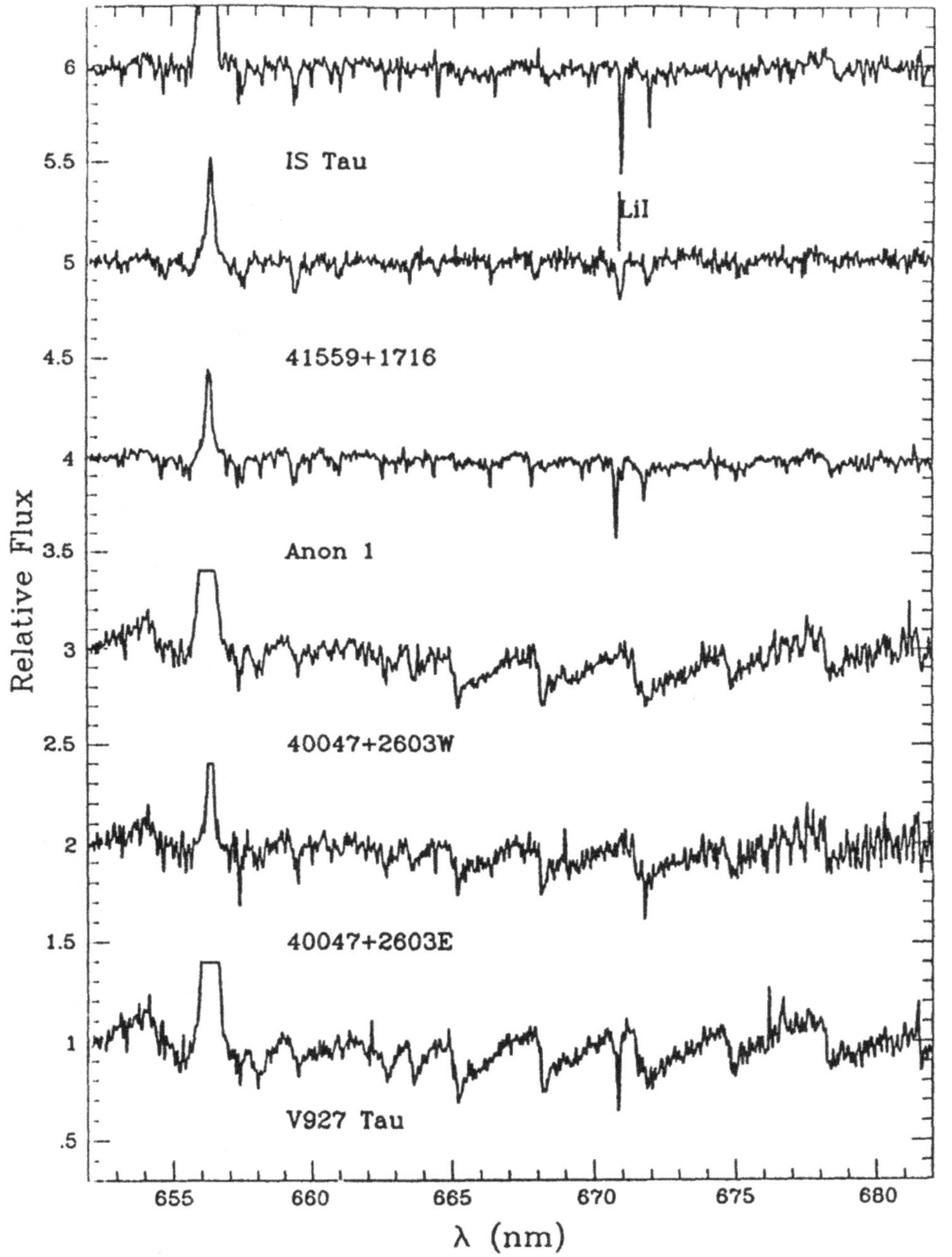

Figure 1.6. Representative spectra of K-type and M-type stars observed with resolving power $R = 9000$ with the 4.2 m William Herschel Telescope. The position of the Li I resonance doublet is marked. Note the clear Li I feature in IS Tau, 41559+1716, Anon 1, and V927 Tau, and the non-detection in the visual binary 40047+2603 East and West. Figure credit: Martín, E. L. et al. 1994 "Pre-main sequence lithium burning. I. Weak T Tauri stars" A&A 282 503, reproduced with permission © ESO.

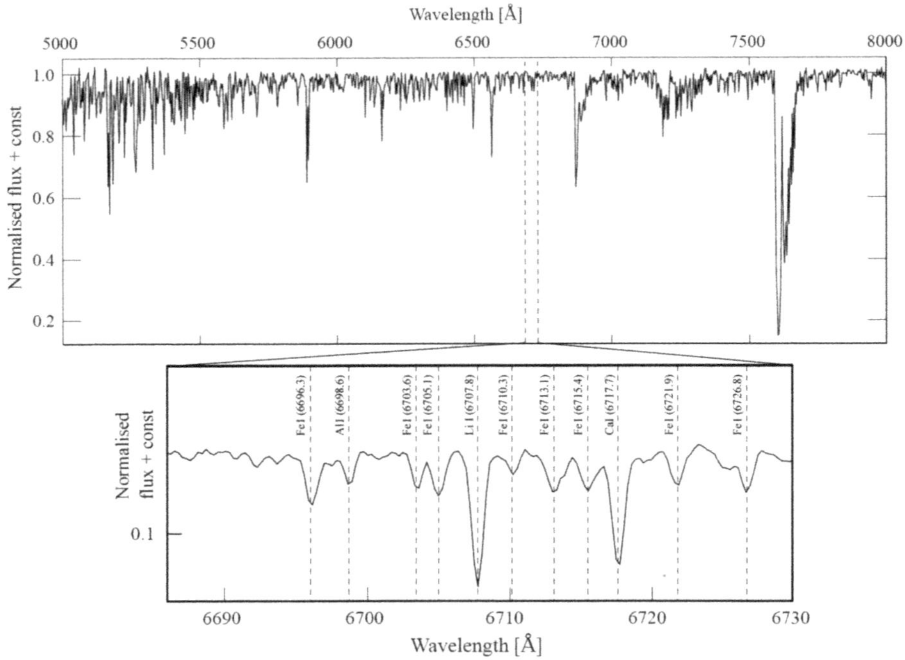

Figure 1.7. Normalized spectrum of the runaway G8V-type star HIP 46843 obtained with the FLECHAS spectrograph. A zoom into the Li I resonance feature is displayed in the bottom panel. The strong Li I resonance doublet in this star stands out from the other lines. The identification of each line is labeled. Figure credit: Bischoff, R. et al. 2020 "Identification of young nearby runaway stars based on Gaia data and the lithium test" AN 341 908.

1.3 The Lithium Isotope Ratio

Two stable lithium isotopes are found on Earth with an isotope ratio $R_I = {}^6Li/{}^7Li = 0.08$, which means that the abundance of ^{6}Li is only 8% that of ^{7}Li. In other words, for every 100 atoms of the most abundant lithium isotope, namely ^{7}Li, there are 8 atoms of the rarer isotope ^{6}Li in the Earth's crust. This value is important as a reference to compare with other values of the lithium isotope ratio that have been estimated in a variety of cosmic environments.

Owing to the almost perfect coincidence between the doublet hyperfine splitting (0.0152 nm) and the isotopic shift (0.0158 nm), the resonance doublets of the two isotopes conspire to blend in the Li I resonance feature at 670.8 nm. Furthermore, line broadening effects in stellar atmospheres such as convective motions, magnetic fields, rotation, and turbulence make it impossible to resolve the lines, no matter how high the spectral resolution can be.

An example of high-resolution (R = 100,000) spectra of bright nearby stars is shown in Figure 1.11. Note that the lines are completely blended. It is therefore a formidable challenge to determine the Li isotope ratio in stars and the ISM.

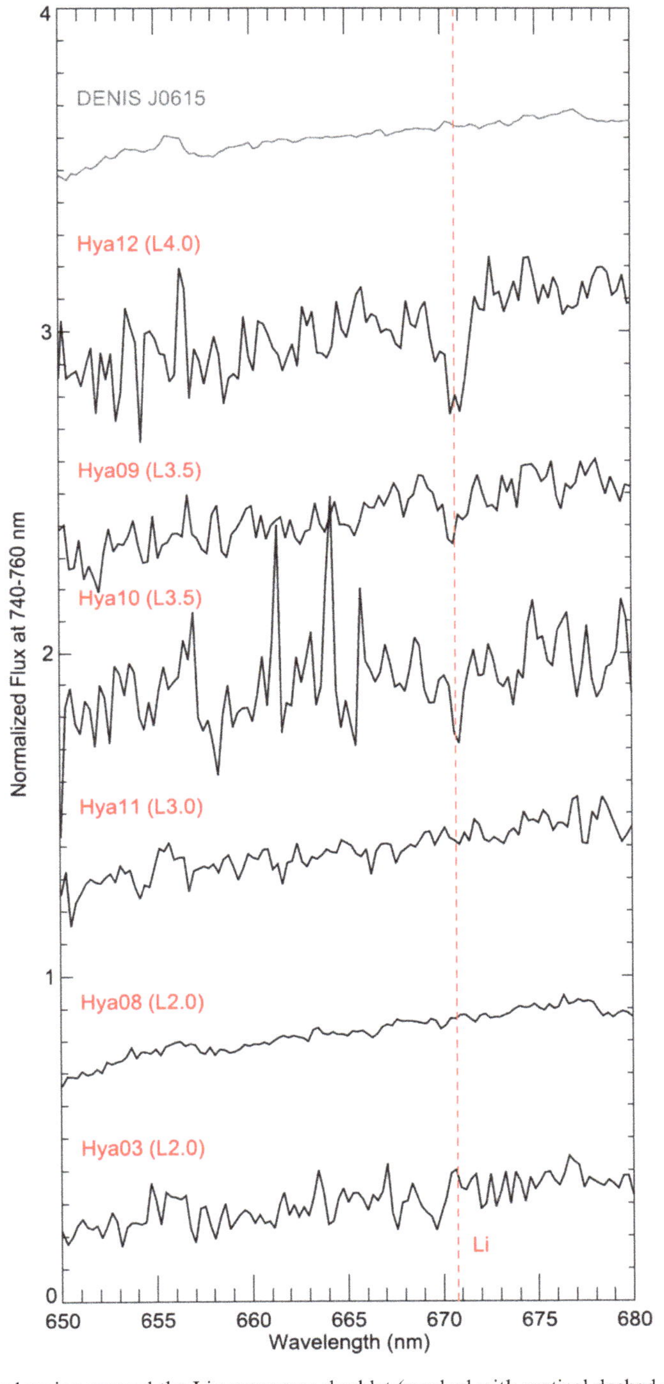

Figure 1.8. Spectral region around the Li I resonance doublet (marked with vertical dashed red line) of L-type dwarfs observed with the long-slit mode of OSIRIS at the GTC. The two objects with strong Li I detections are confirmed members of the Hyades cluster. Figure credit: Martín, E. L. et al. 2018 "The lithium depletion boundary and the age of the hyades cluster" ApJ 856 40.

Figure 1.9. From left to right: the spectra of the R and N-type stars HD 156074, HD 52432, 19 Piscium, laboratory spectra of the CN molecule, U Hydrae, VX Andromedae, U Cygni, and WX Cygni. Note that the spectrum of the C9 star WX Cygni has no signal at the wavelength corresponding to the Li I resonance feature at 6708 Å (670.8 nm). Figure credit: Sanford, R. E. 1950 "An atlas of spectra of six stars of classes R and N" ApJ 111 262.

It has been proposed that the center of gravity of the ^{6}Li + ^{7}Li resonance doublet blend, and the shape of the line profile could be sensitive to the isotope ratio as illustrated in Figure 1.11. The higher the Li isotope ratio is, the more redshifted becomes the mean wavelength of the blend of the two doublets.

Herbig (1964), Merchant et al. (1965), and Smith et al. (1998) provided analytical expressions for the relation between the mean wavelength of the Li I resonance doublet and the isotope ratio. This simple relationship led to estimates of high values for the isotope ratio (Feast 1966), but it has been proved to be naïve because there are several factors that introduce significant line shifts in the Li I feature.

One well-known problem in the determination of the Li isotope ratio in stellar atmospheres is the strong dependence of the Li I doublet on convective motions, temperature effects, and magnetic fields. The Li I lines are much stronger in cool magnetic spots than in the quiet photosphere. This is illustrated in Figure 1.12.

Determination of the lithium isotope ratio requires higher resolution and higher S/N than those needed for obtaining just the bulk lithium abundance, and more sophisticated line formation analysis. As a result, attempts to obtain the isotope ratio have been scarcer than those dedicated to measuring lithium abundances, and the results have not been conclusive.

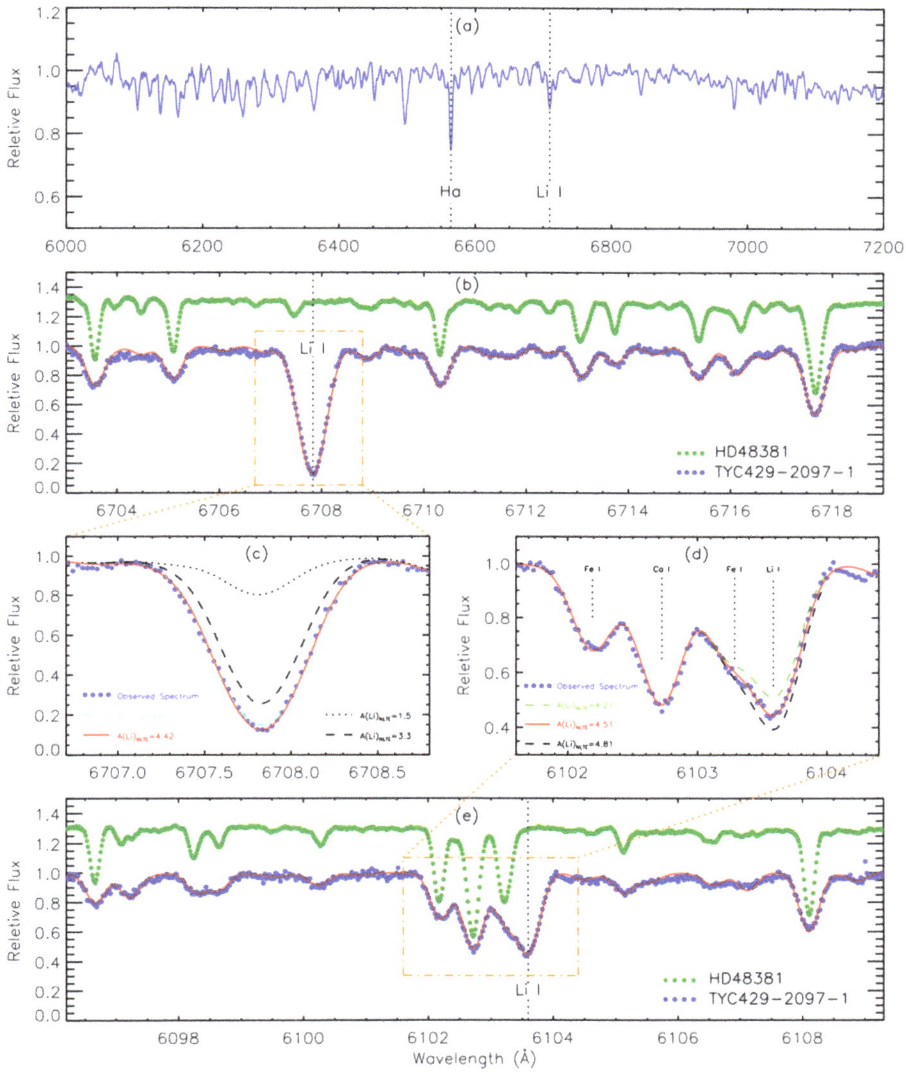

Figure 1.10. Observed spectra for TYC 429-2097-1, an object with one of the highest surface lithium abundance ever detected. Panel (a) displays the LAMOST data. Panels (b)–(e) show the high-resolution APF spectra for TYC429-2097-1 (blue dots) and for HD48381 (green dots) for comparison. HD48381 has similar parameters to those of TYC429-2091-1 (mass = 1.5 $M_{\odot}$, $T_{\rm eff}$ = 4700 K, surface gravity = log g = 2.2, metallicity = Fe/H = −0.36) but no detectable lithium abundance. Synthetic spectra for different Li abundances are shown in panels (c) and (d), which have been used to estimate the best fits to the observed spectrum. Figure credit: Yan, H.-L. et al. 2018 "The nature of the lithium enrichment in the most Li-rich giant star" NatAs 2790.

The detection of ^{6}Li has been attempted in solar-type stars, metal-poor stars, and red giants, but it is still unclear whether or not it has been achieved.

Rebolo et al. (1986) reported ^{6}Li measurements in main-sequence stars and subgiants using the coudé Echelle spectrograph fed by the ESO 1.4 m telescope that

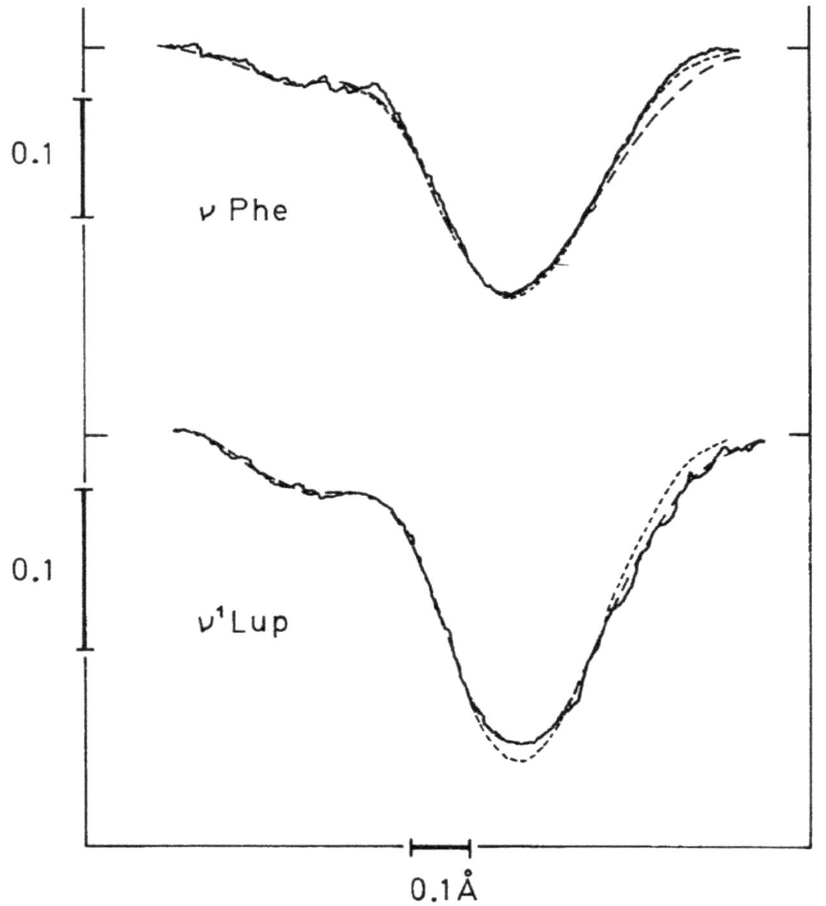

Figure 1.11. Observed (solid line) profile of the Li I resonance doublet at 6708 Å in two late-F-type main-sequence stars. The data were taken with the Echelle spectrograph at the ESO 1.4 m coudé auxiliary telescope in La Silla and the resolving power is 10^5. The spectrum of the upper star is best fitted with a theoretical line profile using a Li isotope ratio = R_I = 0.02 (short-dashed line), whereas the spectrum of the lower star is best fitted with R_I = 0.10 (long-dashed line). Note that these subtle effects in the line profiles can be ascribed to other phenomena that are independent of the Li isotopic ratio (see text). Figure credit: Andersen, J. et al. 1984 "The lithium isotopic ratio in F and G stars" A&A 136 65, reproduced with permission © ESO.

provided $R = 10^5$ spectra. However, Pavlenko et al. (2018) failed to confirm the detection of ^{6}Li in high-quality spectra obtained with HARPS at the ESO 3.6 m telescope of solar-type stars with $R = 10^5$ and S/N = 100.

Mott et al. (2017) used very high spectral resolution (R = 120,000) and high S/N (400 per resolution element) spectra obtained with the coudé Echelle spectrograph Gecko at the CFHT to derive the lithium isotope ratio in the active K0-type red giant branch star HD123351.

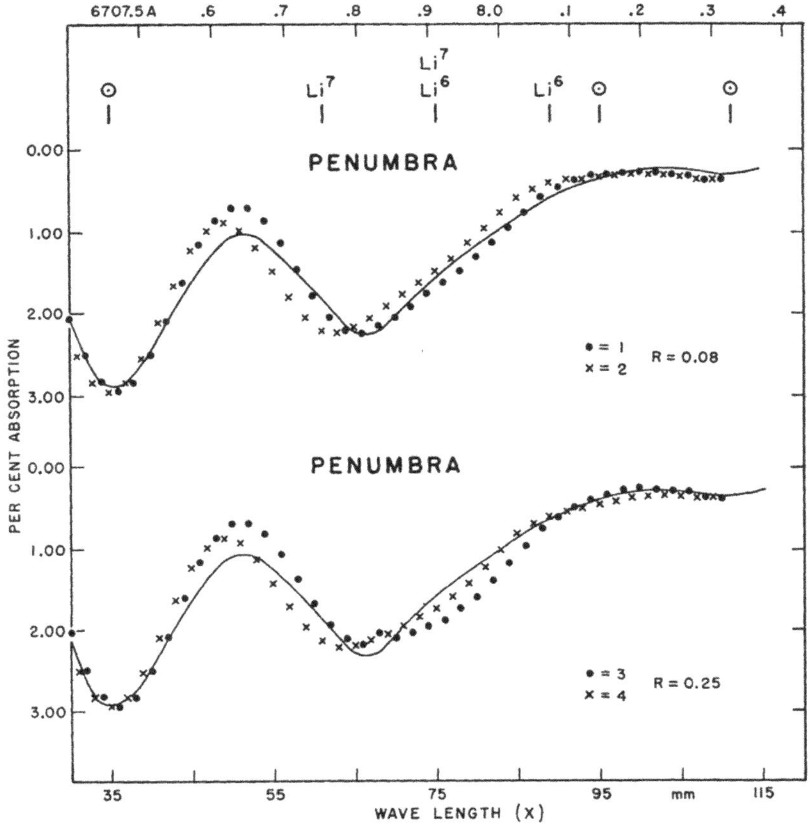

Figure 1.12. Solar spectra in sunspot penumbra (upper curve) and in the quiet photosphere (lower curve). The Li I resonance lines present in the region and other solar lines are labeled. Computed profiles of the Li I lines are denoted by dots and crosses for different assumptions for the Li isotopic ratio = R_I = 0.00, 0.08, and 0.25. Figure credit: Greenstein, J. L., & Richardson, R. S. 1951 "Lithium and the internal circulation of the Sun" ApJ 113 536.

An advantage of the red giant over the solar-type stars is that it has very sharp spectral lines due to low rotational broadening ($V \sin i = 1.8$ km s^{-1}). Moreover, HD123351 has a clear Li I resonance doublet with an equivalent width of 73 mÅ.

In their detailed analysis of HD123351, Mott et al. considered different line lists for the spectral region around the Li I resonance doublet. They also made computations of synthetic spectra using one-dimensional model atmosphere and three-dimensional hydrodynamical code.

They concluded that the best agreement between the observational and theoretical spectra were reached for the computations that used the line list from Meléndez et al. (2012) and the 3D non-LTE code. Even with the impressive dataset and computational effort carried out by Mott et al., their detection of ^{6}Li has only a modest significance of 2σ.

Despite many attempts and contradictory reports, the presence of ^{6}Li in stellar spectra of metal-poor stars is still not confirmed. Cayrel et al. (2007) noticed that convective Doppler shift in stellar spectra could mimic line asymmetries that had been ascribed to the contribution of ^{6}Li to the Li I resonance doublet at 670.8 nm.

State of the art efforts to detect ^{6}Li were reported in the detailed work of Wang et al. (2022) who used data of exceptional high quality (R = 146,000; S/N > 10^3 per pixel) obtained with the ESPRESSO high-resolution spectrograph (Pepe et al. 2021) at the VLT in Cerro Paranal Observatory between 2019 April and 2019 June for three well-known halo stars.

For one of them, namely HD 84937, there were consistent claims of detections of ^{6}Li at the level of 5% (R_{I} = 0.05) reported by several teams (Smith et al. 1993, 1998; Hobbs & Thorburn 1994; Cayrel et al. 1999; Asplund et al. 2006). However, Lind et al. (2013) carried out three-dimensional (3D) NLTE line formation analysis and obtained results in conflict with the previous authors because they were unable to confirm the presence of ^{6}Li in HD 84937, or in any other of the stars in their study, thus confirming the previous suspicions by Cayrel et al. (2007).

The ESPRESSO spectra in the Li I resonance doublet region of HD 84937 and two other metal-poor stars are shown in Figure 1.13, together with synthetic spectra obtained from 3D calculations. The best-fitting spectra were calibrated with other spectral regions containing K I and Fe I lines. Wang et al. (2022) concluded that there is not a significant contribution of ^{6}Li in these spectra.

The presence of ^{6}Li in solar-type stars is not confirmed as demonstrated by the detailed 3D NLTE analysis of two stars observed with HARPS (Harutyunyan et al. 2018).

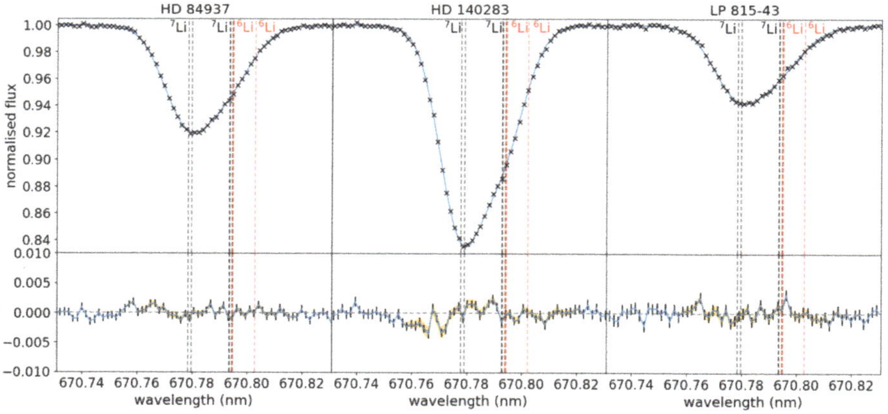

Figure 1.13. ESPRESSO spectra of three metal-poor stars in the spectral region around the Li I resonance doublet. Synthetic spectra are overplotted on top of the observational data. Residuals obtained from the comparison of observational and theoretical data are plotted at the bottom. The wavelengths of the ^{6}Li lines are marked in red and those of the ^{7}Li lines are marked in black. Upper limits on the lithium isotope ratio of R_{I} = 0.007, 0.006 and 0.017 were obtained for these three stars, respectively. These results do not confirm previous claims of ^{6}Li detections in metal-poor stars, and imply a much lower than terrestrial lithium isotopic ratio. Figure credit: Wang, E. X. et al. 2022 "Non-detection of ^{6}Li in Spite plateau stars with ESPRESSO" MNRAS 509 1521.

These authors find that 1D LTE analysis tends to overestimate the contribution of ^{6}Li in the atmospheres of solar-type stars.

An attempt to detect ^{6}Li in very high-resolution spectra (R = 130,000) of a bright visual binary with cool components obtained with PEPSI on the LBT reported null results on both stars (Strassmeier & Steffen 2022).

The detection of the lithium isotopic ratio in the ISM is easier than in stars because the Li I lines are narrower. The presence of clouds at different velocities along the line of sight may complicate the analysis, but in some favorable lines of sight it has been possible to obtain quite convincing measurements of lithium isotopic ratios (Lemoine 1995; Howk et al. 2012).

1.4 Lithium in Molecules

The spectra of objects cooler than about 4000 K (spectral classes M, L, T, Y) are so rich in molecular bands that it is not possible to determine the true continuum as a reference to measure absolute equivalent width.

A theoretical example is shown in Figure 1.14, where it can be seen that the effect of switching TiO absorption on and off for a $T_{\rm eff}$ of 2000 K has dramatic consequences in the appearance of the Li I feature for a constant Li abundance. Very cool objects do not have any real continuum regions. Therefore, only pseudo-equivalent widths (pEW) can be measured in very cool spectra, meaning that the

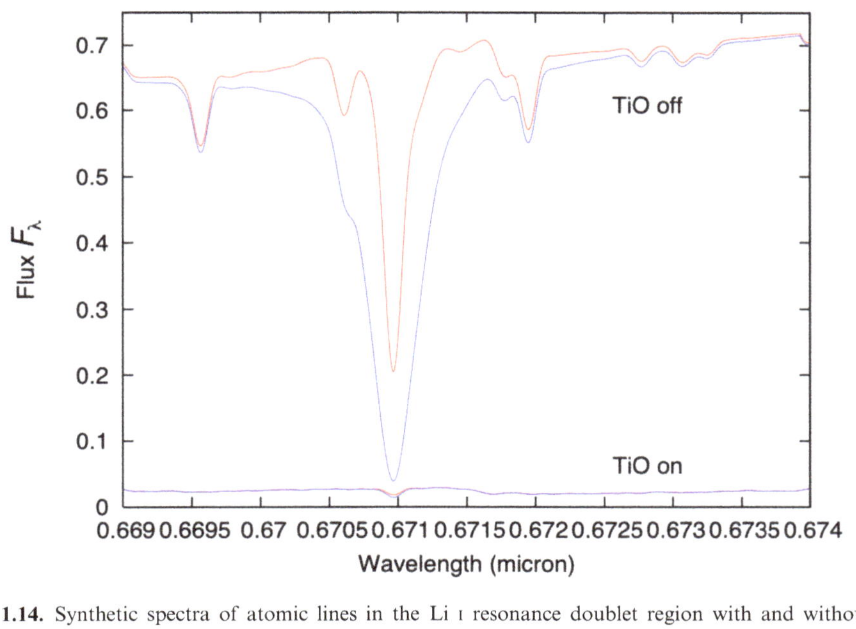

Figure 1.14. Synthetic spectra of atomic lines in the Li I resonance doublet region with and without TiO absorption for a brown dwarf model atmosphere with 2000 K effective temperature, projected rotational velocity ($V \sin i$) of 30 km s^{-1} and solar metallicity. Blue and red lines denote lithium abundances of log N(Li) = 3.2 and 2.0, respectively. The equivalent width of the Li I feature for log N(Li) = 3.2 is 13.5 Å and 0.65 Å for the TiO off and on cases, respectively. Figure credit: Pavlenko Ya et al. 2007 "Lithium in LP944-20" MNRAS 380 1285.

EW of atomic lines are measured with respect to a local pseudo-continuum, which is strongly affected by the presence of molecular band absorption.

Thermochemical equilibrium calculations have shown that at $T_{\rm eff}$ = 1700 K lithium starts converting into LiOH (Burrows & Sharp 1999).

At a temperature of 1641 K and a pressure of 21.1 bar, there is a triple point where lithium is equally divided into Li, LiCl, and LiOH (Lodders 1999).

Below 1525 K, LiCl becomes the dominant Li-bearing molecule, but between 1076 K and 944 K, LiOH dominates again.

At a temperature of 918 K and a pressure of 0.065 bar there is another triple point where lithium is equally distributed into LiCl, LiF, and LiOH. At even cooler temperatures (< 700 K), lithium is predicted to get removed from the gas phase as it condenses into solids (grains of $LiAlO_2$, LiF, Li_2S).

Below 1200 K, less than 1% of lithium is expected to reside in the neutral atomic form in the gas phase. It is therefore thought that the Li I resonance doublet should weaken in cooler and cooler brown dwarfs at constant lithium abundance.

The complex chemistry of lithium in ultracool atmospheres is illustrated in Figure 1.15. A promising way of observing lithium in ultracool brown dwarfs and giant planets with $T_{\rm eff}$ < 1200 K is to detect the transitions from Li-bearing molecules.

Calculations of the thermal spectrum of brown dwarfs and giant planets with $T_{\rm eff}$ in the range from 2400 K to 500 K indicate that the strongest spectral features from LiF, LiCl and LiH are located in the spectral range from 9 to 18.5 μm (Gharib-Nezhad et al. 2021), which could be observed with the JWST.

The first JWST spectrum from 1 to 20 μm of a young planetary-mass wide companion, namely VHS 1256-1257b (Gauza et al. 2015), has recently been reported by Miles et al. (2023).

Absorption feature from CH_4, CO, CO_2, and H_2O have been identified on it. This substellar object is expected to have preserved its initial lithium content because it has a very low mass (<20 $M_{\rm J}$), and its $T_{\rm eff}$ is low enough (1100 K) for it to have Li-bearing molecules in the atmosphere.

A search for the imprint of these molecules in the JWST spectrum of VHS 1256-1257b and an optical measurement of the Li I atomic lines would be worthwhile in order to understand how Li atoms get distributed in the atmosphere of this benchmark substellar object.

1.5 Lithium in Solids

Lithium was first isolated in chemical analysis of the mineral castorite (also known as petalite), which is a tabular crystal feldspar of molecular composition $LiAlSi_4O_{10}$.

This mineral was discovered by the Brazilian mineralist José Bonifácio de Andrada e Silva in 1800 in an iron-ore mine located on Utö, a small island that belongs to the Stockholm archipelago in Sweden. Using the private laboratory of Jöns Jakob Berzelius, one of the founders of modern chemistry, the element lithium was isolated in 1817 by Johan August Arfwedson.

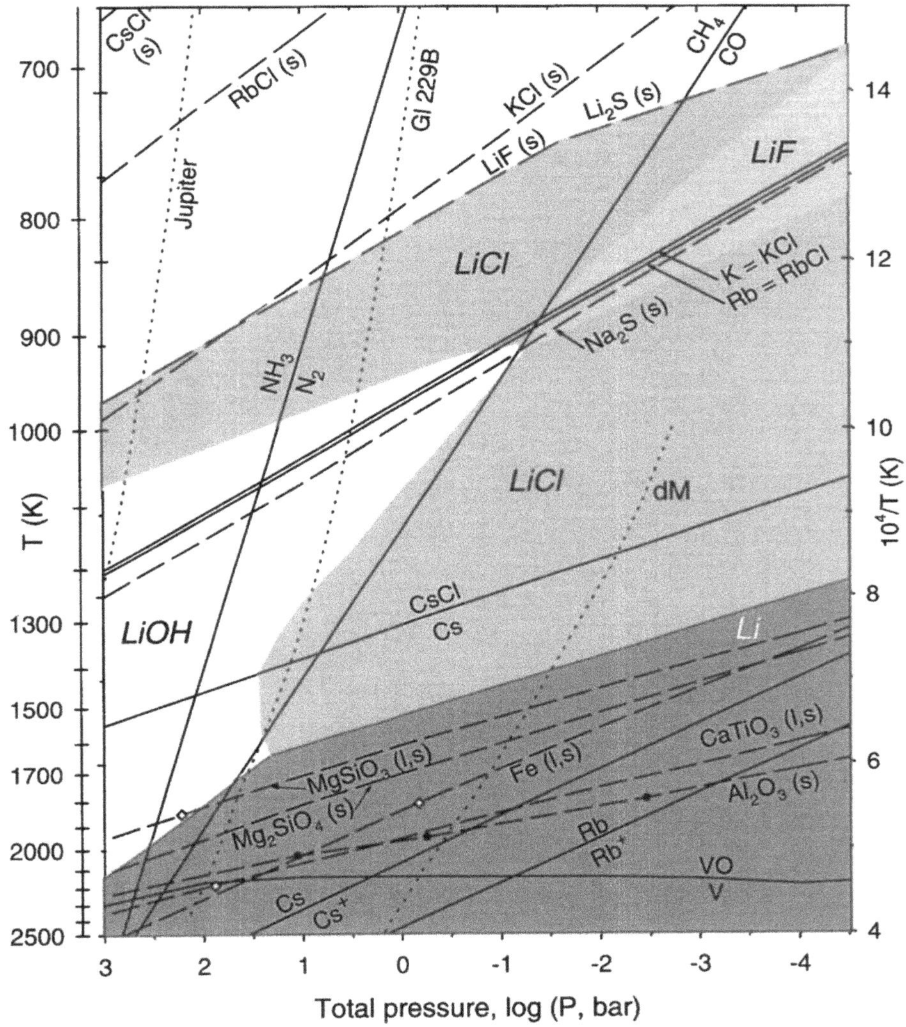

Figure 1.15. A diagram of temperature versus pressure conditions showing the region where most of the lithium is present as neutral Li atoms in the gas form (dark shaded area) and as molecules of LiCl and LiF (light shaded area). Solid lines denote boundaries where pairs of different species have the same abundance. Dashed lines mark the boundaries of condensation into solid grains of different species. Solid grains of LiF (lithium fluoride) and Li_2S (lithium sulfidolithium) condense in the upper part of the diagram and are labeled (s). Dotted lines represent the atmospheric conditions in a late-M dwarf (T_{eff} = 2200 K, Tsuji et al. 1996), a mid T-dwarf (Gl 229B, T_{eff} = 960 K, Marley et al. 1996) and Jupiter. Figure credit: Lodders, K. 1999 "Alkali element chemistry in cool dwarf atmospheres" ApJ 519 793.

In the 1920s industrial production of lithium started in Germany and has increased ever since. In the 1990s the Japanese company Sony produced the first rechargeable battery with lithium ions. Nowadays, lithium batteries are widely used. For example, the electric car Tesla Model S carries 63 kilos of Li in it, an amount comparable to 10^4 cell phones. It is currently estimated that the lithium reserve in the world may last for no more than a few centuries.

Concentrations of lithium and the isotope ratio has been studied in meteorites (Urey 1952, 1964; Balsiger et al. 1968; Nichiporuk & Moore 1970; Anders & Ebihara 1982) and in lunar samples from Apollo 11 (Tera et al. 1970) using atomic absorption spectroscopy in the laboratory, and the values have been found to be consistent with those measured in terrestrial rocks (Taylor 1960).

As for most other chemical elements, the standard concentration of lithium in the solar system comes from CI chondrites, which are widely regarded as the meteorite most representative of the chemical abundances in the protosolar nebula.

The updated compilation of Anders & Grevesse (1989) is frequently used. Nevertheless, it has been noted that there is evidence that CI chondrites have been altered inside their parent bodies, and consequently their role as the solar system abundance standard has been questioned (Ebihara et al. 2000).

The behavior of lithium in the Earth's crust and oceans is complicated. Variations in lithium content among different types of volcanic rocks have been discussed by various authors (e.g. Ryan & Langmuir 1987; Marschall et al. 2017). The possibility of using Li isotopes to reconstruct paleo-seawater composition has been discussed by Song et al. (2021).

The presence of solid Li in extrasolar planetesimals and/or icy exomoons falling onto an old (age > 7 Gyr) white dwarf has been inferred from optical spectroscopy of WDJ1644-0449 obtained with the Goodman spectrograph at the 4.1 m Southern Astrophysical Observatory (SOAR) telescope by Kaiser et al. (2021). These authors interpret the presence of calcium, lithium, potassium, and sodium in the spectrum of the white dwarf as evidence for accretion of a rocky planetesimal onto the white dwarf. A similar argument has been made to explain the excess of Be (and by inference Li) on the surface of the white dwarfs GALEX J2339-0424 and GD 378 (Klein et al. 2021; Doyle et al. 2021; Trierweiler et al. 2022).

With a boiling point at 1615 K (1342°C, 2448°F), lithium is a solid on all objects of the solar system except for the Sun. It is also highly reactive, and hence on Earth it is found associated with other elements in mineral compounds.

The boiling point of lithium closely corresponds to the maximum strength of the Li I resonance line in the atmospheres of L dwarfs, which have effective temperatures in the range from 1700 K to 1500 K (Martín et al. 2022).

The behavior of the Cs I resonance line (atomic cesium) at 852.1 nm is similar to that of the Li I resonance line at 670.8 nm but the maximum is shifted to a lower effective temperature (later spectral type, Lodieu et al. 2015). Cesium has a lower boiling point (944 K, 671°C, 1240°F) than lithium.

The comparison between the behavior of the Cs I and Li I resonance lines confirm that in the coolest population of the Milky Way, the brown dwarfs and planetary mass objects, these elements become locked into molecules and dust grains. Direct detection of Li-bearing molecules or grains outside the solar system has not been achieved yet.

In the solar system the collision of comet D/Shoemaker-Levy 9 with Jupiter produced transient emission lines arising from lithium and other metals (Na, Fe, Ca,

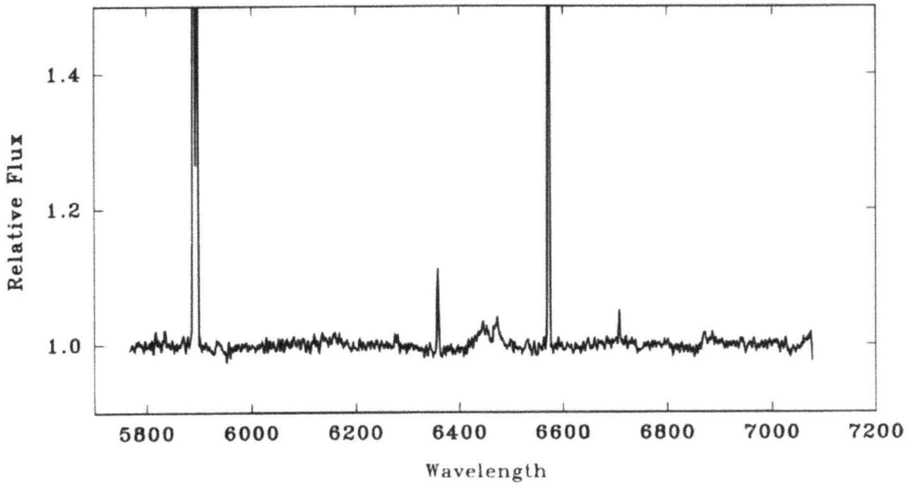

Figure 1.16. Emission spectrum obtained at Pico dos Dias of the hot plume ($T > 3000$ K; Carlson et al. 1997) originating from the collision of comet D/Shoemaker-Levy 9 with Jupiter. Emission features from the Na I resonance doublet at 589.00 and 589.59 nm, Fe I at 635.87 nm, Ca I at 657.28 nm and Li I at 670.78 nm are detected. Figure credit: Costa, R. D. D. et al. 1997 "Atomic emission lines in the spectrum observed after the impact between the L-fragment of Comet D/Shoemaker-Levy 9 and Jupiter" ApJ 485 382.

and K) that were observed with high-resolution optical spectroscopy at the 2 m telescope in Pic-du-Midi Observatory (Pyrenees, France) by Roos-Serote et al. (1995), the WHT by Moreno et al. (1995), and the 1.6 m telescope in Pico dos Dias (Brazil) by Costa et al. (1997).

The Pico dos Dias emission spectrum of the hot plume produced by the cometary impact is shown in Figure 1.16. All the constituents of the impacting body were vaporized. The atomic transitions observed in the spectrum are thought to come from the material ejected in the plume. Incomplete ablation of small plume grains can explain the observed emission lines (Moses 1997; Deming & Harrington 2001).

Observations of other comets have resulted in Na I and K I detections, but have failed to find Li I, such as the report by Preston (1967) on the Sun-grazing comet Ikeya-Seki (C/1965 S1) and the report by Fulle et al. (2013) on comet C/2011 L4 (Panstarrs).

A significant lithium depletion was inferred from the observations of comet C/2011 L4 by Fulle et al. (2013). These authors noted that more high-resolution spectra of comets are needed to understand whether comet nuclei are depleted in lithium, or the ejection mechanism of lithium from the nucleus and into the coma is less efficient than that of sodium and potassium.

Bibliography

Abia, C., et al. 1991, A&A, 245, L1
Anders, E., & Ebihara, M. 1982, GeCoA, 46, 2363
Anders, E., & Grevesse, N. 1989, GeCoA, 53, 197

Andersen, J., Gustaffson, B., & Lambert, D. L. 1984, A&A, 136, 65
Asplund, M., Lambert, D. L., Nissen, P. E., Primas, F., & Smith, V. V. 2006, ApJ, 644, 229
Balsiger, H., Geiss, J., Groegler, N., & Wyttenbach, A. 1968, E&PSL, 5, 17
Basri, G. 2021, An Introduction to Stellar Magnetic Activity (Series AAS-IOP Astronomy)
Bischoff, R., Mugrauer, M., Torres, G., et al. 2020, AN, 341, 908
Boesgaard, A. 2023, ApJ, 943, 40
Brown, H. 1949, RvMP, 21, 625
Burrows, A., & Sharp, C. M. 1999, ApJ, 512, 843
Campbell-White, J., Manara, C. F., Sicilia-Aguilar, A., et al. 2023, A&A, 673, A80
Carlsson, M., Rutten, R. J., Bruls, J. H. M. J., & Shchukina, N. G. 1994, A&A, 288, 860
Cayrel, R., Cayrel de Strobel, G., & Spite, M. 1988, The Impact of Very High S/N Spectroscopy on Stellar Physics IAU Symp. 132, 345
Cayrel, R., Spite, M., Spite, F., et al. 1999, A&A, 343, 923
Cayrel, R., Steffen, M., Chand, H., et al. 2007, A&A, 473, L37
Costa, R. D. D., de Freitas Pacheco, J. A., & Singh, P. D. 1997, ApJ, 485, 380
Deliyannis, C. P., Pinsonneault, M. H., & Duncan, D. K. 1993, ApJ, 414, 740
Deming, D., & Harrington, J. 2001, ApJ, 561, 468
Doyle, A. E., Desch, S. J., & Young, E. D. 2021, ApJ, 907, L35
Ebihara, M., Shinotsuka, K., Ozaki, H., & Oura, Y. 2000, Origins of Elements in the Solar System: Implications of Post-1957 Observations, ed. O. Manuel (Dordrecht: Kluwer Academic/Plenum) 289
Eddington, A. S. 1933, Natur, 132, 639
Feast, M. W. 1966, MNRAS, 134, 321
Franciosini, E., Randich, S., de Laverny, P., et al. 2022, A&A, 668, A49
Fulle, M., Molaro, P., Buzzi, L., & Valisa, P. 2013, ApJ, 771, L21
Gamow, G., & Landau, L. 1933, Natur, 132, 567
Gao, Q., Shi, J.-R., Yan, H.-L., et al. 2021, ApJ, 914, 116
Gauza, B., Béjar, V. J. S., Pérez-Garrido, A., et al. 2015, ApJ, 804, 96
Gharib-Nezhad, E., Marley, M. S., Batalha, N. E., et al. 2021, ApJ, 919, 21
Greenstein, J. L., & Richardson, R. S. 1951, ApJ, 113, 536
Harutyunyan, G., Steffen, M., Mott, A., et al. 2018, A&A, 618, A16
Herbig, G. H. 1964, ApJ, 140, 702
Hernández, J., Calvet, N., Pérez, A., et al. 2014, ApJ, 794, 36
Hobbs, L. M., & Thorburn, J. A. 1994, ApJ, 428, L25
Houtgast, J., & Hubenet, H. 1966, IAU Symp., (New York: Academic) 26, 38
Howk, J. C., Lehner, N., Fields, B. D., & Grant, M. J. 2012, Natur, 489, 121
Kaiser, B. C., Clemens, J. C., Blouin, S., et al. 2021, Sci., 371, 168
Kenyon, M. J., Jeffries, R. D., Naylor, T., Oliveira, J. M., & Maxted, P. F. L. 2005, MNRAS, 356, 89
King, A. 1916, ApJ, 44, 169
Klein, B. L., Doyle, A. E., Zuckerman, B., et al. 2021, ApJ, 914, 61
Kowkabany, J., Ezzeddine, R., Charbonnel, C., et al. 2022, arXiv:2209.02184
Kurucz, R. L. 1993, ATLAS 9 Stellar Atmosphere Program and 2 km s^{-1} grid (Kurucz CD Rom No. 13)
Lemoine, M., & Crane, P. 1995, The Light Elements Abundances (ESO/EIPC Workshop) 132, 350
Lind, K., Melendez, J., Asplund, M., Collet, R., & Magic, Z. 2013, A&A, 554, A96

Lodders, K. 1999, ApJ, 519, 793
Lodieu, N., Zapatero Osorio, M. R., Rebolo, R., et al. 2015, A&A, 581, A73
Lodieu, N., Rebolo, R., & Pérez-Garrido, A. 2018, A&A, 615, L12
Marley, M. S., Saumon, D., Guillot, T., et al. 1996, Sci., 272, 1919
Marschall, H. R., Wanless, V. D., Shimizu, N., et al. 2017, GeCoA, 207, 102
Martín, E. L., Rebolo, R., Magazzu, A., & Pavlenko, Ya. V. 1994, A&A, 282, 503
Martín, E. L., Lodieu, N., Pavlenko, Y., & Béjar, V. J. S. 2018, ApJ, 856, 40
Martín, E. L., Lodieu, N., & del Burgo, C. 2022, MNRAS, 510, 2841
Meléndez, J., Bergemann, M., Cohen, J. G., et al. 2012, A&A, 543, A29
Merchant, A. E., Bodenheimer, P., & Wallerstein, G. 1965, ApJ, 142, 790
Miles, B. E., Biller, B. A., Patapis, P., et al. 2023, ApJ, 946, L6
Moreno, F., Muñoz, O., Molina, A., et al. 1995, GeoRL, 22, 1609
Morton, D. C. 2003, ApJS, 149, 205
Moses, J. I. 1997, JGR, 102, 21619
Mott, A., Steffen, M., Caffau, E., et al. 2017, A&A, 604, A44
Mugrauer, M., Avila, G., & Guirao, C. 2014, AN, 335, 417
Nichiporuk, W., & Moore, C. B. 1970, E&PSL, 9, 280
Pavlenko, Y. V., & Magazzù, A. 1996, A&A, 311, 961
Pavlenko, Ya V, Jones, H. R. A., Martín, E. L., et al. 2007, MNRAS, 380, 1285
Pavlenko, Ya V., Jenkins, J. S., Ivanyuk, O. M., et al. 2018, A&A, 611, A27
Pepe, F., Cristiani, S., Rebolo, R., et al. 2021, A&A, 645, A96
Preston, G. W. 1967, ApJ, 147, 718
Rebolo, R., Crivellari, L., Castelli, F., Foing, B., & Beckman, J. E. 1986, A&A, 166, 195
Roos-Serote, M., Barucci, A., Crovisier, J., et al. 1995, GeoRL, 22, 1621
Ryan, J. G., & Langmuir, C. H. 1987, GeCoA, 51, 1727
Sanford, R. F. 1950, APJ, 111, 262
Smith, V. V., Lambert, D. L., & Nissen, P. E. 1993, ApJ, 408, 262
Smith, V. V., Lambert, D. L., & Nissen, P. E. 1998, ApJ, 506, 405
Soderblom, D. R., Jones, B. F., Balachandran, S., et al. 1993, AJ, 106, 1059
Song, Y., Zhang, X. (Yvon), Bouchez, J., et al. 2021, GeCoA, 313, 340
Strassmeier, K. G., & Steffen, M. 2022, AN, 343, e20220036
Urey, H. C. 1952, PhRv, 88, 248
Urey, H. C. 1964, RvGSP, 2, 1
Taylor, S. R. 1960, GeCoA, 20, 85
Tera, F., Eugster, O., Burnett, D. S., & Wasserburg, G. J. 1970, GeCAS, 1, 1637
Trierweiler, I. L., Doyle, A. E., Melis, C., Walsh, K. J., & Young, E. D. 2022, ApJ, 936, 30
Tsuji, T., Ohnaka, K., Aoki, W., & Nakajima, T. 1996, A&A, 308, L29
Wang, E. X., Nordlander, T., Asplund, M., et al. 2021, MNRAS, 500, 2159
Wang, E. X., Nordlander, T., Asplund, M., et al. 2022, MNRAS, 509, 1521
Yan, H.-L., Shi, J-R, Zhou, Y.-T., et al. 2018, NatAs, 2, 790

Lithium Across the Universe

Eduardo Martín

Chapter 2

Creation, Destruction and Preservation of Lithium

This chapter introduces the reader to the processes that create, destroy, and preserve lithium in the cosmos.

Lithium stands out from the other elements in the rich variety of processes and environments that play a role in shaping its behavior over cosmic distance and time.

A first glance at the diversity of lithium's origins as compared to other chemical elements is illustrated in Figure 2.1. This simple diagram already suggests that lithium has the most complex and diverse history of all chemical elements. In this chapter the different sources of lithium shown in Figure 2.1 will be presented.

The most abundant lithium isotope, ^{7}Li, is a direct decay product of the beta-unstable ^{7}Be atom by the nuclear reaction ^{7}Be(e^{-},n)^{7}Li. The half-life on Earth of the ^{7}Be decay by electron capture is 53 days.

Net production of ^{7}Li occurs when ^{7}Be is made by the nuclear reaction ^{3}He(^{4}He, γ) ^{4}He at a temperature of about 10^{7} K, the material then cooling down faster than the ^{7}Be decay timescale to a temperature below 3×10^{6} K so that destruction of lithium by the nuclear reactions ^{7}Li(p, γ)^{8}Be and ^{7}Li(p, α)^{7}He is avoided.

Such rapid cooling requires efficient mass transport, which may take place during explosive events (the Big Bang and novae) or during unstable phases in stellar evolution (such as the helium flash).

Other nuclear reactions that lead to lithium production include the collisions of protons and α particles (^{4}He nuclei) with ^{12}C and ^{13}C (carbon isotopes), ^{14}N (nitrogen), ^{16}O (oxygen) and ^{20}Ne (neon) nuclei, and the α-α reaction. These nuclear reactions may happen in high-energy transients, X-ray outbursts in binaries, stellar flares, active galactic nuclei, and supernovae.

Lithium existence had a quick start with the Big Bang. It was the heaviest of the light elements created at the origin of the universe.

doi:10.1088/2514-3433/acd8bbch2

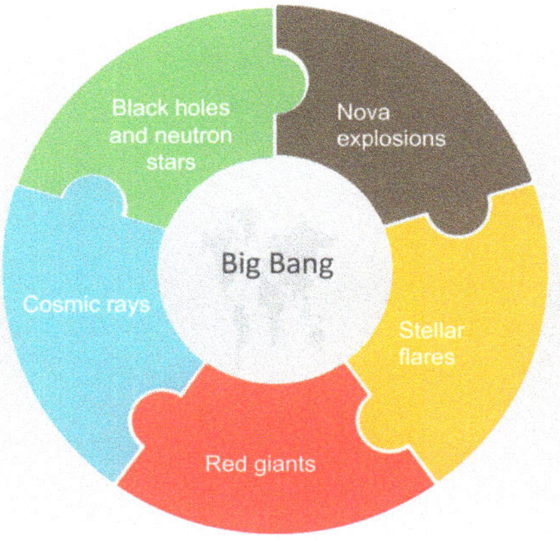

Figure 2.1. Lithium has the most diverse origins of all the chemical elements, most of which are created by just one production mechanism. In contrast, lithium has at least six, and possibly more, such mechanisms. The size of the pieces of the puzzle does not reflect the relative contribution of each lithium source.

The abundance of lithium in the oldest stars of the Galaxy is systematically lower than the primordial lithium abundance predicted by Big Bang nucleosynthesis. This discrepancy is known as the cosmological lithium problem.

The abundance of lithium in the solar system and in the youngest stars of the Galaxy is systematically higher than the primordial one, indicating that there has been lithium enrichment during the evolution of the Milky Way.

Sources of Galactic lithium enrichment that are considered in this chapter include red giants via the Cameron–Fowler mechanism, explosions such as novae and supernovae, compact relativistic objects such as black holes and neutron stars, and solar and stellar flares.

The lithium abundance on the surface of the Sun is much lower than that in the solar system, suggesting that there are processes of lithium destruction inside stars that can observed on the surface.

Lithium is very efficiently depleted by convection in the most numerous stars in the Galaxy (red dwarfs), but there is a limit to lithium destruction in the interiors of very low-mass objects. This limit is located just below the substellar frontier that separates stars from brown dwarfs. Lithium gets preserved forever in brown dwarfs with masses below the limit for thermonuclear lithium burning.

2.1 Big Bang Lithium Synthesis

The standard cosmological scenario that explains the origin of the universe is the Big Bang, a nickname given by Sir Fred Hoyle, one of its foremost detractors.

In this framework, the universe started so hot (10^{32} K) and small that all four fundamental forces (electromagnetic, weak and strong nuclear interactions, and gravity) were intimately bound and acted together (Grand Unified Theory).

The beginning of the nuclear reactions after the Big Bang was delayed by about 1s by the D bottleneck. Then, during the first few minutes nuclear reactions took place producing the two main isotopes of helium (^{3}He and ^{4}He) and traces of lithium nuclei (only the ^{7}Li isotope), see Peebles (1966), Wagoner (1967, 1973), Kajino & Boyd (1990), Bertulani & Kajino (2016), and Pitrou et al. (2018).

The chain of nuclear reactions was halted when the infant universe (just about 3 minutes old) expanded and cooled fast because its temperature was inversely proportional to its size. Atoms formed when the universe recombined at a redshift of 1000.

Standard Big Bang nucleosynthesis (SBBN) is a very successful and elegant theory that is based on the standard model of particle physics and the theory of general relativity. The main simplifying assumptions are: kinetic equilibrium (the Maxwell–Boltzmann distribution of particles), the cosmological principle (homogeneous and isotropic universe), and the non-relevance of both dark energy and dark matter (because of the weakness of gravity with respect to the other forces).

The ratio of baryons (the most familiar of which are the neutrons and protons that contain most of the atomic mass) to photons ($\eta = N_{\text{baryon}}/N_{\text{photon}}$) is constant everywhere in the universe, not only at the time of SBBN, but also throughout the adiabatic expansion (inherent to an isolated system that does not exchange heat with its surroundings) and up to the present day. Photons are created by stars but that is not a significant contribution in a cosmic scale.

The nuclear reactions that took place in SBBN produce a small but significant amount of the dominant ^{7}Li isotope, but not of the rarer ^{6}Li isotope. The lithium isotopic ratio coming from SBBN is predicted to be $R_{\text{I}} < 10^{-4}$.

The ^{7}Li produced in the SBBN came from fusion of tritium atoms with ^{4}He atoms and from the production of ^{7}Be that recombined with free electrons or neutrons. The ^{7}Be atoms came from fusion of the main two isotopes of helium (^{3}He and ^{4}He).

The ^{7}Li atoms can also recombine with free protons to produce two ^{4}He atoms, but the nuclear reactions stopped suddenly because the universe quickly cooled to a temperature below the threshold needed to penetrate the Coulomb barriers of the nuclei.

It has been estimated that 1.1 seconds after the Big Bang, the temperature was 10^{10} K, dropping to $<10^7$ K only 5 hours later. The chemical composition of the primordial universe was frozen after the initial moments of the Big Bang.

The prevalence of H and He originating from SBBN has remained almost unchanged until the present time. The ratio of ^{4}He to H has increased in the Galaxy from 0.25 to 0.28 by mass.

The chain of nuclear reactions in the primordial universe that were involved in primordial nucleosynthesis includes over 400 reactions, but only a few were really important (Coc & Vangioni 2010; Coc et al. 2012; Bertulani & Kajino 2016; Pitrou et al. 2018). Some radioactive isotopes of lithium such as ^{8}Li, ^{9}Li, and ^{11}Li were produced but these have very short lifetimes and did not survive. The main reactions are shown in Figure 2.2.

It took about 400,000 yr after the Big Bang for the first chemical reactions to start taking place, including the formation of lithium-bearing molecules such as LiH (Lepp & Shull 1984; Galli & Palla 2013).

The critical reaction cross sections have been measured and compiled by several teams in different laboratories such as Lawrence Berkeley Laboratory, Indiana

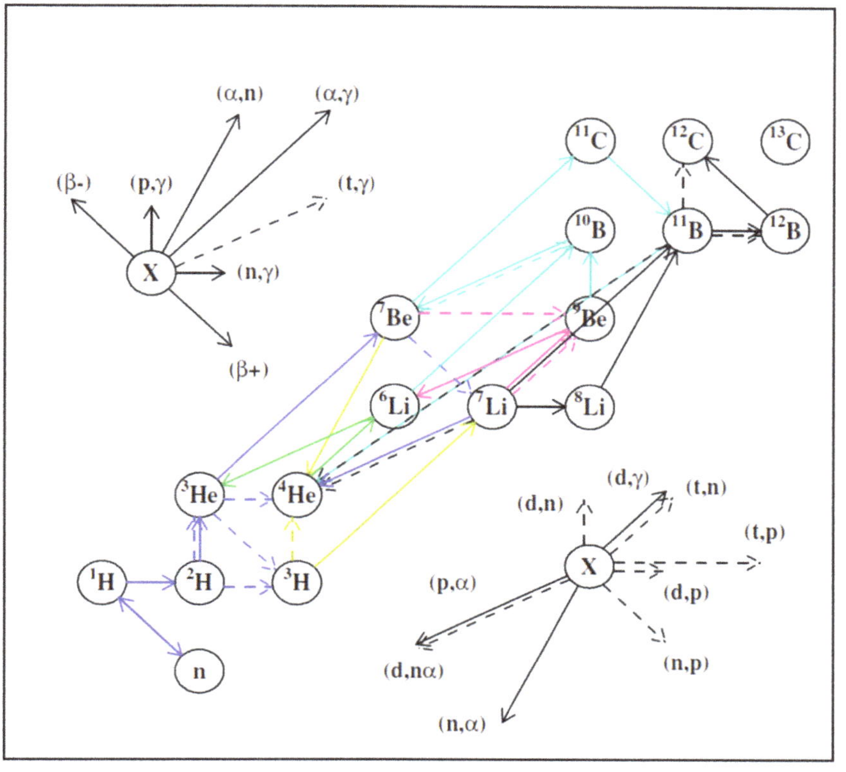

Figure 2.2. The most important nuclear reactions for SBBN. Blue arrows indicate the reactions leading to ^{7}Li. Figure credit: Pitrou, C. et al. 2018 "Precision big bang nucleosynthesis with improved Helium-4 predictions" PhR 754 1.

University, Michigan State, Orsay, and the University of Washington (Austin 1981; Read & Viola 1984). The production rates of lithium and other light elements were calculated by Walker et al. (1985).

The net results of SBBN are illustrated in Figure 2.3. As can be seen, the predicted abundances of the main chemical elements other than H produced in the Big Bang (^{3}He, ^{4}He, D and ^{7}Li) are tightly constrained by the value of η measured by the Planck space mission. The abundance of ^{4}He is very sensitive to the expansion rate too.

While the SBBN values for ^{3}He, ^{4}He, and D coincide within the uncertainties with the observed measurements, ^{7}Li is the exception.

SBBN predicts a constant lithium abundance in the early universe that is dubbed **primordial**. This abundance is about a factor of three higher than the upper envelope of Li abundances measured in the oldest and most metal-poor stars in the Galaxy (the so-called Spite plateau). This disagreement is known as the cosmological lithium problem.

The lithium abundance in meteorites and in the youngest stars (T Tauris) of the Milky Way is dubbed the **cosmic** lithium abundance, and it is about four times larger than the primordial one. This difference between the cosmic and primordial lithium

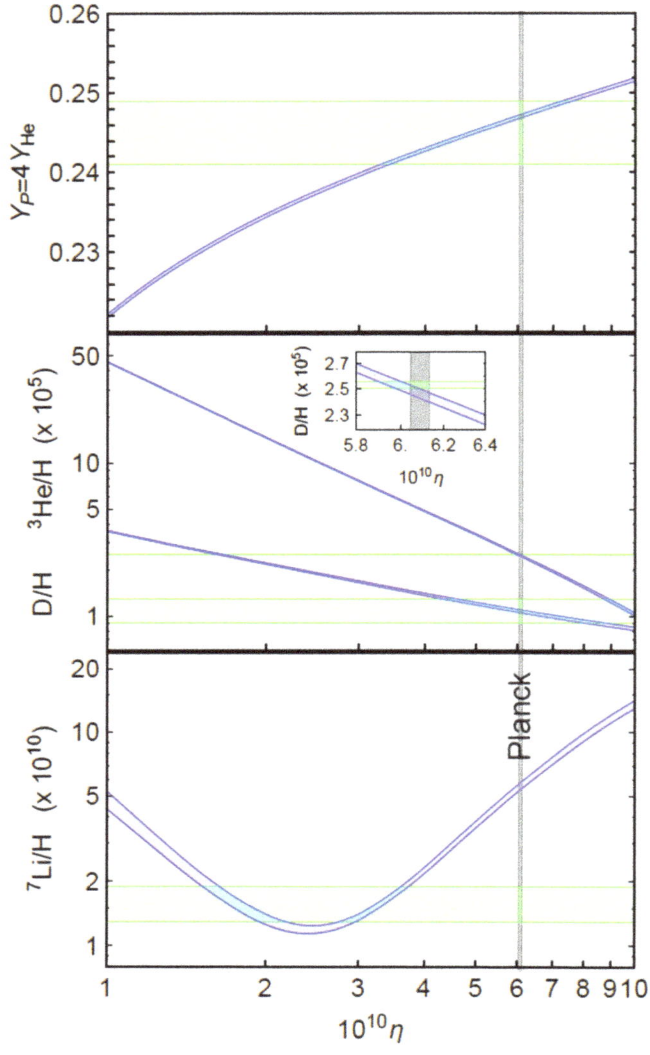

Figure 2.3. From top to bottom: ^{4}He, ^{3}He, D, and ^{7}Li yields from SBBN as a function of η (the ratio of baryons to photons in the universe). Note that as the density of baryonic matter increases to the right of the figure, the amount of ^{4}He and ^{7}Li increase, while that of ^{3}He and D decrease. The thickness of the blue curves represents 1σ uncertainties from neutron decay lifetimes and nuclear reaction rates. A vertical gray line indicates the value of η determined by analysis of the cosmic microwave background (CMB) data taken by the Planck space mission. Horizontal green bands denote the observational constraints on each element. Figure credit: Pitrou, C. et al. 2018 “Precision big bang nucleosynthesis with improved helium-4 predictions” PhR 754 1.

abundances indicates that there are sources of net lithium enrichment of the ISM throughout the evolution of the Galaxy.

The primordial abundances of the elements obtained from the Big Bang could have changed had there been are injections of extra particles during the first minutes of the universe. Such particles could have come from primordial black holes small and hot enough for significant high-energy emission of Hawking radiation, or if inhomogeneous stochastic magnetic fields were able to produce significant deviations from the Maxwell–Boltzmann distribution of particles (Luo et al. 2019).

It has been argued that the solution to the cosmological lithium problem is likely to come from lithium depletion in halo stars (Fields & Olive 2022). These authors have attempted to correct the lithium abundances in extremely metal-poor stars using estimates for lithium depletion obtained from recent limits on the lithium isotopic ratio.

Figure 2.4 summarizes their main result. They obtain corrected lithium abundances that tend to be higher that the prediction from SBBN, and there is a large

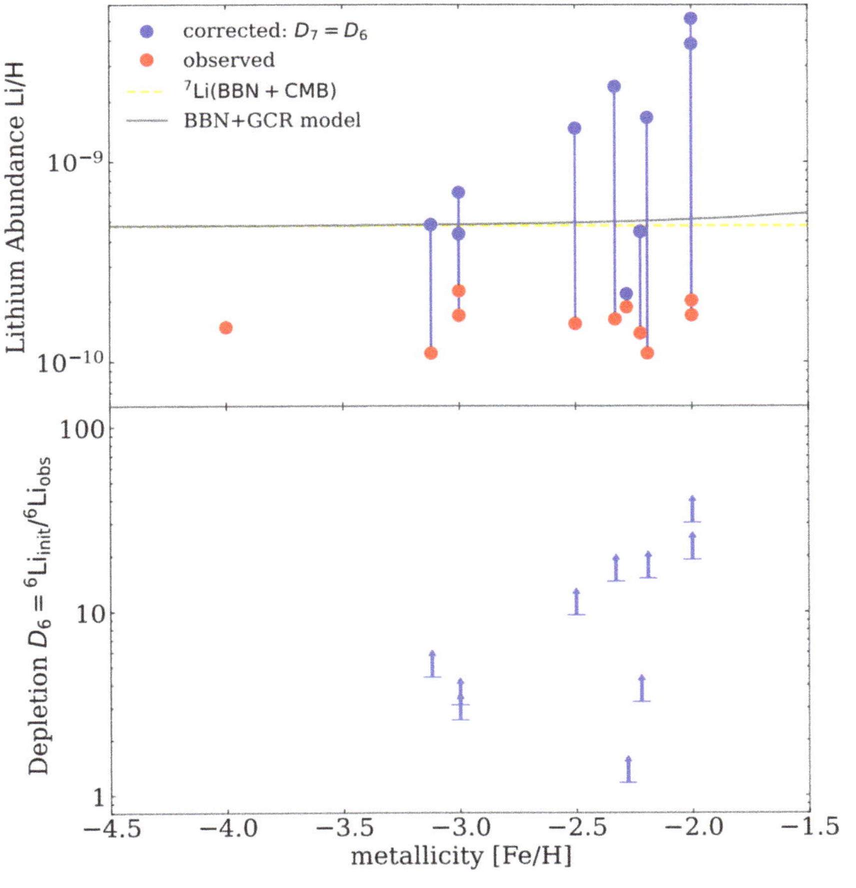

Figure 2.4. Lithium abundance as a function of iron abundance in very metal-poor stars. Red circles denote observed data, blue circles denote corrected values using the ^{6}Li depletion values shown in the lower panel. Primordial lithium abundance level from SBBN is shown as a yellow horizontal dashed line. A model that incorporates the contribution of nucleosynthesis by galactic cosmic rays is shown as a black curve. Figure credit: Fields, B. D., & Olive, K. A. 2022 "Implications of the non-observation of ^{6}Li in halo stars for the primordial ^{7}Li problem" JCAP 10 78.

dispersion in the results. This work does not provide conclusive evidence of depletion of ^{7}Li in halo stars. The jury is still out on the resolution of the cosmological lithium problem.

2.2 Galactic Sources of Lithium

Since ^{6}Li is not expected to have been created in SBBN, and the primordial abundance of ^{7}Li predicted by SBBN is about a factor of four below what is found in the Earth's crust and meteorites, there is a need to identify sources of net lithium production in the Galaxy.

Of every 108 lithium atoms in the Earth, 33 are thought to be ^{7}Li atoms originating from the Big Bang, 67 are ^{7}Li atoms coming from other creation processes in the Galaxy and 8 are ^{6}Li atoms.

There is no shortfall of possible sources of lithium creation in the Galaxy, and this section provides an overview of them.

2.2.1 The Cameron–Fowler Mechanism

Thermonuclear production of lithium is possible under non-equilibrium conditions in stellar interiors, such as those that may happen during the latest stages in the evolution of He-burning giant stars (Cameron 1955).

Very strong Li I features were detected in some evolved stars (McKellar 1940; Sanford 1950; Wallerstein & Conti 1969; Boesgaard 1970; Wallerstein & Sneden 1982), suggesting lithium abundances as high as, or even exceeding, those found in meteorites and in the youngest (T Tauri) stars, and providing evidence for the presence of unusual events that could lead to lithium production.

Cameron & Fowler (1971) proposed that a mechanism of Li production could operate during the helium shell flashes that take place after the He in the stellar core has been exhausted and the star draws energy from H- and He-burning shells. Efficient mixing takes place during this evolutionary stage via deep circulation of matter (Sackmann & Boothroyd 1999).

This lithium production process is widely known as the Cameron–Fowler mechanism, and is thought to be responsible for the high Li abundances observed in giant stars with masses >2 $M_{\odot}$ (Kumar et al. 2020; Singh et al. 2021; Deepak & Lambert 2021). For low-mass giants in the post-He-flash phase or in the lithium flash phase, lithium enhancement has been studied by Kumar & Reddy (2020) and Palacios et al. (2001).

Yan et al. (2018) developed simulations to check the plausibility that the lithium superabundance that they detected in the red giant TYC429-2097-1 using LAMOST could originate from the Cameron–Fowler mechanism. They coupled models of the red giant branch evolutionary phase using the MESA-code with a parametrization of the meridional circulation in the radiative zone. They showed that at the end of a meridional circulation cycle lasting 450 Myr the surface lithium abundance in the red giant can be enhanced to the level observed in TYC429-2097-1. Nevertheless, this evolutionary phase of Li production must be short compared with the overall

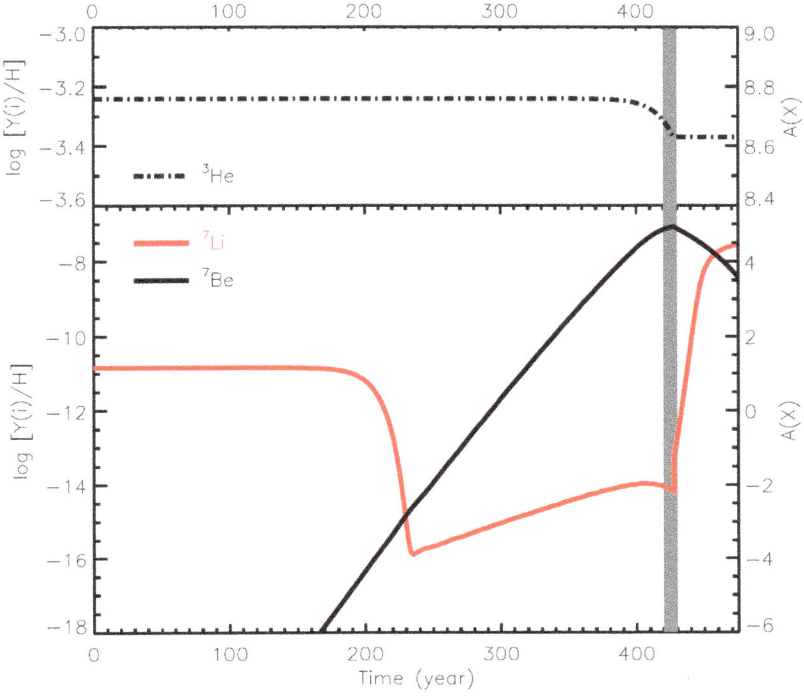

Figure 2.5. Fall and rise of lithium abundance coming from ^{3}He thermonuclear burning and asymmetric mass flow in the interior of a red giant. The thick vertical band separates the downward from the upward motion. The superlithium surface abundance is achieved quickly after the onset of the upward motion. Figure credit: Yan, H.-L. et al. 2018 "The nature of the lithium enrichment in the most Li-rich star" NatAs 2790.

lifetime of red giants, because super-Li-rich giants (defined as those with log N(Li) > 3.5) are rare (<0.1%) among the total population of red giants (Figure 2.5).

2.2.2 Explosive Nucleosynthesis

At the end of their lives, one of the components of binaries of intermediate mass becomes a white dwarf and accretes mass from the other star via an accretion disk.

Recurrent outbursts take place during the so-called symbiotic nova phase. Lithium has been observed in one of these systems (T Coronae Borealis; Woodward et al. 2020).

Eventually, when the accretion disk around the white dwarf reaches a critical mass and becomes unstable, there is an energetic blast known as a classical nova explosion, in which the white dwarf ejects its outer envelope and enriches the insterstellar medium with freshly synthesized nuclei, some of which are predicted to be ^{7}Li atoms produced by the fusion of ^{3}He with ^{4}He nuclei (Hernanz et al. 1996; Jose 2016).

Since recurrent and classical nova explosions are quite frequent events, they held promise of a significant source of lithium enrichment in the Galaxy.

Spectroscopic monitoring of several novae has been carried out with high-resolution spectrographs at the 8 m class telescopes VLT and Subaru at MKO. The spectral analysis of nova high-resolution spectra strongly supports the presence of clear ^{7}Be features that are displaced in wavelength in accordance with the velocity of the ejected material (Tajitsu et al. 2015; Izzo et al. 2015; Molaro et al. 2016, 2020, 2022; Selvelli et al. 2018).

The ^{7}Be atoms are expected to capture e^- and yield ^{7}Li atoms on a timescale of about two months, but no fresh lithium has been detected yet in the nova spectra, even though some events remain bright enough over longer timescales. The problem might be that most of the lithium could be ionized owing to the high temperatures in the nova ejecta, and the Li I resonance doublet might not be strong enough (Molaro et al. 2016; Shore & De Gennaro Aquino 2020).

Modeling of the line formation in nova ejecta using the PHOENIX stellar atmosphere code (Hauschildt et al. 1997), has suggested that the Li and Be overabundances in the spectra of novae should be treated with caution (Shore & De Gennaro Aquino 2020). These authors computed synthetic spectra of novae that do not confirm overabundances of beryllium and lithium in novae spectra. Clearly more observational and theoretical investigations on this topic are required. Detection of gamma-ray emission lines expected from the decay of freshly produced ^{7}Be and ^{7}Li would allow a precise determination of how much lithium is made in the nova explosions.

Not as frequent as nova explosions, but a lot more energetic, supermassive stars and supernovae have been postulated to produce lithium (Colgate 1973; Bodansky et al. 1975; Meneguzzi & Reeves 1975; Norgaard & Fricke 1976; Domogatskii et al. 1978; Starrfield et al. 1978; Boffin et al. 1993; Moriya 2018; Kusakabe et al. 2019; Romano et al. 2021).

Explosive nucleosynthesis events could leave long-lasting imprints on the chemical composition of supernova remnants or on the surface of stars that were sufficiently close to the explosion, but their role as the main source of lithium in the Galaxy has been questioned by models of galactic chemical evolution (Prantzos 2012).

In 1572 Tycho Brahe observed a new star as bright as Venus, which is now known to have been due to a Type I supernova. Such explosions are the end point of a stellar remnant called a white dwarf, which has accreted mass from a companion to the point that it exceeds its limiting mass and is completely destroyed in the blast.

A G-type subgiant named Tycho G has been suggested as the culprit companion of the 1572 explosion (Ruiz-Lapuente et al. 2004) because of its proximity to the location of Tycho's supernova and its abnormally high velocity.

The surface chemical abundances of Tycho G were obtained by González Hernández et al. (2009) using high-resolution optical spectra obtained with the HIRES spectrograph at the 10 m Keck I telescope in Hawaii.

They reported slight overabundances of Li and Ni with respect to what might be expected for a normal star in the disk of the Galaxy. The abundances of many other elements such as Al, Cr, Mg, O, Si, and Ti were consistent with the expected contamination by the supernova explosion.

Ruiz-Lapuente et al. (2004) concluded that their chemical abundance analysis supported the scenario in which Tycho G was the companion that donated mass to the white dwarf until it exploded, and that it is the survivor that carries the imprint of the chemical elements that were ejected by the supernova. However, the argument that lithium was actually created in the supernova is not compelling because the abundance in Tycho G is high, but not higher than those seen in other subgiant stars (Randich et al. 1999).

Lux et al. (2021) searched for lithium in runaway stars in a dozen Galactic supernova remnants, including historic ones such as Cassiopeia A and the Crab Nebula, but found no case of an overabundance with respect to stellar associations and open clusters (see Figure 2.6).

Some supernova explosions result from binary evolution and may leave compact remnants that become neutron stars or black holes. When these remnants are gravitationally bound to stars they form X-ray binary systems. Searches for lithium in the stellar components of these systems have produced a variety of results, consisting of mostly positive detections implying abnormally high lithium abundances.

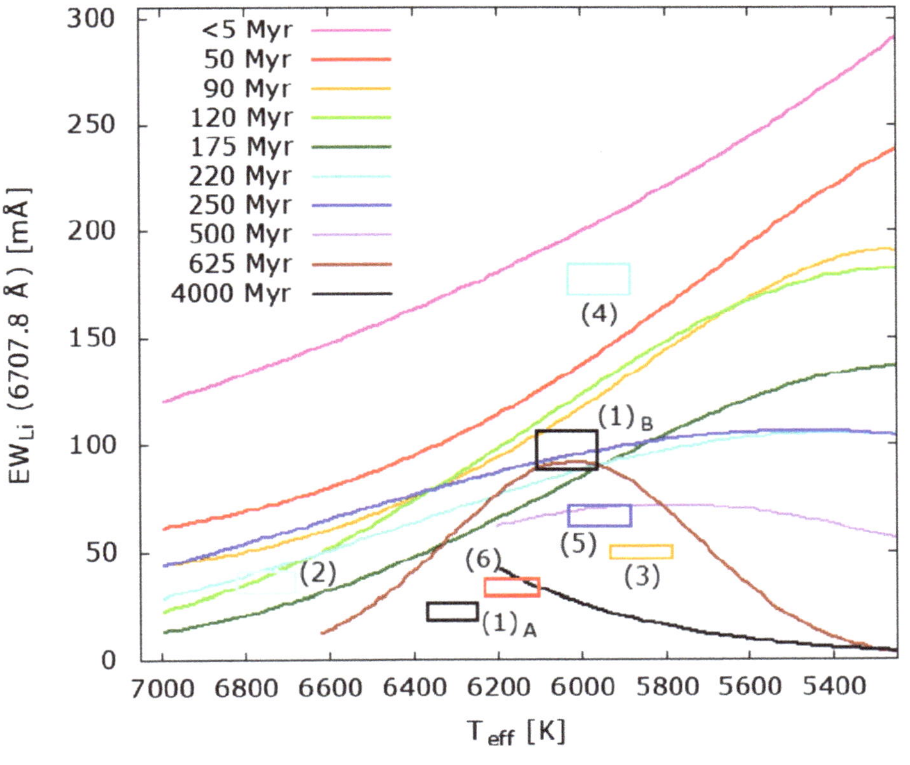

Figure 2.6. Li I equivalent widths versus effective temperatures for runaway star candidates possibly associated with supernova remnants (boxes). The curves for different ages are fits to the Li I data in stellar associations and open clusters. Figure credit: Lux, O. et al. 2021 "A search for runaway stars in 12 Galactic supernova remnants" AN 342 553.

Strong Li I resonance doublets were detected in the black hole binaries V404 Cyg (Martín et al. 1992), A0620-00 (Marsh et al. 1994), GS2000+25 (Harlaftis et al. 1996), Nova Muscae (Martín et al. 1996), and Nova Scorpii 1994 (González Hernández et al. 2008), and in binaries that contain neutron stars such as Cen X-4 (Martín et al. 1994a; González Hernández et al. 2005), PSR J1740-5340 in the globular cluster NGC6397 (Sabbi et al. 2003), and PSR J1023+0038 (Shahbaz et al. 2022).

The lithium abundances inferred from the spectroscopic analysis of all those stellar companions to compact relativistic objects are higher than those of typical stars of the same metallicity, spectral type and surface gravity, but only one case (PSRJ1023+0038) is higher than the cosmic value.

No lithium was detected in the F8IV secondary of the neutron star binary Cyg X-2 (Suárez-Andrés et al. 2015).

While lithium production mechanisms in supernovae ejecta or in the accretion disks around the compact relativistic objects are possible from the energetic point of view (Martín et al. 1994b), the high lithium abundances in some of these stars may be explained by the inhibition of lithium depletion due to tidally locked rotation evolution (Maccarone et al. 2005).

The very high lithium abundance in the companion of the pulsar PSR J1023 +0038 (Shahbaz et al. 2022) is the smoking gun of lithium production in the vicinity of black holes and neutron stars. In the next subsection we take a closer look at the lithium abundances in stars orbiting compact relativistic objects.

2.2.3 Compact Relativistic Objects

The term compact relativistic object includes black holes and neutron stars, both of which share similar properties in terms of how they behave under high accretion rates (Vincentelli et al. 2023). Their gravitational wells are so powerful that they can launch relativistic jets and accelerate particles to energies high enough for nuclear reactions to take place in their surroundings.

A summary of the lithium abundances measured in stars orbiting compact relativistic objects is shown in Figure 2.7. It is noteworthy that some of these stars have high lithium abundances, which suggests that either they are very young (unlikely given that they belong to evolved binary systems), or that they have been enriched by some mechanism.

The most lithium-rich of these systems is the early G-type star in the binary millisecond pulsar PSRJ 1023+0038 (Shahbaz et al. 2022), although it has been noted that other systems show lithium enrichment with respect to what could be expected for a normal star or for binary systems where the compact object is a white dwarf instead of a black hole or a neutron star (Martín et al. 1995; Sabbi et al. 2003).

The chain of nuclear reactions that could take place in the accretion disk around a black hole has been considered in detail by Guessoum & Kazanas (1999) and it is reproduced below (Table 2.1).

These authors note that reactions 4.1, 5.1, 5.2, 6.1, and 6.2 (breakup of Be and Li atoms by neutrons and protons) have no threshold energy to overcome because they

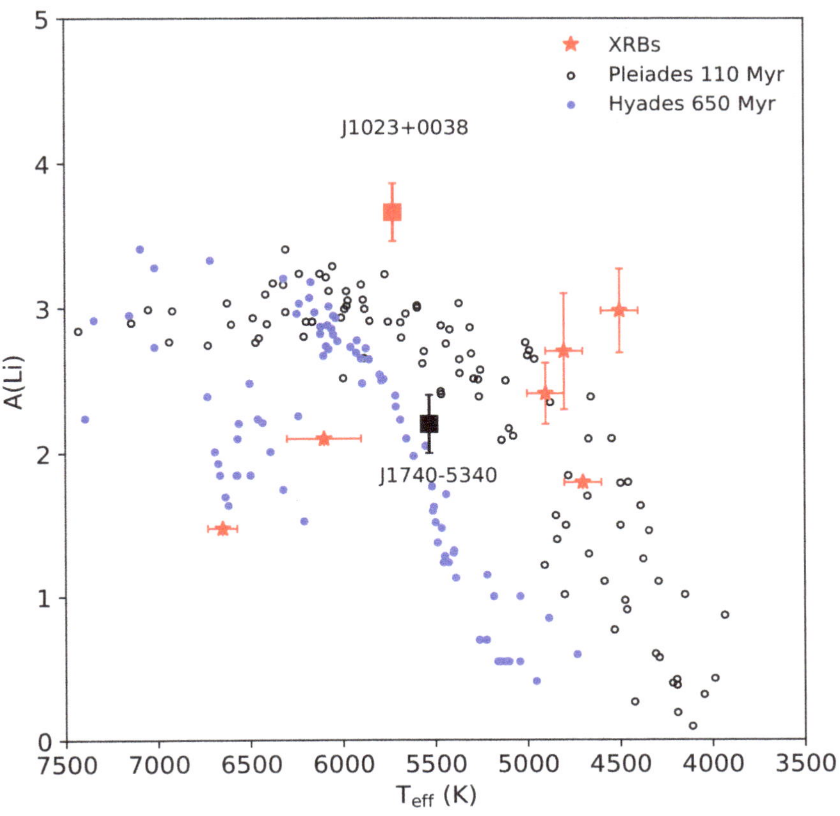

Figure 2.7. Lithium abundance versus effective temperature in X-ray binaries (XRB) compared with two well-known open clusters (the Hyades, age = 650 Myr; and the Pleiades, age = 110 Myr). Note the lithium overabundance in PSR J1023+0038. Figure credit: Shahbaz, T. et al. 2022 "The peculiar chemical abundance of the transitional millisecond pulsar PSR J1023+0038—Li enhancement" MNRAS 513 71.

are exothermic nuclear reactions, and that taken altogether they are more likely to destroy those light elements rather than create them in the disks around black holes, in opposition to what was proposed by Yi & Narayan (1997).

Spallation reactions, rather than fusion reactions of α particles (reaction 2.1a), are considered by Guessoum & Kazanas (1999) as a more likely source of fresh lithium synthesis in the surface of the stellar companions of black holes and neutron stars.

In the next subsection we take a closer look at spallation reactions.

Both the de-excitation of freshly produced Be and Li atoms by α-α reactions and the collisions of those atoms with protons produce gamma-ray lines.

Using SIGMA/GRANAT, a transient detection of a gamma-ray emission line at 481 ± 22 keV during an outburst of black hole X-ray binary Nova Muscae was reported by Goldwurm et al. (1992) and Sunyaev et al. (1982), which is very close to the energy expected from the de-excitation of ^{7}Li atoms (478 keV). The mass-donor

Table 2.1. Nuclear Reactions in an Accretion Disk Around a Black Hole According to Guessoum & Kazanas (1999).

Reaction Number	Nuclear Reactions	Threshold Energies (MeV)
1.1	$p + \alpha \rightarrow {}^3\text{He} + d$	18.35
1.2	$\rightarrow {}^3\text{H} + p + p$	19.81
1.3	$\rightarrow {}^3\text{He} + p + n$	20.58
2.1a	$\alpha + \alpha \rightarrow {}^7\text{Be} + n$	18.99
2.1b	$\rightarrow {}^7\text{Be}^* + n$	19.42
2.2a	$\rightarrow {}^7\text{Li} + p$	17.35
2.2b	$\rightarrow {}^7\text{Li}^* + p$	17.83
2.3a	$\rightarrow {}^6\text{Li} + p + n$	24.60
2.3b	$\rightarrow {}^6\text{Li}^* + p + n$	28.16
2.4	$\rightarrow \alpha + {}^3\text{He} + n$	20.58
2.5	$\rightarrow \alpha + {}^3\text{H} + n$	19.81
2.6	$\rightarrow \alpha + d + d$	23.85
3.1	$\alpha + n \rightarrow {}^3\text{H} + d$	17.59
3.2	$\rightarrow {}^3\text{H} + p + n$	19.81
3.3	$\rightarrow {}^3\text{He} + n + n$	20.58
4.1	$p + {}^7\text{Li} \rightarrow \alpha + \alpha$	−17.35
4.2	$\rightarrow {}^7\text{Be} + n$	1.65
5.1	$n + {}^7\text{Be} \rightarrow \alpha + \alpha$	−18.99
5.2	$\rightarrow {}^7\text{Li} + p$	−1.65
6.1	$p + {}^6\text{Li} \rightarrow {}^4\text{He} + {}^3\text{He}$	−4.02
6.2	$n + {}^6\text{Li} \rightarrow {}^4\text{He} + {}^3\text{H}$	−4.78
7.1	$p + {}^3\text{He} \rightarrow d + p + p$	5.49
7.2	$\rightarrow 3p + n$	7.72
8.1a	$\alpha + {}^3\text{He} \rightarrow {}^6\text{Li} + p$	4.02
8.1b	$\rightarrow {}^6\text{Li}^* + p$	7.58
8.2	$\rightarrow {}^3\text{He} + {}^3\text{He} + n$	19.81
8.3	$\rightarrow {}^3\text{He} + {}^3\text{H} + p$	20.58
8.4	$\rightarrow {}^3\text{He} + d + d$	23.85
9.1	$p + {}^3\text{H} \rightarrow {}^3\text{He} + n$	0.76
9.2	$\rightarrow d + d$	4.03
10.1a	$\alpha + {}^3\text{H} \rightarrow {}^6\text{Li} + n$	4.78
10.1b	$\rightarrow {}^6\text{Li}^* + n$	8.35
10.2	$\rightarrow {}^3\text{He} + {}^3\text{H} + n$	20.58
10.3	$\rightarrow {}^3\text{H} + {}^3\text{H} + p$	19.81
10.4	$\rightarrow {}^3\text{H} + d + d$	23.85
11.1	$\alpha + d \rightarrow {}^3\text{He} + {}^3\text{H}$	14.32
11.2	$\rightarrow \alpha + p + n$	2.22
12.1	$n + {}^3\text{He} \rightarrow {}^3\text{H} + p$	−0.76
13.1	$p + n \rightarrow d + \gamma$	−2.22

Negative threshold energies indicate exothermic nuclear reactions.

star in the Nova Muscae system has a high surface lithium abundance (Martín et al. 1996).

An emission gamma-ray feature was detected in an energy range compatible with a blend of ^{7}Be (429 keV) + ^{7}Li (478 keV) de-excitation lines during an outburst of the black hole binary V404 Cygni that was observed using the SPI spectrometer on the International Gamma-ray Astrophysical Laboratory (INTEGRAL), although it was interpreted as redshifted electron positron pair production or positronium three photon annihilation (Siegert et al. 2016).

Siegert et al. recognized that their interpretation of the gravitational redshift of a positron annihilation feature at 511 keV is expected to produce a much broader line than was actually observed. This inconsistency between the energy of the line and its width may be solved in terms of a scenario involving gamma-ray emission produced by freshly synthesized lithium during the outburst.

Enhanced lithium production by large fluxes of energetic particles in the vicinity of the supermassive black holes in active galactic nuclei (AGN) has been predicted (Boyd & Fencl 1991; Crosas & Weisheit 1996).

A search for the hyperfine lines of Li I in absorption at 732.153 MHz and at 803.504 MHz toward the Galactic Center was carried out using the 43 m radio telescope at Green Bank Observatory (Lubowich et al. 1998). The lines were not detected and the authors reported upper limits on the lithium column density toward the dense Sgr A molecular clouds, which they interpreted as evidence of a lack of accretion activity in the supermassive black hole at the center of the Galaxy.

2.2.4 Spallation Reactions

When high-energy massive particles, such as neutrons, protons or α particles hit C, N, O, and Ne (carbon, nitrogen, oxygen and neon) nuclei, they can break them apart and produce lighter elements. These nuclear reactions are thought to take place in low-density high-energy astrophysical environments such as cosmic ray interaction with the ISM (Reeves et al. 1970; Meneguzzi et al. 1971).

Clayton & Dwek (1976) considered possible lithium production around pulsars. Guessoum & Kazanas (1999) considered in detail the spallation of C, N, O, and Ne nuclei by high-energy neutrons (Table 2.2).

Accretion disks around black holes and neutron stars produce a flow of high-energy neutrons that impact the surface of the stars orbiting them. The observed abundances in the companions of some X-ray binaries could be accounted for if a fraction of only about 10^{-7} of the C, N, O, and Ne nuclei available on the stellar surface is converted into lithium via spallation nuclear reactions induced by the flux of neutrons on a timescale of one month.

Shahbaz et al. (2022) have not considered neutron-induced spallation as a viable mechanism for lithium production in the millisecond binary PSRJ1023+0038. Instead, they propose that protons and gamma rays in the pulsar wind are the most likely sources of spallation reactions that can synthesize fresh lithium in this system.

Table 2.2. Spallation Reactions Produced by High-energy Neutrons Colliding with Carbon, Nitrogen, Oxygen and Neon Atoms.

Nuclear Reactions	Threshold Energies (MeV)
$n + {}^{12}C \rightarrow n + 3\alpha$	7.28
$\rightarrow \alpha + {}^{9}Be$	5.70
$\rightarrow {}^{3}He + {}^{10}Be$	19.47
$\rightarrow {}^{3}H + {}^{10}B$	18.93
$\rightarrow {}^{7}Li + \alpha + {}^{2}H$	22.40
$n + {}^{14}N \rightarrow \alpha + {}^{11}B$	0.16
$\rightarrow 2\alpha + {}^{7}Li$	8.82
$n + {}^{16}O \rightarrow n + 4\alpha$	14.44
$\rightarrow \alpha + {}^{13}C$	2.22
$\rightarrow 2\alpha + {}^{9}Be$	12.86
$\rightarrow {}^{7}Li + {}^{10}B$	23.63
$n + {}^{20}Ne \rightarrow \alpha + {}^{17}O$	0.59
$\rightarrow 2\alpha + {}^{13}C$	6.98
$\rightarrow 3\alpha + {}^{9}Be$	17.60
$\rightarrow {}^{7}Li + \alpha + {}^{10}B$	28.36
$\rightarrow {}^{7}Li + {}^{14}N$	16.74

Table taken from Guessoum N and Kazanas D 1999 "Lithium production in companions of accreting X-ray binaries by neutron spallation of C, N, O elements" *ApJ* 512 332.

Jets launched from misaligned accretion disks around microquasars and quasar systems could accelerate neutrons and protons to velocities nearing that of light, which could impact the surface of the stellar companions close to the disk and synthesize lithium by spallation reactions (Butt et al. 2003).

Spallation reactions can also be induced by a very high flux of neutrinos such as those coming from core-collapse supernovae (Type II). Domogatskii et al. (1978) showed that the most likely site for the neutrino-produced synthesis of both isotopes of lithium is the He-rich layers in the expelled envelope. These authors predicted lithium isotopic ratios ($R_I = {}^{6}Li/{}^{7}Li$) between 0.1 and 0.05, which are consistent with the terrestrial value of $R_I = 0.08$, although they note that these theoretical isotopic ratios are very sensitive to uncertain factors such as the spectral shape of the high-energy neutrino flux and the ratio of antineutrinos to neutrinos.

The possible detection of ${}^{6}Li$ on the surface of the cool (4500 K) low-mass (0.2 $M_\odot$) stellar companion of the neutron star in the Centaurus X-4 binary system (Casares et al. 2007) with a Li isotopic ratio of $R_I = 0.12$ could be consistent with neutrino-induced spallation process of Li production, but not with neutron-induced spallation of CNO atoms, which is expected to yield much higher isotopic ratios ($R_I = 0.7$) according to the calculations made by Fujimoto et al. (2008).

The claim of the detection of the presence of ${}^{6}Li$ in Cen X-4 must be taken with a pinch of salt because the spectral resolution ($R = 43,000$) and S/N of the data are lower than for the solar-type stars and metal-poor stars where claimed detections of ${}^{6}Li$ could not be confirmed. Casares et al. (2007) carried out a one-dimensional LTE analysis that is known now to overestimate the lithium isotopic ratio in solar-type stars. A more detailed analysis of the Li I resonance doublet in Centaurus X-4, and other subordinate lines, using three-dimensional NLTE calculations is desirable.

Cosmic rays have been regarded as a very likely culprit for light-element production in the Galaxy (Reeves 1971). As can be seen in Figure 2.8, lithium, along with beryllium and boron, has cosmic-ray abundance that is much higher than that found in the solar system (Rauch 2014; Tatischeff & Gabici 2018).

The leading explanation of these overabundances is that cosmic rays are accelerated by supernova explosions in the disk of the Milky Way (Baade & Zwicky 1934), and that they are mainly composed of high-energy protons that collide with and break up C, N, and O atoms in the ISM and produce lighter elements.

There are alternative explanations for the acceleration of cosmic rays, such as explosions in the supermassive black hole at the Galactic center (Ginzburg & Syrovatskii 1964; Ptuskin & Khazan 1981) and stellar winds (Cassé & Paul 1980).

Direct evidence of a link between a hypernova, a short gamma-ray burst and a binary neutron star merger has been provided with the detection of gravitational waves, high-energy electromagnetic radiation and optical glow in GW170817 (see Savchenko et al. 2017 and references therein).

A fireball model indicates that neutron capture nucleosynthesis could be important in the baryon-rich outflows probably connected with gamma-ray bursts. Lithium is one of the chemical elements that is likely to be produced in such events (Inoue et al. 2020).

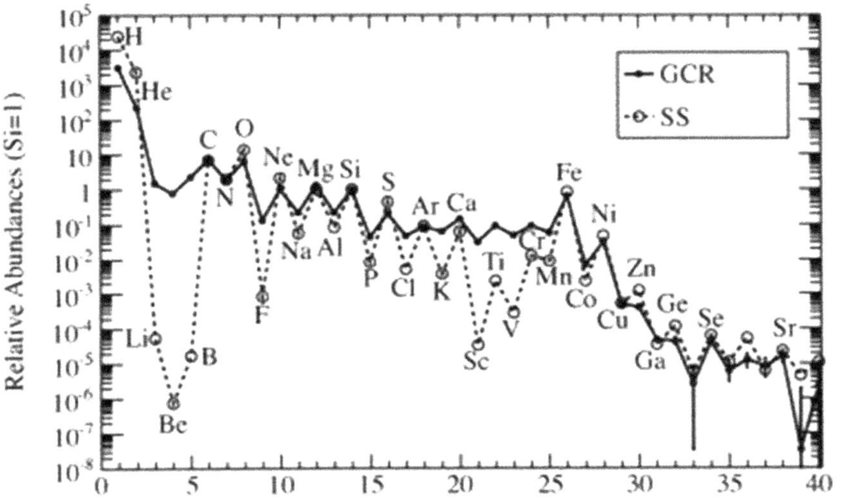

Figure 2.8. Abundances of chemical elements with atomic numbers ranging from $Z = 1$ (H) to $Z = 40$ (Zr) in cosmic rays (solid black line) compared with those in the solar system (dashed line: Lodders, Palme and Gail 2009). Note in particular that the abundances of Li, Be, and B are more than three orders of magnitude higher in cosmic rays than in the solar system. The abundances are normalized to that of Si ($Z = 14$). Figure credit: Rauch, B. F. for the CALET collaboration 2014 AdSpR 53 1444.

2.2.5 Solar Flares

The first suggestions that lithium, together with other light elements (deuterium, beryllium and boron) could be created in non-thermonuclear reactions on the surface of magnetic stars is due to Fowler et al. (1955). The idea that lithium could be created in magneto-hydrodynamic shock waves on stellar surfaces was pursued by Gold (1960) and Bashkin & Peaslee (1961).

A model for the early solar system was constructed by Fowler et al. (1962) that considered a detached circumstellar disk with a small mass of a hundredth (1%) that of the Sun made the following considerations regarding lithium:

When the Sun itself was very young, it rotated rapidly and had a strong magnetic field that connected the solar surface to the proto-planetary disk. During a phase that lasted 10 million years, preceding the formation of the Earth but posterior to the formation of the giant planets and the removal of most of the gas from the disk, a sequence of powerful flares launched from the Sun could have induced spallation reactions on planetesimal material that produced lithium.

The lithium isotope ratio was reduced by reactions of ^{6}Li with neutrons that took place simultaneously with the spallation process. The same process was proposed to be happening in the very young objects called T Tauri stars.

The Fowler et al. model of the early solar system is still being used to account for the abundances of some chemical elements in modern studies of the chemical composition of charbonaceous chondrites (Fukuda et al. 2021).

The feasibility of lithium creation in stellar flares was criticized by Ryter et al. (1970) because the energy requirements for spallation reactions of protons with CNONe nuclei are too high for stellar surfaces. Nevertheless, Canal (1974) showed that the energy required for α–α production of lithium and α-induced spallation of C, N, O, and Ne nuclei is lower and could be consistent with the physical conditions that are present in stellar flares.

In the framework proposed by Ramón Canal, detailed calculations of the lithium production cross sections and yields in spallation reactions were made by Canal et al. (1975), and have been updated by Hultqvist (1977), Glagola et al. (1982), Woo et al. (1985), Livshits (1997), and Kelly et al. (2020).

Evidence for lithium creation by α–α reactions in 20 solar flares was reported from observations collected by the Gamma Ray Spectrometer onboard the Solar Maximum Mission (SMM) satellite (Forrest 1983).

A careful analysis of the 1981 April 27 solar flare was carried out by Murphy et al. (1990), who considered the interaction of α particles accelerated in the flare with ambient He atoms in the solar surface, which had been proposed as a possible source of gamma-ray emission lines at 429 keV and 478 keV (Kozlovsky & Ramaty 1974, 1977; Wang & Ramaty 1975; Murphy et al. 1988). One of the best-fitting models is shown in Figure 2.9.

Since the ^{7}Be nuclei decay into ^{7}Li with a mean lifetime of only about 2 months, they estimated that a total of approximately 2×10^{29} ^{7}Li nuclei were produced in that flare alone.

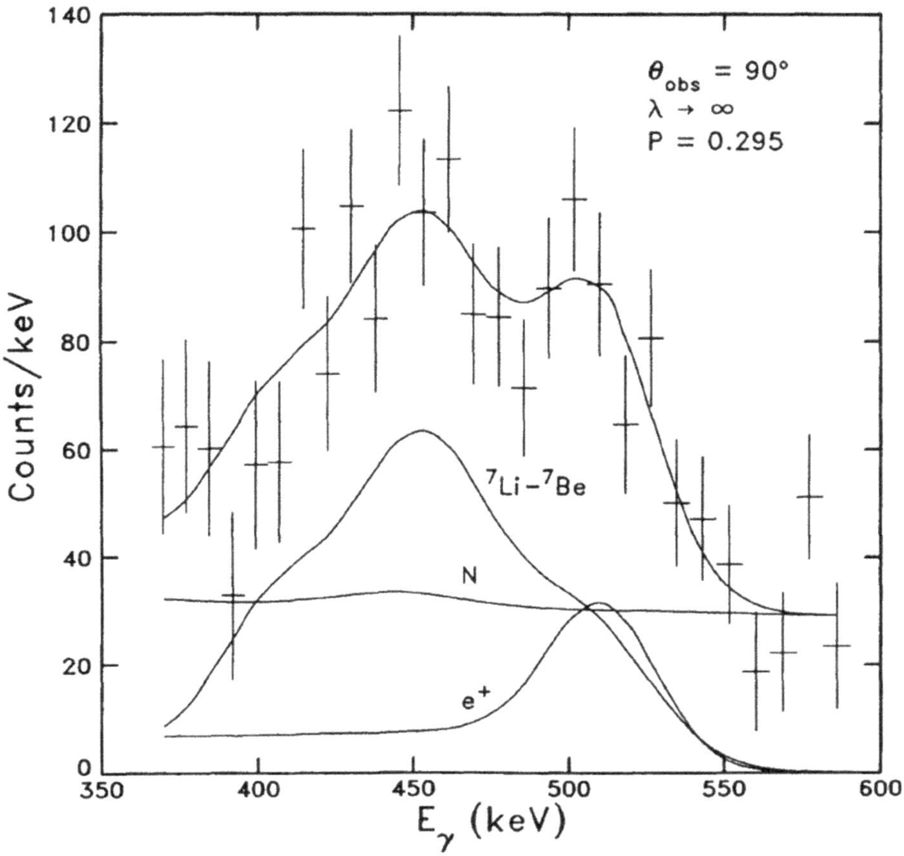

Figure 2.9. Energy (in keV) dependence of the photons counted by the detector of the Gamma Ray Spectrometer onboard the Solar Maximum Mission satellite during the flare observed on 1981 April 27. Datapoints are denoted with error bars on the number of counts and the energy. The contribution of a power-law continuum due to electron bremsstrahlung has been subtracted out. Three theoretical emission curves are shown. The one peaking at around 450 keV is due to de-excitations of ^{7}Be (429 keV) and ^{7}Li (478 keV) nuclei produced by α– α reactions for a flare observed at the limb. The one peaking at 511 keV is due to a narrow positron (e^+) annihilation line. The broad one is a nuclear spectrum determined from fitting data at higher energies. The blend of these three contributions provides a good fit to the data. Figure credit: Murphy, R. J. et al. 1990 "^{7}Li and ^{7}Be de-excitation lines: probes for accelerated particle transport models in solar flares" ApJ 351 299.

Significant flare to flare variability of the gamma-ray emission lines was reported by Share & Murphy (1997). The energy spectrum of the 1981 April 27 flare can be seen compared with the SMM spectra of four other flares in Figure 2.10. Sometimes, the blend of the ^{7}Be (0.429 MeV) and ^{7}Li (0.478 MeV) lines has more counts than the positron annihilation line (0.511 MeV), but at other times the relative intensities of these lines is reversed. In one case, only a very strong positron line is seen, whereas the α–α lines may be veiled by the continuum emission.

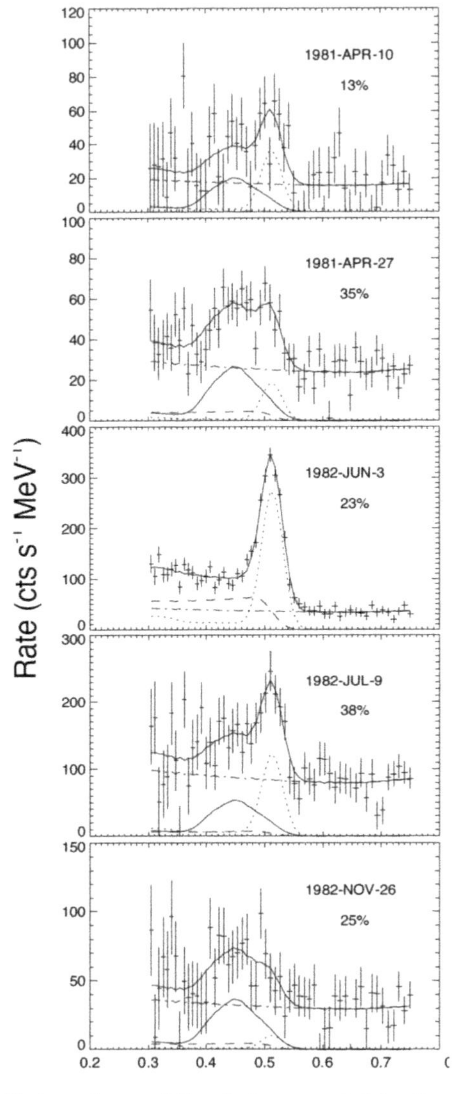

Figure 2.10. Energy (in MeV) dependence of the photons counted by the detector of the Gamma Ray Spectrometer onboard the Solar Maximum Mission satellite during four solar flares observed between 1981 and 1982. Datapoints are denoted with error bars on the number of counts at each energy. The contribution of a power-law continuum due to electron bremsstrahlung has been subtracted out. Three theoretical emission curves are shown. The one peaking at around 0.450 MeV (solid line) is due to de-excitations of ^{7}Be (0.429 MeV) and ^{7}Li (0.478 MeV) nuclei produced by α–^{4}He fusion reactions. The one peaking at 0.511 MeV is due to positron (e^{+}) annihilation (dotted line). The flat one (dot dashed line) is a nuclear spectrum determined by fitting data at higher energies. The blend of these three contributions provides a good fit to the data. Note the very strong flare to flare variability of the relative intensities of the gamma-ray emission lines. Figure credit: Share, G. H., & Murphy, R. J. 1997 "Intensity and directionality of flare-accelerated α particles at the Sun" ApJ 485 409.

Additional support for the production of lithium in solar flares was presented by the analysis of data provided by the hard X-ray spectrometer onboard of the Yohkoh satellite.

Kotov et al. (1996) selected 10 large solar flares with energies extending above 1 MeV and found excess in two of them (1991 October 27 and November 15), which they interpreted as coming from the ^{7}Be and ^{7}Li gamma-ray emission lines. They estimated that about 0.3 T of ^{7}Li were produced in the 1991 November 15 event alone. They proposed that powerful flares could account for the lithium abundance observed in the solar photosphere.

It has been suggested that searches for photospheric lithium abundance variations with time and with solar latitude might be worthwhile, but an early preliminary search for these spectroscopic lithium variability effects in the photosphere of the Sun did not yield positive results (Hultqvist 1974). It could be worthwhile to revisit this question with modern solar telescopes.

The Earth may be accreting some of the lithium produced in solar flares. After over five years orbiting in space at an altitude of 310 km, the Long Duration Exposure Facility (LDEF) was returned to our planet in 1990 January. The concentration of ^{7}Be nuclei on the leading edge of the spacecraft was found to be 2000 times higher than the measurements obtained using balloons in the stratospheric regions of the atmosphere (at altitudes between 20 and 50 km; Lal & Peters 1967) where ^{7}Be is believed to be produced by cosmic-ray induced spallation of N and O nuclei (Fishman et al. 1991).

According to Share & Murphy (1997), the origin of the lithium enhancement found with the LDEF could well be a flux of ^{7}Be reaching the Earth as a result of α–α reactions in several strong solar flares that were observed with the SMM satellite during 1989.

Motivated by the LDEF results, collections of Be^7 nuclei were made using stainless steel foils attached to the Russian spacecrafts COSMOS and RESURS F1 from 1996 to 1999 (Phillips et al. 2001).

A strong correlation between the high orbital concentration of Be^7 nuclei and the solar energetic particle (SEP) fluence was found, but no correlation with the solar X-ray fluence. Phillips et al. considered local spallation of N and O nuclei in the Earth atmosphere induced by SEP and cosmic rays as the most likely explanation of the high ^{7}Be concentrations at satellite altitudes.

The ^{7}Be production rate due to SEP associated with solar great coronal mass ejection is higher than that for cosmic rays, but their occurrence rate is low (Yoshimori et al. 2003). A search for ^{7}Be in interplanetary space is needed to assess the possible direct contribution of α–^{4}He reactions in solar flares to the concentration of these nuclei in the Earth's atmosphere.

2.2.6 Stellar Flares

Magnetically active stars can have flares that are orders of magnitude stronger than solar flares. Pioneering detections of strong Li I resonance doublet in magnetic stars include those reported by Faraggiana & Hack (1963) and Wallerstein & Hack (1964).

A search for lithium among flare stars was performed by de La et al. (1981) who identified one remarkable lithium-rich nearby M dwarf. They attributed the presence of lithium to the star's being young rather than to production of Li nuclei in flares.

An enhancement of the Li I resonance line during a long-duration flare in the very active K-type star 2RE J0743+224 was attributed to spallation reactions by Montes & Ramsey (1998). The star was observed for 10 nights in 1998 with the Sandiford Cassegrain Echelle Spectrograph on the 2.1 m telescope at the McDonald Observatory in Texas. A time series of high-resolution (R = 42,000) spectra is shown in Figure 2.11.

The EW of the Li I resonance doublet in 2RE J0743+224 was measured to change by a factor of between two and five times more than the EW of neighboring Ca I, Fe I and Ni I lines. The maximum of the Li I strength happened just after the maximum of chromospheric emission from a long duration flare. Possible contributions from the unresolved secondary star in the system and from large cold spots and faculae were considered to be an unlikely explanation for the observed variability in the Li I resonance doublet.

Montes & Ramsey (1998) indicated that the changes in the Li I resonance doublet happened mainly in the line core and that the lithium isotopic ratio appeared to be variable, both effects in agreement with theoretical predictions for spallation reactions by Livshits (1997).

Variability of the Li I resonance feature in cool magnetic stars was reported by Polosukhina et al. (1999) and Shavrina et al. (2001). This variability was attributed to the presence of lithium-rich cool spots in atmospheres with cosmic lithium abundance (Shavrina et al. 2003).

The nearby (distance = 10.36 pc) K1-type main-sequence star GJ 117 (HD 17925) has been the target of detailed very high-resolution (R = 130,000) spectroscopic studies with the Echelle spectrograph on the Harlan J. Smith telescope at McDonald Observatory, and ultraviolet photometric monitoring with the Extreme Ultraviolet Explorer (EUVE).

The presence of ^{6}Li was inferred from line profile analysis of the spectra, and ultraviolet flares were observed. This observing team concluded that there is enough high-energy activity in GJ 117 to produce detectable amounts of ^{6}Li by spallation reactions on the stellar surface (Christian et al. 2008). However, as with all ^{6}Li detections based on observations of the Li I resonance doublet in stars, particularly among active ones, a cautionary note about this detection is due here.

Another nearby (distance = 18.29 pc) active lithium-rich K1 main-sequence star that has been monitored for lithium variability is LQ Hya (GJ 355, HD 82558). A time series of 199 high-resolution spectra (R=30,000) were obtained from 2011 December to 2012 April using the STELLA Echelle Spectrograph (SES) coupled to the 1.2 m STELLA-1 robotic telescope at Teide Observatory (Flores Soriano et al. 2015). This star is known to display frequent high-energy flares (Covino et al. 2001).

These authors found that the side of the star with stronger magnetic activity is 30 K cooler and has a slightly, but statistically significantly higher lithium

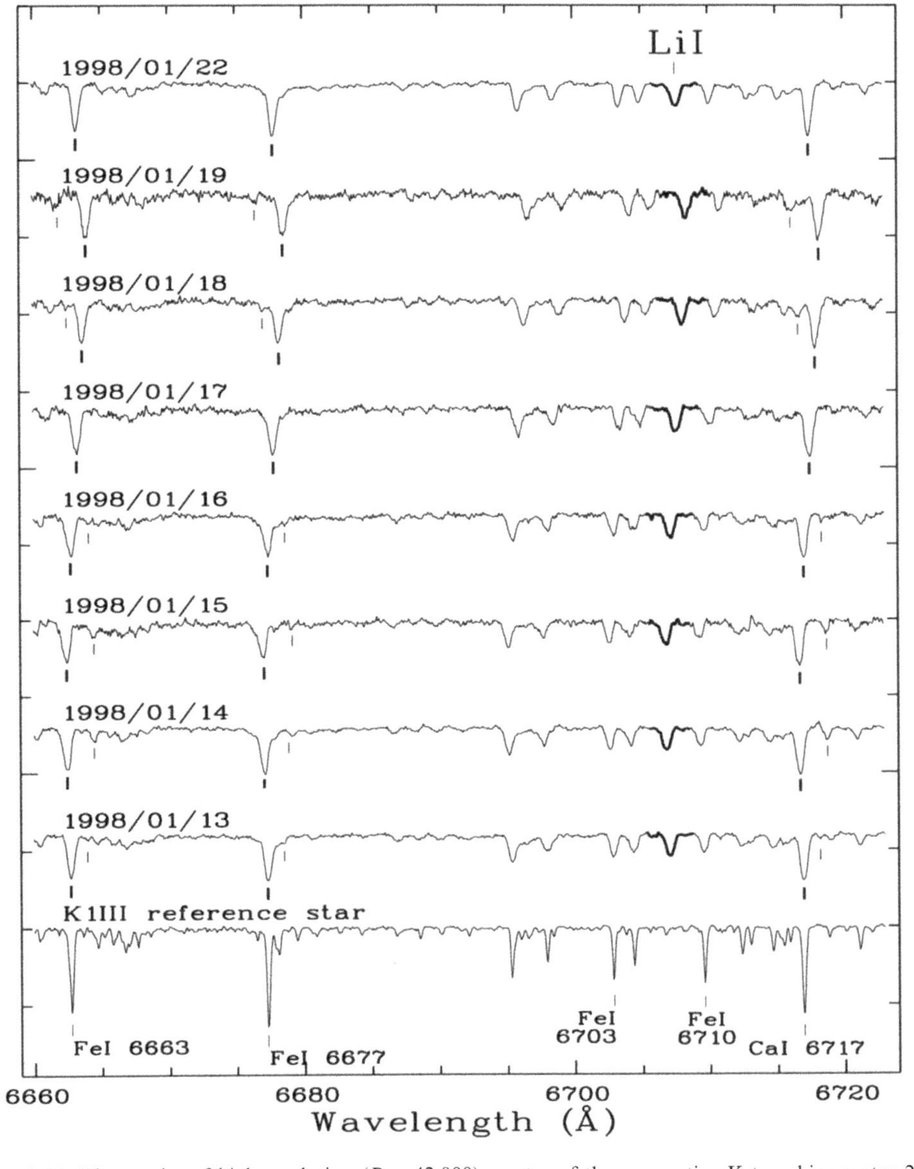

Figure 2.11. Time series of high-resolution (R = 42,000) spectra of the very active K-type binary star 2RE J0743+224 obtained at McDonald Observatory. Note that the relative strength of the Li I resonance doublet (thick line) changes with respect to the neighboring absorption features of Ca I and Fe I. The spectrum of the lithium-free K-type giant star HR 5340 is plotted at the bottom for comparison. Figure credit: Montes, D., & Ramsey, L. W. 1998 "Li I enhancement during a long-duration stellar flare" A&A 340 L5, reproduced with permission © ESO.

abundance. The rotational modulation of the lithium abundance is shown in Figure 2.12. The lithium isotopic ratio also increased from $R_I = 0.08 \pm 0.02$ on the quieter side to $R_I = 0.14 \pm 0.03$ on the more active side. Flores Soriano et al. concluded that their data supported a scenario of lithium production, mostly ^{6}Li, by magnetically driven spallation reactions. However, as with all claims of ^{6}Li detections in stars, this result must be taken with a pinch of salt.

Some magnetically active red giants have been reported to show indications of high lithium abundance and isotope ratio. For instance, the star HD 123351 is a

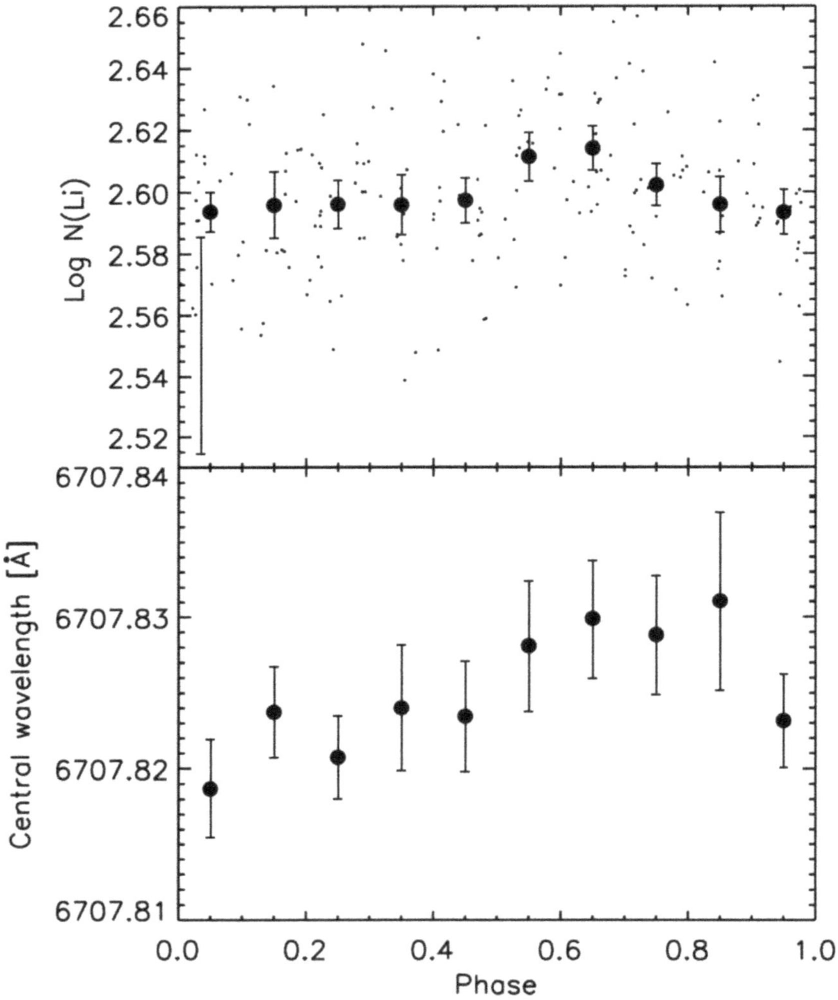

Figure 2.12. Rotational modulation of the lithium abundance and the central wavelength of the Li I resonance doublet, which might be connected to the isotope ratio. Higher lithium abundances and isotopic ratios are obtained for the stellar side that shows stronger magnetic activity. Figure credit: Flores Soriano, M., Strassmeier, K. G., & Weber, M. 2015 "Chromospheric activity and lithium line variations in the spectra of the spotted star LQ Hydrae" A&A 575 A57, reproduced with permission © ESO.

K0-type 1.3 $M_{\odot}$ star that has evolved away from the main-sequence and has an age of about 5 Gyr. According to standard stellar evolution models, it should have a ^{7}Li abundance of log N(Li) = 1.2 and no ^{6}Li at all.

A detailed analysis of high-quality (S/N = 400) high-resolution spectra (R = 120,000) of HD 123351 obtained with the Gecko Coudé Echelle spectrograph at the CFHT was performed by Mott et al. (2017) who noted that there still does not exist a complete atlas of the atomic and molecular lines that contribute to the spectral region around the Li I doublet. Consequently, they compared the effects of using four different line lists. They used the 12 lithium lines that include hyperfine and isotopic splittings (Kurucz 1995), and a model atom with 17 levels and 34 radiative transitions (Cayrel et al. 2007; Sbordone et al. 2010). A more complete model atom (Klevas et al. 2016) did not have any impact on the calculations.

Mott et al. (2017) used 3D hydrodynamical (CO5BOLD code, Freytag et al. 2012) as well as 1D (LHD model, Caffau & Ludwig 2007) models to perform NLTE spectral synthesis calculations with the Linfor3D package (Steffen et al. 2015). The best fit that they achieved with the theoretical spectra to the CHFT spectrum of HD 123351 is shown in Figure 2.13.

The lithium abundance obtained from the 3D calculations is log N(Li) = 1.687 ± 0.114 and the isotopic ratio is $R_{\rm I} = 0.07984 \pm 0.0431$. The 1D calculations provided similar values within the uncertainties. The authors recognized that the claimed detection of ^{6}Li should be treated with caution because it depends on the details of the line lists (particularly the V I line at 670.81096, Lawler et al. 2014) and it could also be affected by physical effects, such as Zeeman magnetic broadening, not included in the models.

Three possible lithium enhancement mechanisms in HD 123351 were discussed by Mott et al. (2017), all of them being considered as highly unlikely. Should the detection of ^{6}Li not be real, then the most likely explanation is the Cameron–Fowler mechanism. The arguments given by Mott et al. for and against each scenario are summarized below:

(1) The first one is the Cameron–Fowler mechanism that could be enhanced by tidal interactions with a highly eccentric binary companion to HD 123351 and could account for the lithium excess in this star with respect to other giants at a similar evolutionary stage. However, the Cameron–Fowler mechanism is not expected to produce any significant amount of ^{6}Li. If the lithium isotopic ratio of $R_{\rm I}$ = 0.08 in HD 123351 is not a fluke, then the Cameron–Fowler mechanism could not be the main source of lithium production in this star.

(2) The second mechanism is stellar flares. Even though these could produce significant amounts of ^{6}Li, it was deemed to be unlikely to account for the lithium isotopic ratio in HD 123351 because the convection zone in this star has about 1 $M_{\odot}$. Too many superflares during the lifetime of the star would be required to produce a detectable amount of ^{6}Li distributed over the entire convection zone.

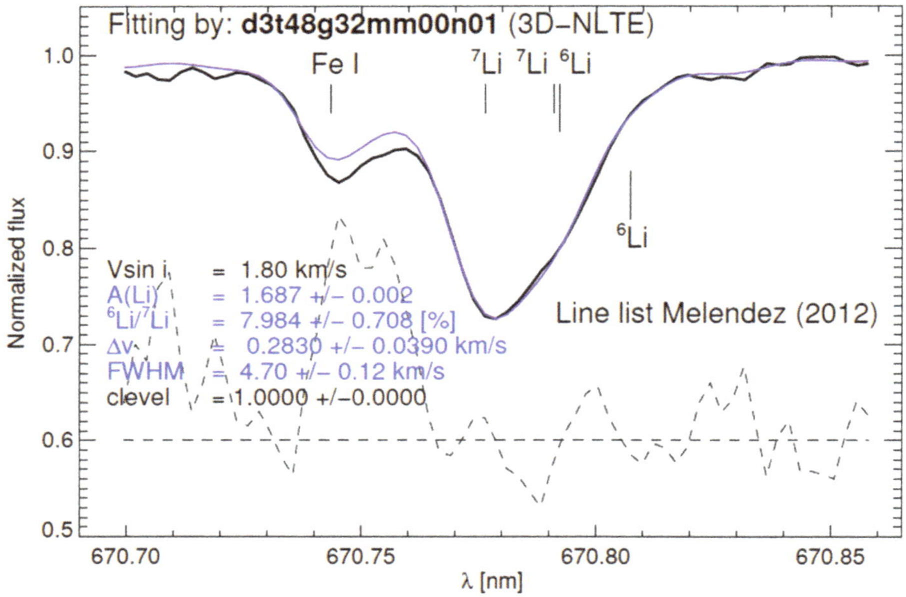

Figure 2.13. The Gecko CFHT spectrum of HD 123351 (black line) compared with the best-fit 3D synthetic spectrum (blue line). The line list from Meléndez et al. (2012) was used. Highlighted in blue are the values of the four free parameters (including the photospheric lithium abundance and the Li isotopic ratio), that enter into the computation of the best fit. Note that the Li isotopic ratio is given as a percentage, and corresponds to $R_I = 0.07984$. The formal 1σ error bars obtained from the fit are provided. The $V \sin i$ and the continuum level were fixed. Dashed lines denote the residuals of the fit, which are multiplied by a factor of 10 and shifted by 0.6 in flux. Figure credit: Mott, A. et al. 2017 "Lithium abundance and ^{6}Li/^{7}Li ratio in the active giant HD 123351. I. A comparative analysis of 3D and 1D NLTE line-profile fits" A&A 604 A44, reproduced with permission © ESO.

(3) External lithium enrichment by ingestion of brown dwarfs, planets, or rocky circumstellar material is the third hypothesis. This scenario will be discussed in some detail later in this book in the context of lithium and exoplanets. It has the advantage of producing a lithium isotope ratio similar to that of meteorites in the solar system (R_I = 0.08). However, it would need the accretion of 53 M_E of lithium-rich rocky material over the last 100 Myr. Such a very large mass would require the ingestion of a superplanet or a brown dwarf during the expansion of the star to become a red giant. The authors did not enter into any quantitative assessment of the probability that such an event could have happened. Detailed modeling of this process can be found in the work of Aguilera-Gómez et al. (2020).

2.3 Where Is Lithium Destroyed?

The most recent determination of the lithium abundance in the photosphere of the Sun is A(Li)= 0.96 ± 0.05 (Wang et al. 2021) from synthetic spectral fitting to the Hamburg solar disk-center flux atlas obtained by James W. Brault using the Fourier transform spectrometer at the 1.6 m Macmath–Pierce telescope in the KPNO. This atlas was made digitally available by Neckel (1999) following the analysis presented in Neckel & Labs (1984).

The current solar lithium abundance determination considers 3D and NLTE effects (Asplund et al. 2021; Wang et al. 2021) and is slightly lower than previous compilations and determinations, such as those provided by Müller et al. (1975), King et al. (1997), Asplund et al. (2009), and Harutyunyan et al. (2018). The 1D/LTE abundance values are quite similar to the 3D/NLTE ones because the spectral lines are very weak.

Lithium is a rare element in the solar photosphere, with an abundance comparable to those of heavy elements with atomic numbers $Z > 40$ (see Figure 2.14), and more than two orders of magnitude lower than in meteorites (Lodders 2003).

The initial lithium abundance inherited by the Sun from the protosolar nebula must have been much higher than what remains at present in the solar photosphere. Carbonaceous Ivuna (CI) meteorites are relics from the origin of the solar system and they have a Li abundance of $A(\text{Li}) = 3.26 \pm 0.05$ (Lodders et al. 2009). The abundances of most chemical elements and their isotopic ratios are very similar between the Sun and the solar system minor bodies and planets (Brault & Holweger 1981; Scott et al. 2015). Why is lithium an exception?

The process by which lithium is destroyed in the Sun started over 4000 million years ago, before the Sun became a main-sequence star, and it is still continuing. Our star and its planetary system, including the Earth, formed from a large cloud that condensed out from the ISM and collapsed gravitationally.

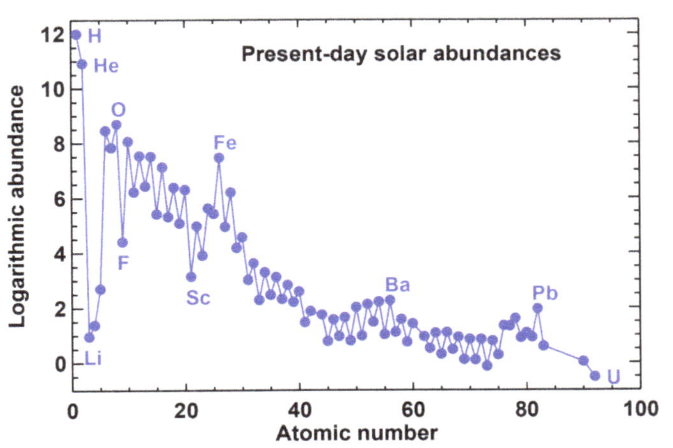

Figure 2.14. Abundances of chemical elements with atomic numbers from Z=1 (hydrogen) to Z = 92 (uranium) at the surface of the Sun. Note that the abundance of lithium (Z = 3) is lower than that of Be (Z = 4), in contrast to what is observed in cosmic rays and in the rest of the solar system. The abundances are normalized to a value of log N(H) = 12. Figure credit: Asplund, M., Amarsi, A. M., & Grevesse, N. 2021 "The chemical make-up of the Sun: A 2020 vision" A&A 653 A141, reproduced with permission © ESO.

The protosun got rid of most of its angular momentum and developed a core at the center where the density and temperature kept rising until stable thermonuclear fusion reactions converting hydrogen into helium stopped the contraction.

The core of the Sun is an enormous H-fusion nuclear reactor that produces just the right amount of energy to balance the pull of gravity. The central temperature of the Sun is kept constant at 15.5×10^6 K (Fowler 1965). At this temperature all the lithium is gone because it is readily destroyed by reactions with protons at temperatures higher than 2.3×10^6 K. Lithium burning in the solar interior must have started during the pre-main-sequence phase of evolution, when the Sun was only a few million years old, it was still contracting and the central temperature was rising (Iben 1965).

All stars destroy lithium in their interiors, but they may be able to retain some of it on their surfaces if there is no inside out mixing of material. Chemical mixing is efficient in stellar interiors when the dominant mechanism of energy transport is convection, and it is inefficient when the dominant mechanism is radiation.

The structure of the present Sun is thought to consist of a radiative interior that is not chemically mixed and rotates as a solid body, and a convective envelope that is thoroughly mixed and experiences differential rotation.

The frontier between the inner radiative region and the convective outer region is called the tachocline. Its location and width have been constrained by measuring vibrations in solar light (helioseismology) that trace changes in the sound speed of seismic waves (Christensen-Dalsgaard et al. 1991).

These authors estimated that the base of the convective zone is located at a depth of $0.287 \pm 0.003\ R_{\odot}$ where the temperature is about 2.2×10^6 K. At those internal temperatures, the rate of lithium burning is so slow that it would take about three times longer than the age of the universe to produce the observed Li depletion of a factor of 200 on the surface of the Sun. The location of the zone where thermonuclear Li destruction is happening must be somewhat deeper than the base of the convective zone as shown in Figure 2.15 taken from Lebreton & Maeder (1987).

Standard models of solar structure and evolution are in remarkably good agreement with helioseismology data (Bachall et al. 1995), but predict a Li depletion of only a factor of three, which is about 50 times lower than observed in the solar photosphere (Pinsonneault 1997).

It is currently accepted that only about one part in a hundred of the initial lithium atoms in the Sun still remains visible in the photosphere. What process is responsible in the Sun for bringing up the lithium-poor material from the radiative interior into the outer convective zone where it gets completely mixed all the way to surface? This is an important topic that remains a challenge for solar models (Buldgen et al. 2023).

Other fragile light elements in stellar interiors are B and Be, but they are destroyed at higher temperatures than Li in deeper regions of the solar interior and there is no evidence for mixing up to the surface (see Boesgaard 2023 for a review).

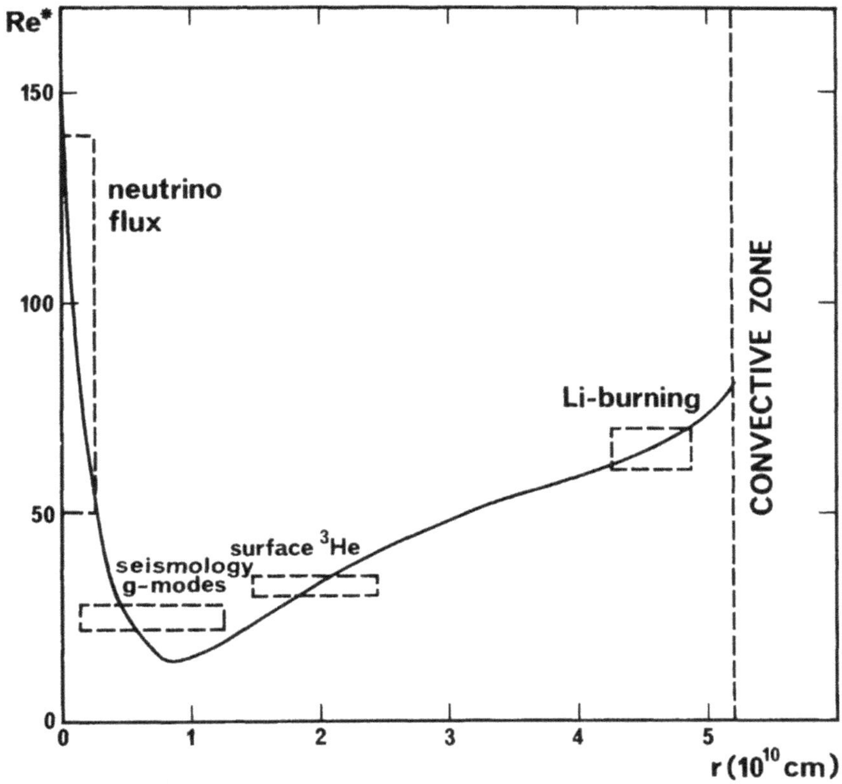

Figure 2.15. Theoretical estimates of the Reynolds number (a dimensionless number that parametrizes how a fluid flows) as a function of the distance from the center of the Sun. Some of the main observational constraints on the Solar model are indicated. Note that thermonuclear lithium burning in the Sun is a unique probe of what happens in a region slightly deeper than the base of the convective zone. The depth of the convection zone has been determined by heliosismology. Figure credit: Lebreton, Y., & Maeder, A. 1987 "Stellar evolution with turbulent diffusion mixing. VI—The solar model, surface ^{7}Li, and ^{3}He abundances, solar neutrinos and oscillations" A&A 175 99, reproduced with permission © ESO.

A recent book about the chemical composition of the Sun and other stars has been published by van den Heuvel (2016).

2.4 Where Is Lithium Preserved?

Lithium in the Milky Way gets recycled from one generation of stars to the next, with a net increase over time in its abundance in the ISM (Rebolo et al. 1988).

Stars with masses and temperatures slightly higher than that of the Sun, may have preserved lithium in their surface for hundreds of Myr, as can be seen in Galactic open clusters such as M48 and the Hyades (Cummings et al. 2017; Sun et al. 2023), but eventually even those stars will experience lithium depletion via post-main-sequence dilution.

It takes a very different type of object to preserve lithium forever.

The lithium abundance at a given time in a particular region of the Galaxy where stars form gets preserved forever in special kinds of objects, which act as cosmic beads of amber.

The objects where lithium gets locked are called lithium brown dwarfs.

Inspired by the detection of a strong Li I resonance feature in UX Tau C (Magazzù et al. 1991), the very low-mass companion of the T Tauri star UXTau A, the first specific calculations of lithium burning versus preservation in very low-mass stars and brown dwarfs were made by Pozio (1991) using the evolutionary code of D'Antona & Mazzitelli (1984, 1985).

She found that 0.06 $M_{\odot}$ could be considered as the theoretical minimum mass for thermonuclear lithium burning (MMLB) because her models showed that Li would never be depleted by more than 90% at that mass; this is just 20% below the H-burning minimum mass of 0.075 $M_{\odot}$ that constitutes the boundary between stars and brown dwarfs with solar chemical composition.

Even though the 0.06 $M_{\odot}$ brown dwarf model reached sufficiently high central ρ and temperature to start burning lithium, it failed to destroy all of it because of the development of electron degeneracy, which eventually dominated and refrigerated the core, thus halting nuclear reactions.

Magazzù et al. (1993) using the simple approximation of a polytrope for the maximum central temperature of a brown dwarf according to a prescription provided by Stevenson (1991) and updated cross sections for the $^{7}Li(p,\alpha)^{4}He$ thermonuclear reaction, confirmed that the theoretical MMLB is located at 0.06 $M_{\odot}$ (see Figure 2.16).

Nelson et al. (1993) obtained similar results and also included the depletion of Be and B in their models. They suggested that a benchmark for the MMLB could be that for 0.062 $M_{\odot}$ there should be a preservation of 10% of the initial Li abundance in the brown dwarf.

They also predicted that there is a narrow range of masses where total depletion of ^{6}Li is expected whereas ^{7}Li is preserved. Future observations of the lithium isotopic ratio in brown dwarfs could explore this possibility, although they probably require the advent of a new generation of high-resolution spectrographs on 40 m class ground-based telescopes.

The models of lithium burning presented by Chabrier et al. (1996) for solar composition predicted that a slow depletion by a factor of 100 would take place in 1 Gyr for 0.060 $M_{\odot}$.

Below 0.055 $M_{\odot}$ all models predict that the initial lithium abundance should remain essentially untouched throughout the evolution of brown dwarfs.

The remarkable coincidence (within about 10%) between the calculations of various groups (Pozio, Magazzù et al., Nelson et al. and Chabrier et al.) indicated that the calculations of lithium burning in brown dwarfs were not very sensitive to the input physics or to the details of the models because the dominant effect was the maximum central temperature attained during the evolution and for how long it was sustained.

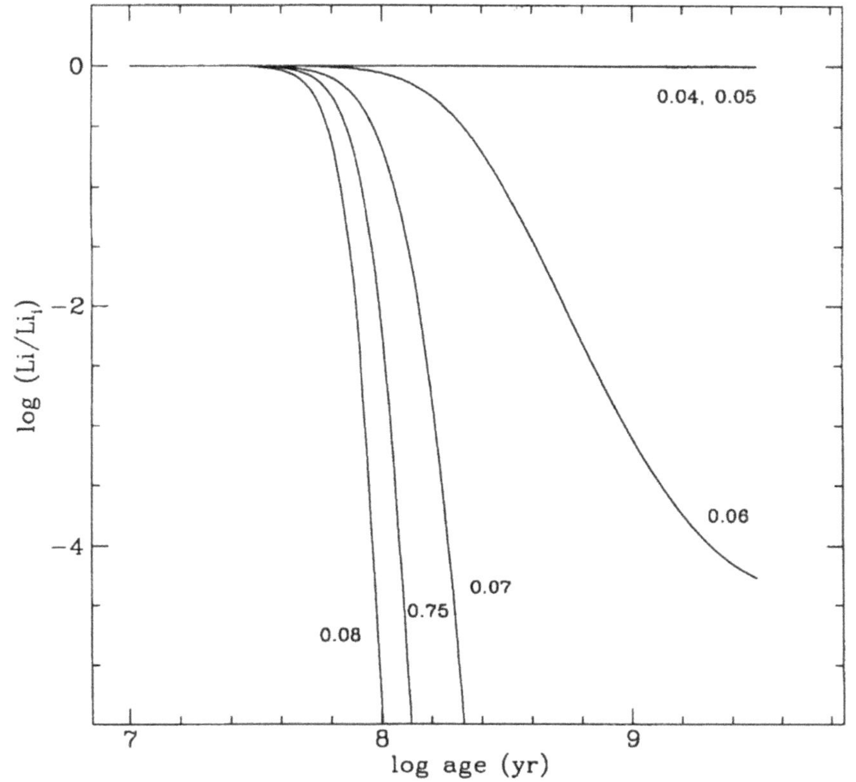

Figure 2.16. Theoretical evolution of lithium abundance versus age for very low-mass stars with 0.08 and 0.075 $M_\odot$ and brown dwarfs with 0.06, 0.05, and 0.04 $M_\odot$. Note that the stars athe most massive brown dwarfs totally destroy lithium whereas the brown dwarfs with masses at and below 0.06 $M_\odot$ do not, which is the limit of lithium burning minimum mass. Figure credit: Magazzù, A. et al. 1993 "A spectroscopic test for substellar objects" ApJ 404 L17.

The model-independent robustness of the lithium depletion method in estimating the ages and masses of fully convective stars and brown dwarfs younger than 100 Myr was reinforced by the analytical calculations of Bildsten et al. (1997) who suggested that it could be a very reliable way of dating young open clusters in the Milky Way.

Magazzù et al. (1993) also carried out the first spectroscopic search for lithium among brown dwarf candidates in binaries. None of the five binaries in their study showed a detectable Li I resonance feature.

The search was extended to 10 single brown dwarf candidates in the solar vicinity, four brown dwarf candidates in open clusters (Hyades and Pleiades) and two brown dwarf candidates in binaries but still with negative results (Marcy et al. 1994; Martín et al. 1994c).

The term "**lithium test**" was coined as a way to check the substellar nature of brown dwarf candidates (Rebolo et al. 1992).

All the brown dwarf candidates to which it was applied between 1993 and 1995 failed to pass the test. The first objects to pass the lithium test were PPl 15 (Basri et al. 1996), Calar 3 and Teide1 (Rebolo et al. 1996), all members of the Pleiades open cluster.

Very recently, the model predictions for the destruction and preservation of lithium in brown dwarfs have been tested using brown dwarf binaries with dynamical masses.

Figure 2.17 shows a schematic representation of the past and present lithium content in the Sun compared with other objects in the solar system, and the brown dwarf binary Reid 1AB. The primary component of this brown dwarf binary has a dynamical mass of 59 ± 5 M_J and has destroyed all of its initial lithium content, while the secondary has a dynamical mass of 41 ± 4 M_J and has preserved it completely (Martín et al. 2022). This is a clear example of lithium depletion and preservation in the same brown dwarf binary system where one component is above the limit for lithium preservation and the other one is below it.

The preservation of lithium in brown dwarfs and the little variation in the cosmic lithium abundance in the Milky Way over the last 5 Gyr make it possible to use the upper envelope of Li I resonance doublet at 670.8 nm pEW measurements as a reference for how the Li I feature depends on effective temperature in ultracool dwarfs (see Figure 2.18). Differential analysis with respect to this reference has been useful to derive lithium abundances and lithium depletion factors in ultracool dwarfs (Martín et al. 2018, 2020, 2022). Knowledge of the lithium abundance in ultracool dwarfs with different metallicities offers a pathway to determine Galactic evolution of lithium and the primordial lithium abundance.

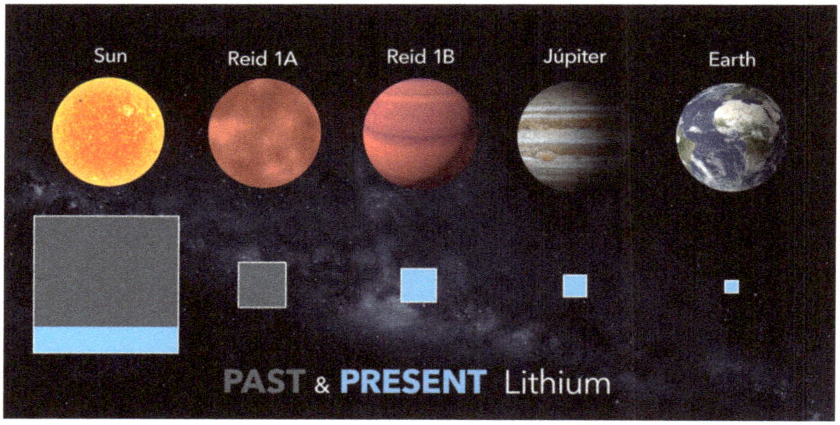

Figure 2.17. Schematic impression of the relative lithium content in the Sun, the two brown dwarf components in the Reid 1 AB binary system, Jupiter and the Earth. The Sun retains only about 1% of its initial lithium abundance. The brown dwarf Reid 1A has a dynamical mass of 59 ± 5 M_J (Dupuy & Liu 2017) and has completely destroyed all of its initial lithium (Martín et al. 2022). The lithium brown dwarf Reid 1B has dynamical mass of 41 ± 4 M_J and, together with Jupiter and the Earth, it has preserved all of its original lithium because it has a mass below the observational lithium thermonuclear burning limit at 51.5 M_J. Figure credit: the author with assistance from Gabriel Pérez Díaz and the IAC science communication and outreach unit IAC press release 24/11/2021.

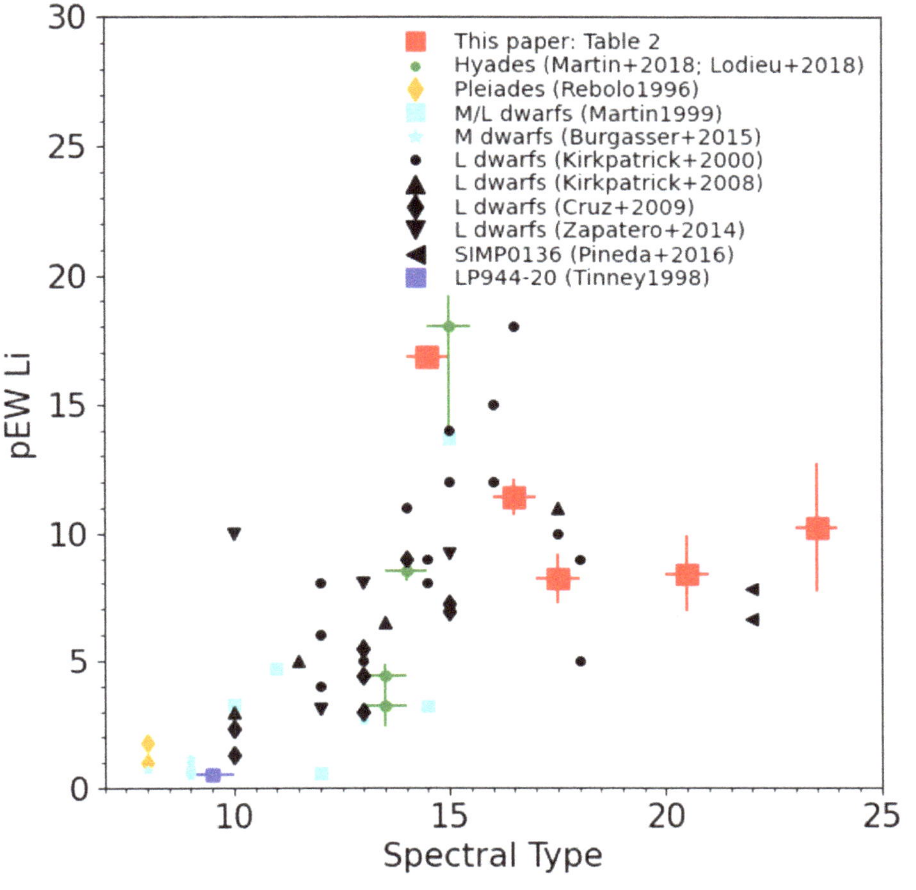

Figure 2.18. Dependence of the pEW of the Li I resonance doublet on spectral type. The coding for spectral type is: 7.5 = M7.5; 10.0 = L0.0; 12.5 = L2.5; 15.0 = L5.0; 17.5 = L7.5; 20.0 = T0.0. Note that the maximum strength of the Li I feature is found around spectral class L5 which corresponds to $T_{\rm eff}$ = 1580 ± 110 K, very close to the boiling point of lithium (1615 K). Figure credit: Martín, E. L. et al. 2022 "New constraints on the minimum mass for thermonuclear lithium burning in brown dwarfs" MNRAS 510 2841.

Bibliography

Aguilera-Gómez, C., Chanamé, J., & Pinsonneault, M. H. 2020, ApJ, 897, L20
Asplund, M., Grevesse, N., Sauval, A. J., & Scott, P. 2009, ARA&A, 47, 481
Asplund, M., Amarsi, A. M., & Grevesse, N. 2021, A&A, 653, A141
Austin, S. M. 1981, PrPNP, 7, 1
Baade, W., & Zwicky, F. 1934, PhRv, 46, 76
Bahcall, J. N., Pinsonneault, M. H., & Wasserburg, G. J. 1995, RvMP, 67, 781
Bashkin, S., & Peaslee, D. C. 1961, ApJ, 134, 981
Basri, G., Marcy, G. W., & Graham, J. R. 1996, ApJ, 458, 600
Bertulani, C. A., & Kajino, T. 2016, PrPNP, 89, 56

Bildsten, L., Brown, E. F., Matzner, C. D., & Ushomirsky, G. 1997, ApJ, 482, 442
Boesgaard, A. M. 1970, ApJ, 161, 1003
Boesgaard, A. 2023, ApJ, 943, 40
Bodansky, D., Jacobs, W. W., & Oberg, D. L. 1975, ApJ, 202, 222
Boffin, H. M. J., Paulus, G., Arnould, M., & Mowlavi, N. 1993, A&A, 279, 173
Boyd, R. N., & Fencl, H. S. 1991, ApJ, 373, 84
Brault, J. W., & Holweger, H. 1981, ApJ, 249, L43
Buldgen, G., Eggenberger, P., Noels, A., et al. 2023, A&A, 669, L9
Butt, Y. M., Maccarone, T. J., & Prantzos, N. 2003, ApJ, 587, 748
Cameron, A. G. W. 1955, ApJ, 121, 144
Cameron, A. G. W., & Fowler, W. A. 1971, ApJ, 164, 111
Caffau, E., & Ludwig, H. G. 2007, A&A, 467, L11
Canal, R. 1974, ApJ, 189, 531
Canal, R., Isern, J., & Sanahuja, B. 1975, ApJ, 200, 646
Casares, J., Bonifacio, P., González Hernández, J. I., Molaro, P., & Zoccali, M. 2007, A&A, 470, 1033
Cassé, M., & Paul, J. A. 1980, ApJ, 237, 236
Cayrel, R., Steffen, M., Chand, H., et al. 2007, A&A, 473, L37
Chabrier, G., Baraffe, I., & Plez, B. 1996, ApJ, 459, L91
Christensen-Dalsgaard, J., Gough, D. O., & Thompson, M. J. 1991, ApJ, 378, 413
Christian, D. J., Mathioudakis, M., & Jevremović, D. 2008, ApJ, 686, 542
Clayton, D. D., & Dwek, E. 1976, ApJ, 206, L59
Coc, A., & Vangioni, E. 2010, JPhCS, 202, 012001
Coc, A., Goriely, S., Xu, Y., Saimpert, M., & Vangioni, E. 2012, ApJ, 744, 158
Colgate, S. A. 1973, ApJ, 181, L53
Covino, S., Panzera, M. R., Tagliaferri, G., & Pallavicini, R. 2001, A&A, 371, 973
Crosas, M., & Weisheit, J. 1996, ApJ, 465, 659
Cummings, J. D., Deliyannis, C. P., Maderak, R. M., & Steinhauer, A. 2017, AJ, 153, 128
de La Reza, R., Torres, C. A. O., & Busko, I. C. 1981, MNRAS, 194, 829
D'Antona, F., & Mazzitelli, I. 1984, A&A, 138, 431
D'Antona, F., & Mazzitelli, I. 1985, ApJ, 296, 502
Deepak, & Lambert, D. L. 2021, MNRAS, 507, 205
Domogatskii, G. V., Eramzhian, R. A., & Nadezhin, D. K. 1978, Ap&SS, 58, 273
Dupuy, T. J., & Liu, M. C. 2017, ApJS, 231, 15
Faraggiana, R., & Hack, M. 1963, PASP, 75, 376
Fields, B. D., & Olive, K. A. 2022, JCAP, 2022, 78
Flores Soriano, M., Strassmeier, K. G., & Weber, M. 2015, A&A, 575, A57
Fishman, G. J., Harmon, B. A., Gregory, J. C., et al. 1991, Natur, 349, 678
Forrest, D. J. 1983, AIPC, 101, 3
Fowler, W. A., Burbidge, G. R., & Burbidge, E. M. 1955, ApJS, 2, 167
Fowler, W. A., Greenstein, J. L., & Hoyle, F. 1962, GeoJ, 6, 148
Fowler, W. A. 1965, Nuclear Astrophysics Jayne Lectures for 1965 (Philadelphia, PA: Memoirs of the American Philosophical Society) 67
Freytag, B., Steffen, M., Ludwig, H. -G., et al. 2012, JCoPh, 231, 919
Fujimoto, S.-I., Matsuba, R., & Arai, K. 2008, ApJ, 673, L51
Fukuda, K., Hiyagon, H., Fujiya, W., et al. 2021, GeCoA, 293, 187

Galli, D., & Palla, F. 2013, ARA&A, 51, 163
Ginzburg, V. L., & Syrovatskii, S. I. 1964, SvA, 8, 342
Glagola, B. G., Viola, V. E. Jr, Breuer, H., et al. 1982, PhRvC, 25, 34
Gold, T. 1960, ApJ, 132, 274
Goldwurm, A., Ballet, J., Cordier, B., et al. 1992, ApJ, 389, L79
González Hernández, J. I., Rebolo, R., Israelian, G., et al. 2005, ApJ, 630, 495
González Hernández, J. I., Rebolo, R., & Israelian, G. 2008, A&A, 478, 203
González Hernández, J. I., Ruiz-Lapuente, P., Filippenko, A. V., et al. 2009, ApJ, 691, 1
Guessoum, N., & Kazanas, D. 1999, ApJ, 512, 332
Harlaftis, E. T., Horne, Keith, F., & Alexei, V. 1996, PASP, 108, 762
Harutyunyan, G., Steffen, M., Mott, A., et al. 2018, A&A, 618, A16
Hauschildt, P. H., Shore, S. N., Schwarz, G. J., et al. 1997, ApJ, 490, 873
Hernanz, M., Jose, J., Coc, A., & Isern, J. 1996, ApJ, 465, L27
Howk, J Ch, Lehner, N., Fields, B. D., & Mathews, G. J. 2012, Natur, 489, 121
Hultqvist, L. 1974, SoPh, 34, 25
Hultqvist, L. 1977, SoPh, 52, 101
Iben, I. 1965, ApJ, 142, 1447
Inoue, A., Tamii, A., Chan, P., et al. 2020, JPhCS, 1643, 012049
Izzo, L., Della Valle, M., Mason, E., et al. 2015, ApJ, 808, L14
Jose, J. 2016, Stellar Explosions: Hydrodynamics and Nucleosynthesis (Boca Raton, FL: CRC Press/Taylor and Francis)
Kajino, T., & Boyd, R. N. 1990, ApJ, 359, 267
Kelly, D. E., Christian, D. J., Mathioudakis, M., & Jevremović, D. 2020, RAA, 20, 104
King, J. R., Deliyannis, C. P., Hiltgen, D. D., et al. 1997, AJ, 113, 1871
Klevas, J., Kučinskas, A., Steffen, M., Caffau, E., & Ludwig, H.-G. 2016, A&A, 586, A156
Kotov, Y. D., Bogovalov, S. V., Endalova, O. V., & Yoshimori, M. 1996, ApJ, 473, 514
Kozlovsky, B., & Ramaty, R. 1974, ApJ, 191, L43
Kozlovsky, B., & Ramaty, R. 1977, ApL, 19, 19
Kumar, Y. B., & Reddy, B. E. 2020, JApA, 41, 49
Kumar, Y. B., Reddy, B. E., Campbell, S. W., et al. 2020, NatAs, 4, 1059
Kurucz, R. L. 1995, ApJ, 452, 102
Kusakabe, M., Cheoun, M.-K., Kim, K. S., et al. 2019, ApJ, 872, 164
Maccarone, T. J., Jonker, P. G., & Sills, A. I. 2005, A&A, 436, 671
Magazzù, A., Martín, E. L., & Rebolo, R. 1991, A&A, 249, 149
Magazzù, A., Martín, E. L., & Rebolo, R. 1993, ApJ, 404, L17
Marcy, G. W., Basri, G., & Graham, J. R. 1994, ApJ, 428, L57
Marsh, T. R., Robinson, E. L., & Wood, J. H. 1994, MNRAS, 266, 137
Martin, E. L., Casares, J., Charles, P. A., & Rebolo, R. 1995, A&A, 303, 785
Martin, E. L., Casares, J., Molaro, P., Rebolo, R., & Charles, P. A. 1996, NewA, 1, 197
Martín, E. L., Lodieu, N., Pavlenko, Y., & Béjar, V. J. S. 2018, ApJ, 856, 40
Martín, E. L., Lodieu, N., & Béjar, V. J. S. 2020, A&A, 640, A9
Martín, E. L., Lodieu, N., & del Burgo, C. 2022, MNRAS, 510, 2841
Martín, E. L., Rebolo, R., Casares, J., & Charles, P. A. 1992, Natur, 358, 129
Martín, E. L., Rebolo, R., Casares, J., & Charles, P. A. 1994a, ApJ, 435, 791
Martín, E. L., Rebolo, R., & Magazzù, A. 1994c, ApJ, 436, 262
Martin, E. L., Spruit, H. C., & van Paradijs, J. 1994b, A&A, 291, L43

McKellar, A. 1940, PASP, 52, 407
Meléndez, J., Bergemann, M., Cohen, J. G., et al. 2012, A&A, 543, A29
Meneguzzi, M., Audouze, J., & Reeves, H. 1971, A&A, 15, 337
Meneguzzi, M., & Reeves, H. 1975, A&A, 40, 91
Molaro, P., Izzo, L., Bonifacio, P., Hernanz, M., Selvelli, P., & della Valle, M. 2020, MNRAS, 492, 4975
Molaro, P., Izzo, L., D'Odorico, V., et al. 2022, MNRAS, 509, 3258
Molaro, P., Izzo, L., Mason, E., Bonifacio, P., & Della Valle, M. 2016, MNRAS, 463, L117
Montes, D., & Ramsey, L. W. 1998, A&A, 340, L5
Moriya, T. J. 2018, MNRAS, 475, L49
Mott, A., Steffen, M., Caffau, E., et al. 2017, A&A, 604, A44
Müller, E. A., Peytremann, E., & de la Reza, R. 1975, SoPh, 41, 53
Murphy, R. J., Kozlovsky, B., & Ramaty, R. 1988, ApJ, 331, 1029
Murphy, R. J., Hua, X -M, Kozlovsky, B., & Ramaty, R. 1990, ApJ, 358, 299
Neckel, H., & Labs, D. 1984, SoPh, 90, 205
Neckel, H. 1999, SoPh, 184, 421
Nelson, L. A., Rappaport, S., & Chiang, E. 1993, ApJ, 413, 364
Norgaard, H., & Fricke, K. J. 1976, A&A, 49, 337
Lal, D., & Peters, B. 1967, Cosmic Rays II Handbuch der Physik (Berlin: Springer) 46, 551
Lawler, J. E., Wood, M. P., Den Hartog, E. A., et al. 2014, ApJS, 215, 20
Lebreton, Y., & Maeder, A. 1987, A&A, 175, 99
Lemoine, M., Ferlet, R., & Vidal-Madjar, A. 1995, A&A, 298, 879
Lepp, S., & Shull, J. M. 1984, ApJ, 280, 465
Livshits, M. A. 1997, SoPh, 173, 377
Lodders, K. 2003, ApJ, 591, 1220
Lodders, K., Palme, H., & Gail, H. -P. 2009, Solar System, Landolt-Börstein—Group IV A&A, 4B, 712
Lubowich, D. A., Turner, B. E., & Hobbs, L. M. 1998, ApJ, 508, 729
Luo, Y., Kajino, T., Kusakabe, M., & Mathews, G. J. 2019, ApJ, 872, 172
Lux, O., Neuhäuser, R., Mugrauer, M., & Bischoff, R. 2021, AN, 342, 553
Palacios, A., Charbonnel, C., & Forestini, M. 2001, A&A, 375, L9
Peebles, P. J. E. 1966, ApJ, 146, 542
Phillips, G. W., Share, G. H., & King, S. E. 2001, GeoRL, 28, 939
Pinsonneault, M. 1997, ARA&A, 35, 557
Pitrou, C., Coc, A., Uzan, J.-P., & Vangioni, E. 2018, PhR, 754, 1
Polosukhina, N., Kurtz, D., Hack, M., et al. 1999, A&A, 351, 283
Pozio, F. 1991, MmSAI, 62, 171
Prantzos, N. 2012, A&A, 538, A80
Ptuskin, V. S., & Khazan, Y. M. 1981, SvA, 25, 547
Randich, S., Gratton, R., Pallavicini, R., Pasquini, L., & Carretta, E. 1999, A&A, 348, 487
Rauch, B. F., & For the CALET collaboration, 2014, AdSpR, 53, 1444
Read, S. M., & Viola, V. E. 1984, ADNDT, 31, 359
Rebolo, R., Molaro, P., Abia, C., & Beckman, J. E. 1988, A&A, 193, 193
Rebolo, R., Martín, E. L., & Magazzù, A. 1992, ApJ, 389, L83
Rebolo, R., Martin, E. L., Basri, G., Marcy, G. W., & Zapatero-Osorio, M. R. 1996, ApJ, 469, L53
Reeves, H., Fowler, W. A., & Hoyle, F. 1970, Nature, 226, 727

Reeves, H. 1971, Topics in Astrophysics and Space Physics (New York: Gordon and Breach)
Romano, D., Matteucci, F., Ventura, P., & D'Antona, F. 2021, A&A, 653, A72
Ruiz-Lapuente, P., Comeron, F., Méndez, J., et al. 2004, Natur, 431, 1069
Ryter, C., Reeves, H., Gradsztajn, E., & Audouze, J. 1970, A&A, 8, 389
Sabbi, E., Gratton, R. G., Bragaglia, A., et al. 2003, A&A, 412, 829
Sackmann, I. J., & Boothroyd, A. I. 1999, ApJ, 510, 217
Sanford, R. F. 1950, ApJ, 111, 262
Savchenko, V., Ferrigno, C., Kuulkers, E., et al. 2017, ApJ, 848, L15
Sbordone, L., Bonifacio, P., Caffau, E., et al. 2010, A&A, 522, A26
Scott, P., Asplund, M., Grevesse, N., Bergemann, M., & Sauval, A. J. 2015, A&A, 573, A26
Selvelli, P., Molaro, P., & Izzo, L. 2018, MNRAS, 481, 2261
Shahbaz, T., González-Hernández, J. I., Breton, R. P., et al. 2022, MNRAS, 513, 71
Share, G. H., & Murphy, R. J. 1997, ApJ, 485, 409
Shavrina, A. V., Polosukhina, N. S., Zverko, J., et al. 2001, A&A, 372, 571
Shavrina, A. V., Polosukhina, N. S., Pavlenko, Ya V, et al. 2003, A&A, 409, 707
Shore, S. N., & De Gennaro Aquino, I. 2020, A&A, 639, L12
Siegert, T., Diehl, R., Greiner, J., et al. 2016, Natur, 531, 341
Singh, R., Reddy, B. E., Campbell, S. W., Kumar, Y. B., & Vrard, M. 2021, ApJ, 913, L4
Starrfield, S., Truran, J. W., Sparks, W. M., & Arnould, M. 1978, ApJ, 222, 600
Steffen, M., Prakapavičius, D., Caffau, E., et al. 2015, A&A, 583, A57
Stevenson, D. 1991, ARA&A, 29, 163
Suárez-Andrés, L., González Hernández, J. I., Israelian, G., Casares, J., & Rebolo, R. 2015, MNRAS, 447, 2261
Sun, Q., Deliyannis, C. P., Steinhauer, A., Anthony-Twarog, B. J., & Twarog, B. A. 2023, ApJ, 952, 71
Sunyaev, R., Churazov, E., Gilfanov, M., et al. 1992, ApJ, 389, L75
Tajitsu, A., Sadakane, K., Naito, H., Arai, A., & Aoki, W. 2015, Natur, 518, 381
Tatischeff, V., & Gabici, S. 2018, ARNPS, 68, 377
van den Heuvel, E. 2016, The Amazing Unity of the Universe: And Its Origin in the Big Bang, Astronomers' Universe (Cham: Springer)
Vincentelli, F. M., Neilsen, J., Tetarenko, A. J., et al. 2023, Natur, 615, 45
Wagoner, R. V. 1967, Sci., 155, 1369
Wagoner, R. V. 1973, ApJ, 179, 343
Walker, T. P., Viola, V. E., & Mathews, G. J. 1985, ApJ, 299, 745
Wallerstein, G., & Hack, M. 1964, Obs, 84, 160
Wallerstein, G., & Conti, P. S. 1969, ARA&A, 7, 99
Wallerstein, G., & Sneden, C. 1982, ApJ, 255, 577
Wang, H. T., & Ramaty, R. 1975, ApJ, 202, 532
Wang, E. Xi., Nordlander, T., Asplund, M., et al. 2021, MNRAS, 500, 2159
Woo, L. W., Kwiatkowski, K., Zhou, S. H., & Viola, V. E. 1985, PhRvC, 32, 706
Woodward, C. E., Pavlenko, Ya V, Evans, A., et al. 2020, AJ, 159, 231
Yan, H.-L., Shi, J.-R., Zhou, Y.-T., et al. 2018, NatAs, 2, 790
Yi, I., & Narayan, R. 1997, ApJ, 486, 363
Yoshimori, M., Hirayama, H., Mori, S., Sasaki, K., & Sakurai, H. 2003, AdSpR, 32, 2691

Eduardo Martín

Chapter 3

Lithium Transport Mechanisms

In this chapter we take a look at different mechanisms of lithium transport inside and outside stars. The most effective mixing mechanism is convection because it entails upward motion of plasma bubbles that transport hot material from the interior and release it at the stellar surface.

It is generally assumed that regions dominated by convective transport are thoroughly mixed on timescales that are very short compared to stellar lifetimes. Very young stars, very low-mass stars and brown dwarfs are fully convective (Hayashi 1961).

The evolutionary models predict that Li depletion could be significant during the convective PMS phase of evolution of stars of 1 $M_\odot$, an effect that may have been important during the early life of the Sun.

The situation is radically different in regions where the energy transport is dominated by radiation. The timescale for mixing becomes much longer and could be comparable to, or longer than, the lifetime of the star (Eddington 1933). The present Sun and solar-type stars on the main sequence have internal structures that are mostly radiative. The presence of Li in the atmospheres of solar-type stars is a probe of internal mixing mechanisms.

The effects of mass loss and accretion on the surface lithium abundances of stars are also discussed in this chapter.

3.1 Convection

Convection is a complex mechanism which is challenging to model in stars. It happens in regions that are very opaque or that have a very large energy flux. In the Sun, the convective zone is located in the outer layers (from 0.961 $R_\odot$ to 0.706 $R_\odot$).

Higher-mass main-sequence stars have convective regions closer to their cores. Lower-mass main-sequence stars have larger convective regions, and when the mass is lower than about a third of the solar mass, they become fully convective.

doi:10.1088/2514-3433/acd8bbch3

Fully convective main-sequence stars are not expected to have any lithium at all on their surface because they destroy it in their cores and it gets mixed downwards very efficiently.

Numerical simulations of convection are very demanding in terms of computational power because they need to calculate the equations of thermodynamics in 3D over a large range of values of the parameters involved (chemical composition, ρ, length, pressure, temperature and time). Thus, full 3D calculations are very expensive and their scope is necessarily limited to snapshots of the stellar evolution.

Generic stellar evolution calculations frequently use the one-dimensional (1D) prescription called mixing length theory (MLT). It was introduced in models of terrestrial convection by Taylor (1920) to study the effects of tidal friction in the Earth–Moon system, and in stellar evolution by Biermann (1948). The MLT is a well-known method in fluid dynamics with applications in many fields of science (Prandtl 1952).

The mixing length is the radial distance that a blob of material travels before getting dissolved in its surroundings. It is considered to be directly proportional to the local pressure, and inversely proportional to the local density. The proportionality is parametrized with a single value for all the convection zone and it can be calibrated by requiring that the solar model evolves to the solar values for the radius and luminosity of the Sun at the solar age (Schwarzschild 1958).

More advanced treatments of convection have been proposed by several authors. Canuto & Mazzitelli (1991), hereafter CM, substituted the single eddy of the MLT by a turbulent spectrum of eddies. The CM theory also requires a free parameter called "a", but the range of values for it is more constrained than for the MLT parameter.

The effects of the CM treatment of convection on the pre-main-sequence (PMS) depletion of lithium as compared with the MLT were presented by D'Antona & Mazzitelli (1994).

Figure 3.1 illustrates different predictions of these models for the evolution of the surface lithium abundance in a star with a mass equal to that of the Sun. It is noteworthy that the PMS Li depletion at a solar mass is much more sensitive to the choice of the MLT free parameter than to the choice of the CM free parameter. However, as we shall see later, all these models predict significant PMS Li depletion for stars of 1 $M_{\odot}$ but the observations indicate that there is no such depletion for this mass during the PMS phase of evolution.

Lydon et al. (1992) incorporated the results of numerical simulations of turbulent convection into a stellar structure code and they were able to reproduce the solar values of luminosity and radius without any adjustable parameter.

They noted that the MLT can be adjusted to almost any combination of input physics and hence cannot be used to test its validity. In other words, the fact that with a single adjustable parameter, the MLT is able to compensate for uncertainties in many fundamental stellar properties indicates how poorly known is the convective transport in stellar interiors.

The impact of considering rotation on the temperature of the base of the convection zone was addressed by Martín & Claret (1996). For low-mass stars

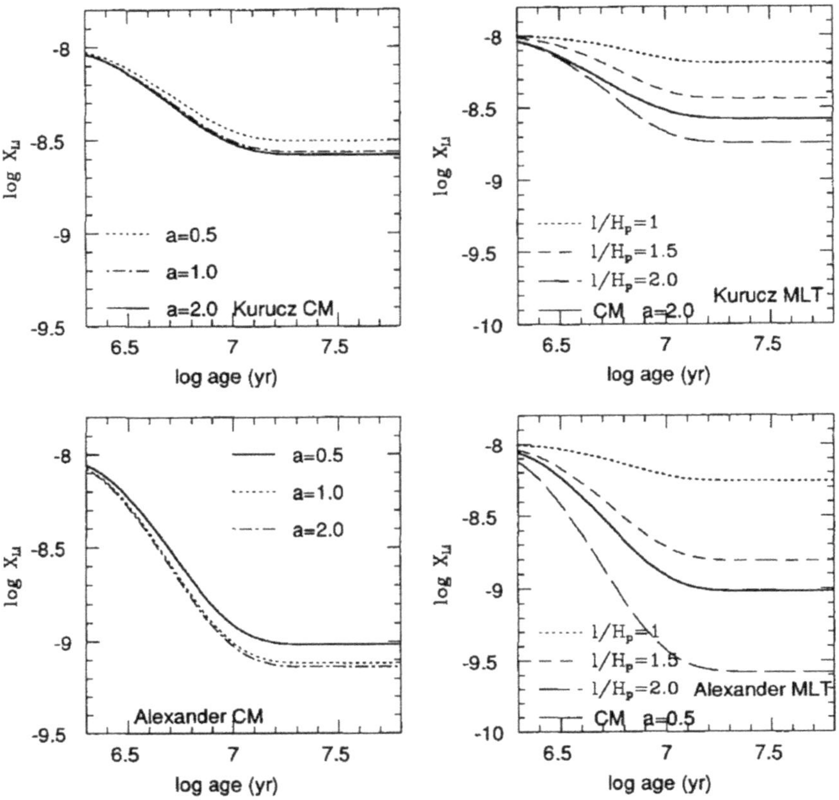

Figure 3.1. Depletion of lithium by convective mixing in the pre-main-sequence (PMS) phase of evolution for models of mass = 1 $M_\odot$. Note that the starting level of undepleted lithium abundance is set at log $X_{\rm Li} = -8$, which is equivalent to log N(Li) = 3.0. The effects of using different opacities (Alexander or Kurucz) and different prescriptions of convection—Canuto and Mazzitelli (CM) or Mixing Length Theory (MLT)—are indicated. Calculations for three different values of the Canuto and Mazzitelli free parameter "a" are shown but the PMS Li depletion is quite insensitive to the adopted value. The mixing length free parameter is denoted as $l/H_{\rm p}$ which is the ratio of the convective length to the local pressure scale height. Results for three different values of $l/H_{\rm p}$ are shown. The higher the $l/H_{\rm p}$ value the deeper the convection and the larger the PMS Li depletion. Figure credit: D'Antona, F., & Mazzitelli, I. 1994 "New pre-main-sequence tracks for $M < 2.5$ $M_\odot$ as tests of opacities and convection model" ApJS 90 467.

with 0.8 $M_\odot$ and 0.7 $M_\odot$, they found that higher rotation rates produce cooler temperatures, both at the stellar surface and at the base of the convection zone, and hence lithium depletion diminishes for fast rotation (Figure 3.2).

Evolutionary models that include the PMS phase have shown that lithium depletion is expected for fully convective low-mass stars and that the amount of surface depletion increases for higher metallicity (Siess et al. 2000; Piau & Turk-Chièze 2002; Amard et al. 2019).

In fully convective very low-mass stars lithium depletion is very efficient. Figure 3.3 shows that lithium is predicted to drop from its initial abundance to zero on a timescale of less than 100 Myr in stars with masses between 0.10 and

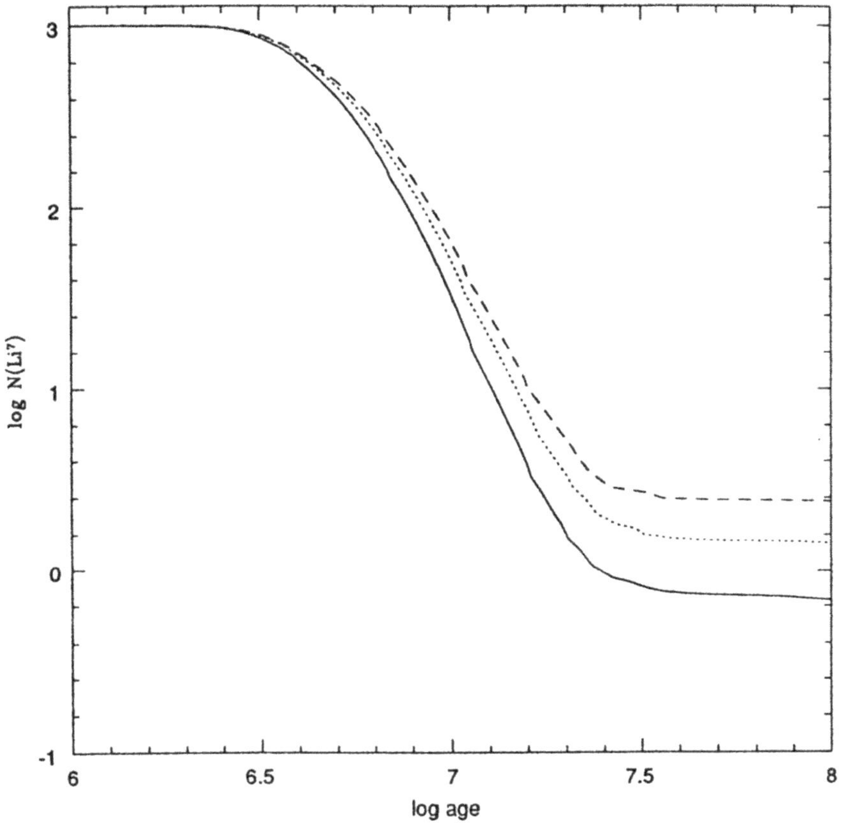

Figure 3.2. Theoretical evolution of the lithium abundance for a 0.8 $M_{\odot}$ star during the first 100 Myr. Solid line denotes a model without rotation. Dotted and dashed lines denote models with increasingly fast rotation that are less efficient at destroying lithium than the non-rotating model. Figure credit: Martín, E. L., & Claret, A. 1996 "Stellar models with rotation: an exploratory application to pre-main sequence lithium depletion" A&A 306 408, reproduced with permission © ESO.

0.09 $M_{\odot}$. This sharp drop is due to thorough convective mixing in the stellar structure. High-mass brown dwarfs also deplete their lithium albeit on a timescale slightly longer than that of 0.10 $M_{\odot}$ stars (Nelson et al. 1993).

Strong magnetic fields are thought to inhibit the efficiency of convective transport and result in cooler stellar structures and diminished lithium depletion rate. These effects may be particularly important in rapidly rotating low-mass young stars (Feiden 2016).

3.2 Overshooting

Convection in stars is a turbulent phenomenon. There is no rigid boundary between the convective zone and the radiative region. Convective bubbles penetrate to some extent into the stable radiative region. If this penetrative effect creates mixing but no

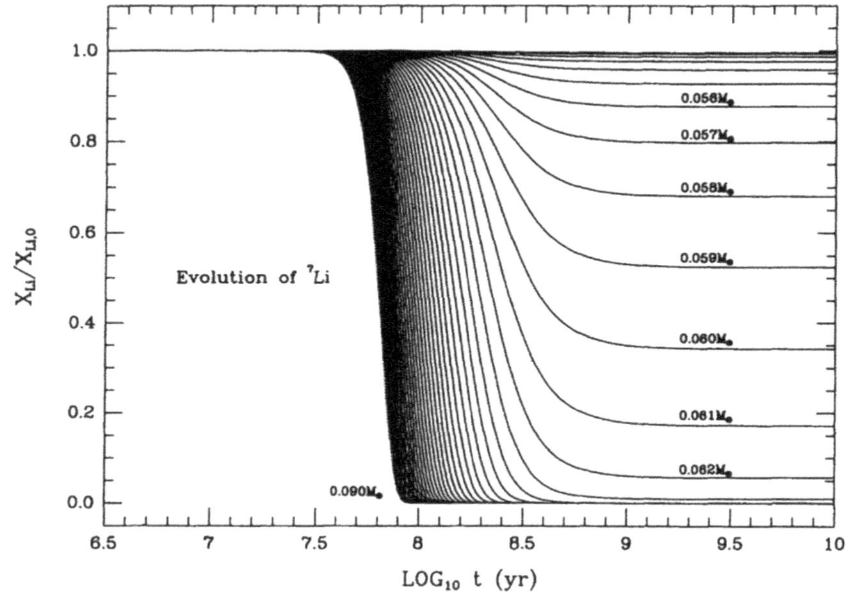

Figure 3.3. A fine grid of evolutionary models of the lithium mass fraction in fully convective brown dwarfs and very low-mass stars covering the region around the lithium burning minimum mass. Figure credit: Nelson, L. A. et al. 1993 "On the Li and Be tests for brown dwarfs" ApJ 413 364.

significant change in the stellar structure, it is known as overshooting and may have important effects on stellar structure and evolution.

Böhm (1963) noted the relevance of overshooting for lithium depletion in solar-type stars. Xiong (1989) suggested that overshooting easily bridges the distance between the convective zone and the base of the photosphere in the Sun.

Singh et al. (1998) solved the full set of Navier–Stokes equations in 3D to calculate the behavior of overshooting in stars.

Overshooting can have dramatic effects on the rate of lithium depletion during the evolution of the Sun because the base of the convective zone is very close to the thermonuclear burning region.

A promising way to improve our understanding of the history of the Sun is to develop models that account for both the lithium depletion and the spinning down. Theoretical efforts have parametrized a diffusion coefficient that is finely tuned to account for the present lithium abundance of the Sun (Denissenkov 2010).

It is thought that the rate of lithium depletion during the Sun's history may be connected to the evolution of its angular momentum. If that is the case, the lithium abundance in the Sun may be indicating something about its rotational evolution, which may be traced back to its formation and early evolution.

During the first 20 Myr the rotation of solar-type stars may be controlled by interactions with the circumstellar protoplanetary disk (Shu et al. 1994). A lithium anomaly in the Sun could suggest that the disk–star interaction was not typical, and that the resulting properties of the solar system may not be common among those of planetary systems in general. This argument is very speculative and lacks clear observational support.

Rotation dependent overshooting has been introduced in stellar evolutionary models by Baraffe et al. (2017) and Constantino et al. (2021).

The later authors fed results from 3D hydrodynamical simulations (Multi-Dimensional Stellar Implicit Code, MUSIC) of convection in the presence of rotation into 1D evolutionary models that calculate the Li depletion in stars with masses ranging from 1.2 to 0.55 $M_{\odot}$. They find that fast rotation severely suppresses lithium depletion.

The theoretical results obtained by Constantino et al. (2021) for the evolution of the surface lithium abundance in a star of 0.75 $M_{\odot}$ are shown in Figure 3.4.

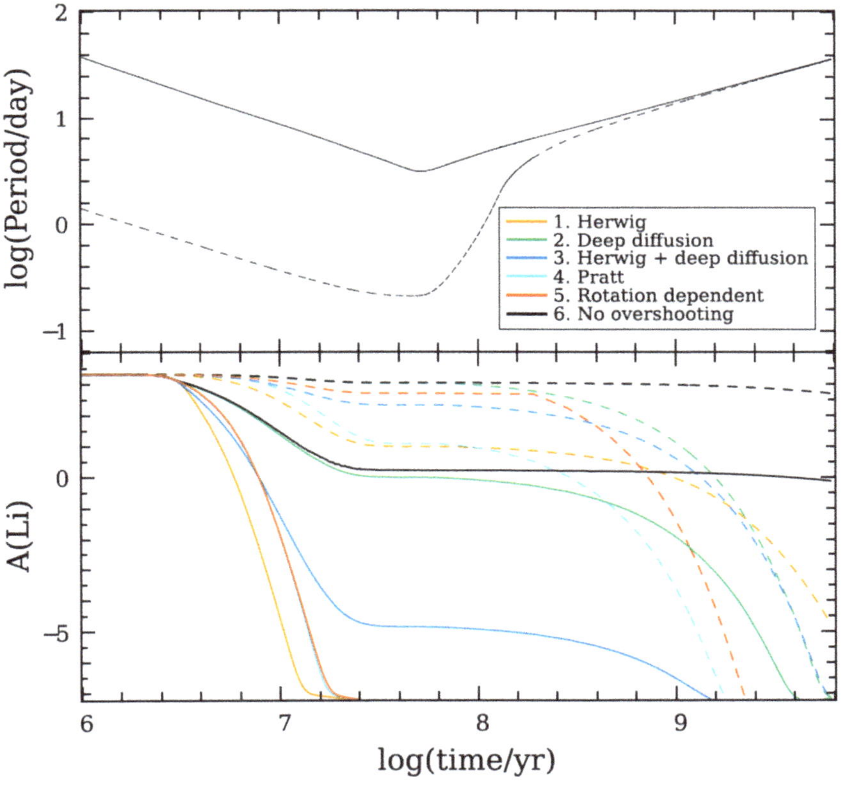

Figure 3.4. Upper panel; two extreme cases for the evolution of rotation in a low-mass star (0.75 $M_{\odot}$). Lower panel: Theoretical evolutionary curves for the evolution of surface lithium abundance in the same star as above. The panel provides the key to identifying the color assigned to each model (see text for more details). Dashed lines denote fast rotating models while solid lines are for the slowly rotating counterparts. Note that all slowly rotating models predict noticeable Li depletion. Figure credit: Constantino, T. et al. 2021 "Suppression of lithium depletion in young low-mass stars from fast rotation" A&A 654 A146, reproduced with permission © ESO.

The six different sets of models considered by Constantino et al. (2021) are the following:

1. Mixing by diffusion that decays exponentially with distance from the convection/radiation boundary (Herwig et al. 1997). These models are shown with orange curves.
2. Deep mixing by diffusion inside the radiative region which is motivated by slow lithium depletion in main-sequence stars, but does not have any physical basis. These model curves are colored in green. They take a longer time to deplete lithium than the previous models (Herwig diffusion scheme) and are less affected by rotation.
3. A combination of the previous two models (dark blue curves) where the diffusion coefficient applied is the larger of the two.
4. A more complicated version of decaying diffusion with distance from the convective/radiation boundary based on 3D MUSIC convection simulations by Pratt et al. (2020). These are called "Pratt" models and are colored in light blue.
5. This is the only model that has a criterion that depends directly on the rotation rate. The diffusion coefficient is the same as in the previous model up to a rotation rate five times solar. For faster rotation, the diffusion length scale is decreased arbitrarily by a factor of 10. This scheme was used by Baraffe et al. (2017). It is labeled as "Rotation dependent" and is shown with a red curve. For slow rotation it coincides with the "Pratt" model. For fast rotation it has significantly reduced lithium depletion owing to the diminished efficiency of convective penetration into the radiative zone.
6. Evolutionary models that do not consider any convective overshooting at all. The effect of increasing rotation in the global properties of the models is that they become more inflated and cooler. The mass of the convective region and the temperature at its base diminish. These results are in good agreement with those reported in Martín & Claret (1996). These models are labeled as "no overshooting" and are shown with black curves.

In the next two sections these models and others will be compared with observational data on lithium in stars belonging to different open clusters.

3.3 Rotationally Induced Mixing

Rotation has important effects on stellar structure and evolution. The possibility that rotation could play a role in the lithium depletion of solar-type stars was already mentioned in the spectroscopic study of lithium abundances in the Hyades cluster by Wallerstein et al. (1965).

Endal & Sofia (1976, 1981) and Zahn (1974) calculated several instabilities induced by rotation in stellar interiors. Stellar winds driven by magnetic fields cause angular momentum loss from the surface (Kawaler 1987, 1988).

Angular momentum redistribution from the outer layers into the interior may carry with it chemical mixing in the process called rotationally induced mixing (Chaboyer 1994; Stephens et al. 1997; Deliyannis et al. 1989).

Pinsonneault et al. (1989) elaborated a very detailed stellar evolution code that included treatments of the following rotation-induced phenomena:

(1) One-dimensional (1D) rotational effects on the stellar structure following the prescriptions given by Kippenhahn et al. (1970) and Endal & Sofia (1976).
(2) Solid body rotation law in all convective zones in agreement with solar data (Brown & Morrow 1987).
(3) Three different choices of initial angular momentum, one in agreement with estimates for the Sun ($J_0 = 1.63 \times 10^{50}$ g cm^2 s^{-1}, Kawaler 1987), one a factor 3 higher and another a factor of 3 lower to explore the dependence of the results on this uncertain parameter.
(4) Surface angular momentum loss with the general parametrization of Kawaler (1988), which combines mass-dependent scaling factors for the magnetic field strength and wind velocity, adjusted to the solar surface rotation velocity at the solar age.
(5) Local conservation of angular momentum in radiative regions combined with gravitational contraction during pre-main-sequence evolution and angular momentum loss generate differential rotation. Two types of mechanisms of angular momentum redistribution were considered: short-time scale (dynamical instabilities) and long-time scale (secular instabilities).
(6) Three dynamical instabilities were included in the models: convection, Solberg–Hoiland, and dynamical shear.
(7) Three secular instabilities were also included: Eddington circulation, Goldreich–Schubert–Fricke instability, and secular shear.

All the different parameters included in the calculations were tuned to find the best fit to the solar data.

The theoretical evolution of the lithium abundance for the best-fitting set of parameters is shown in Figure 3.5.

Models that start with a higher initial angular momentum deplete more lithium at all ages because of enhanced mixing linked to stronger angular momentum loss. If these models are correct, lithium would be a unique tracer of the initial angular momentum of solar-type stars.

The spread in lithium abundances observed among solar-type stars in open clusters could be due to a range of different initial angular momentum values at the time of star formation, which could also be linked to a variety of conditions in the formation of planetary systems.

Consequently, it could be worthwhile to search for connections between surface lithium abundances and the architecture of exoplanetary systems.

The calculations of Pinsonneault et al. (1999), Sills & Deliyannis (2000), Charbonnel & Talon (2005), and Somers & Pinsonneault (2016) suggest that the Sun may have been a moderately fast rotator, which would imply that it may have dissipated the inner circumstellar disk very soon after its formation. On the contrary, the study of PMS Li depletion by Piau & Turk-Chièze (2002) suggests that the Sun may have had a long-lived accretion disk.

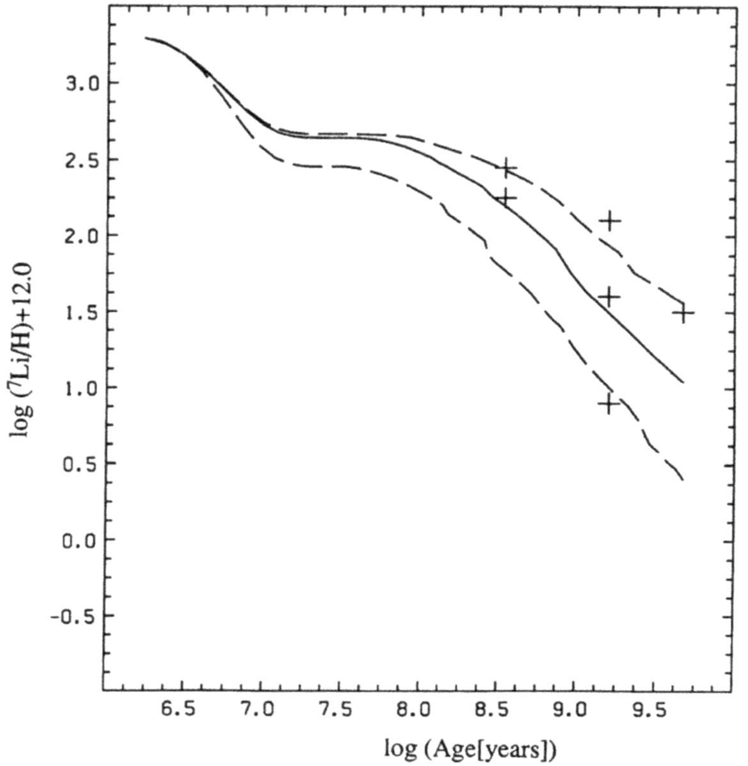

Figure 3.5. Theoretical models of lithium evolution for three different values of the initial angular momentum. The solid curve is for the adopted initial angular momentum of the Sun ($J_0 = 1.63 \times 10^{50}$ g cm^2 s^{-1}) and the present-day solar surface lithium abundance. The upper and lower dashed curves represent the evolution of surface lithium abundance for lower and higher initial angular momentum, respectively. The crosses represent the range of lithium abundances observed for solar analogs in open clusters (Hyades, NGC 752 and M67). Figure credit: Pinsonneault, M. H. et al. 1989 "Evolutionary models of the rotating Sun" ApJ 338 424.

The question of whether or not the Sun is a common star for its age and mass or a rara avis is still open. It is not clear yet whether or not there is a connection between the initial properties of the Sun and the solar system and the properties that we see at the present time.

Theoretical work that has attempted to derive a consistent picture that includes the rotational evolution, helioseismic constraints and lithium depletion in the Sun has so far failed to reach that goal (Amard et al. 2016; Buldgen et al. 2023).

As shown in Figure 3.6, the Amard et al. models tend to predict too much Li depletion. To reduce this effect a parametric turbulent viscosity has been added to couple the interior with the envelope at the age of the Sun (Dumont et al. 2021a).

This parametrization is a simplification of complex physical processes that remain to be identified. It has been employed to reproduce the observed behavior of solar-type stars in 20 Galactic open clusters (Dumont et al. 2021b).

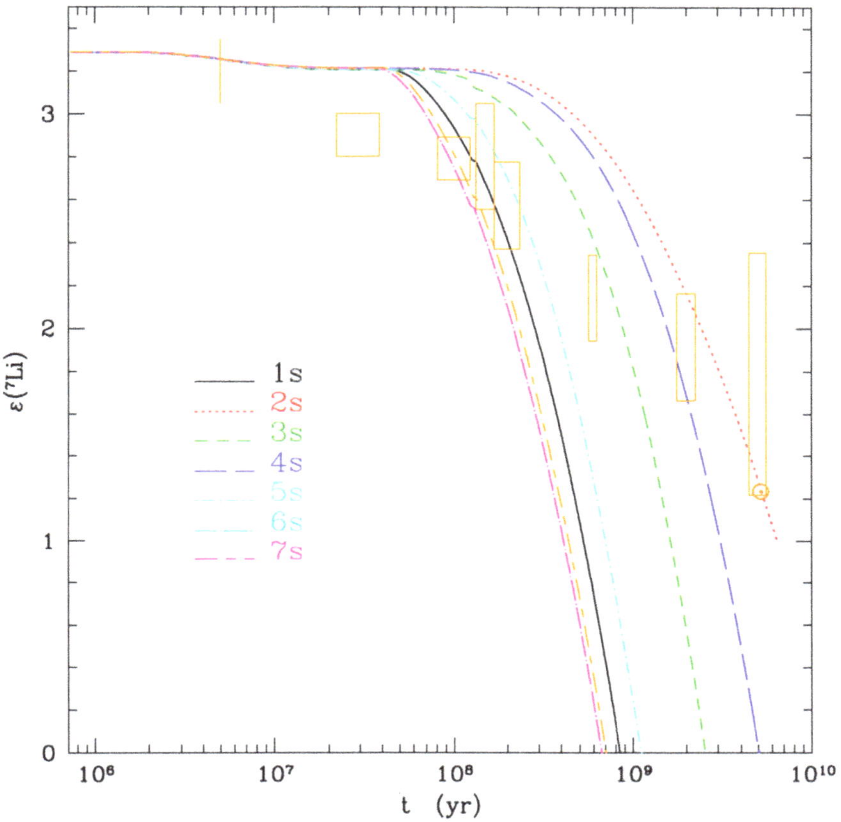

Figure 3.6. Time evolution of lithium abundance in solar-type stars predicted by slow rotating models compared with lithium data in open clusters (orange boxes) from Sestito & Randich (2005). The location of the Sun is marked with the usual symbol. Note that all models but one overpredict the lithium depletion at the Sun's age. Figure credit: Amard, L. et al. 2016 "Rotating models of young solar-type stars. Exploring braking laws and angular momentum transport processes" A&A 587 A105, reproduced with permission © ESO.

3.4 Diffusion

Separation of chemical elements by diffusion below the convective region can take place if the radiative zone is stable enough in main sequence stars. Under some circumstances in stellar interiors, He and Li are not supported by radiation pressure and hence they can sink into the star, leading to a decrease of their abundances in the atmosphere and the convective zone (Michaud et al. 1979; Michaud 1986).

The simplified calculations of Michaud (1986), neglecting the effects of meridional circulation and rotation and using MLT, showed that diffusion can deplete Li from the atmospheres of main sequence stars with $T_{\rm eff}$ below 7000 K in a timescale that has a strong dependence with the stellar mass. For hotter stars, diffusion does not produce Li depletion, but it can lead to chemical separation that has been invoked to explain the abundance anomalies in AmFm stars (Michaud et al. 1983).

The pattern of chemical abundances in A-type stars indicates that processes that counteract diffusion are likely to play a role (Varenne & Monier 1999).

The chemical abundance analysis of evolved stars in the globular cluster NGC 6397 (a factor of 100 less metallicity than the Sun) presented in Korn et al. (2006, 2007) found that models that included atomic diffusion and turbulent mixing were able to provide a satisfactory description of the pattern of Li together with those of other elements (Ca, Fe, Mg, and Ti).

3.5 Gravity Waves

At the interface between convective and radiative regions inside stars, gravity waves can be generated by fluctuations in the pressure that turbulent bubbles exert on the region with stable stratification. These waves may propagate in stellar interiors and carry with them enough material to produce mixing down to the regions where lithium gets destroyed by collisions with protons.

García López & Spruit (1991) were the first to model this mechanism of surface lithium depletion in the context of explaining the lithium dip observed among F-type stars in the Hyades cluster (Boesgaard & Tripicco 1986). As shown in Figure 3.7, their models provided a good account of the data.

However, the main caveat of these models was that they had to assume an ad hoc increase of the gravity wave intensity of a factor of 15 over those obtained from analytical calculations of the energy flux produced at the base of the convective region in the approximation of the mixing length theory.

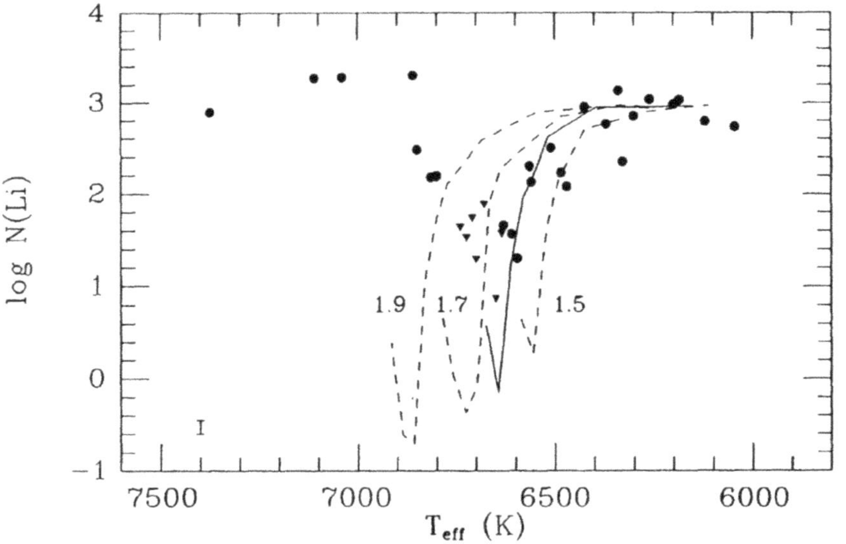

Figure 3.7. Theoretical lithium depletion curves for different values of the convection mixing length parameter as a function of effective temperature for a fixed age of 600 million years. Hyades cluster data for F-type stars are shown as filled circles and inverted triangles (upper limits). Figure credit: García López, R. J., & Spruit, H. C. 1991 "Li depletion in F stars by internal gravity waves" ApJ 377 268.

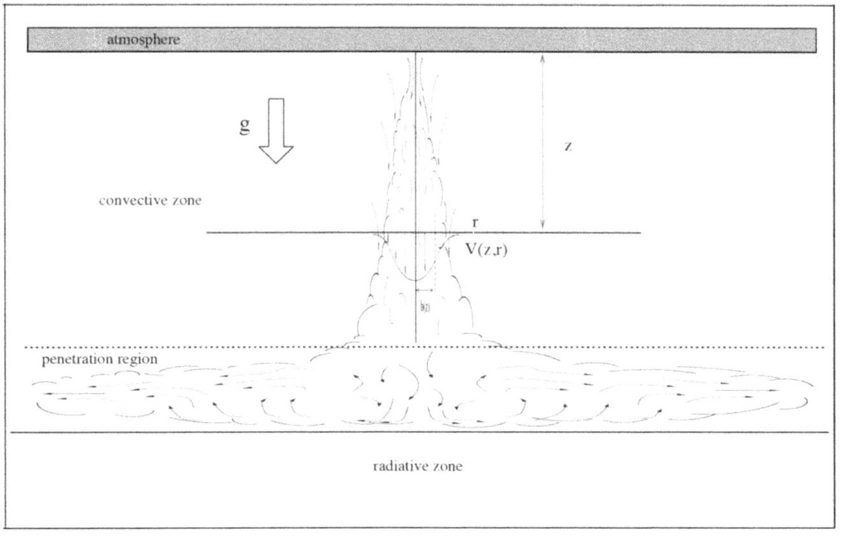

Figure 3.8. Schematic representation of a downward convective plume transporting angular momentum and mixing lithium from the convective zone into the radiative region. $V(z,r)$ is the velocity field at a depth z and a radius r from the perturbation axis for a 2D model. Figure credit: Montalbán, J., & Schatzman, E. 2000 "Mixing by internal waves. III. Li and Be abundance dependence on spectral type, age and rotation" A&A 354 943, reproduced with permission © ESO.

Evry Schatzman (1993) and Josefina Montalbán (1994) developed two-dimensional (2D) analytical models of internal transport of angular momentum and lithium by gravity waves in solar-type stars. The need for an ad hoc increase in gravity wave intensity was found to be unnecessary when the model of convective transport by plumes was adopted by Montalbán & Schatzman (2000) using the prescription of Rieutord & Zahn (1995).

A phenomenological representation of a convective plume is shown in Figure 3.8. These models were able to reproduce the observed behavior of lithium in stars with masses between 0.7 $M_{\odot}$ and 1.2 $M_{\odot}$, and ages between 800 and 5000 Myr. They also noted that fast rotation decreases the efficiency of the transport mechanism induced by gravity waves.

3.6 Mass Loss

The Sun loses mass continuously by means of the solar wind. The removal of the outer layers may bring about surface lithium depletion as lithium-poor material from the interior gets mixed with the convective envelope.

The first detailed calculations of the effects of mass loss on the lithium depletion in the Sun were carried out by Weymann & Sears (1965). They found that the present mass loss rate in the Sun is a factor of 1000 too low to account for its lithium depletion. Nevertheless, the mass loss rate may have been higher when the Sun was younger.

Boothroyd et al. (1991) demonstrated that a mass loss of about 0.1 $M_\odot$ could account for the observed lithium depletion in the Sun. Swenson & Faulkner (1992) revisited the lithium depletion caused by mass loss in the Sun and in F- and G-type stars in the Hyades open cluster. They found that a total mass loss of about 0.1 $M_\odot$ concentrated in the first 400 Myr of the solar evolution could account for the observed lithium depletion and at the same time maintain most of the original beryllium abundance in agreement with the observations of Boesgaard & Budge (1989).

An amount of early mass loss of about 0.1 $M_\odot$ could help to explain the Young Sun Paradox (Graedel et al. 1991), but is incompatible with calculations of the solar oscillation frequencies (p modes) that probe the rate of hydrogen burning in the solar interior (Guzik & Cox 1995).

Mass loss decay with age could help to explain the pattern of lithium abundances seen among F-type stars in the Hyades, but it does not work for G-type stars. Figure 3.9. illustrates one of the comparisons of mass-losing models with lithium

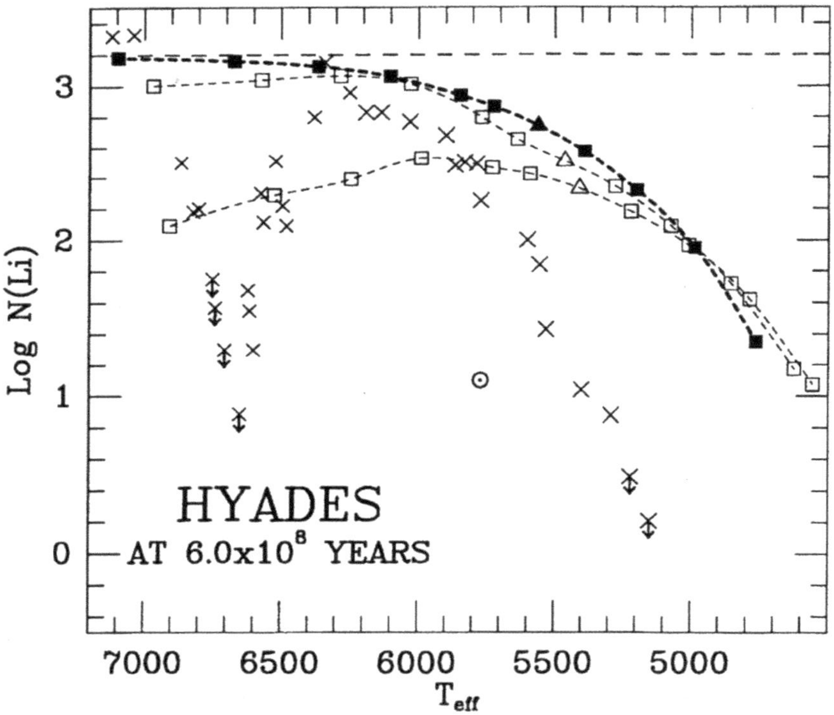

Figure 3.9. Theoretical lithium isochrones for a constant age of 600 Myr compared with lithium abundances from observations of members in the Hyades cluster (crosses) coming from Duncan and Jones (1983), Cayrel et al. (1984), and Boesgaard & Tripicco (1986). The thick short dashed line joins the theoretical points (filled squares) for fixed masses. Two thin dashed lines join the theoretical points (empty squares) for different mass loss amounts (0.03 and 0.045 $M_\odot$, respectively). The triangles denote the position of the models for the mass of the Sun. The horizontal long-dashed line denotes the assumed initial lithium abundance. Figure credit: Swenson, F. J., & Faulkner, J. 1992 “Lithium dilution through main-sequence mass loss” ApJ 395 654.

data for the Hyades clusters (Swenson & Faulkner 1992). The maximum amount of total mass loss that stars can sustain in the first 600 Myr of their evolution without significant lithium depletion is 0.03 $M_{\odot}$. For greater mass loss amounts, the surface lithium abundance is significantly reduced.

3.7 Mass Accretion

Stars accrete mass from circumstellar disks when they are very young (classical T Tauri phase) and also at later stages if they engulf infalling material such as asteroids, comets, or even entire planets.

Evidence for overabundance of refractory chemical elements (Al, Co, Mg, etc) that are expected to condense and survive the dissipation of a protoplanetary disk has been found in a PMS star of about 1.2 $M_{\odot}$, which belongs to the very young (15–21 Myr) Gamma Velorum cluster (Spina et al. 2014; Jeffries et al. 2017).

Pre-main-sequence evolutionary models that include planetary engulfment were computed by Tognelli et al. (2016). They considered accretion events of planets between 180 and $2 \times M_{\mathrm{E}}$ at ages 10, 12 and 15 Myr. The impact of such events on the chemical composition of the star and its position in the CMD were found to be significant.

In the case of lithium-rich and iron-rich main-sequence stars, late ingestion of planets has been considered to explain the chemical enrichment (Gonzalez 1997; Laughlin & Adams 1997; Israelian et al. 2001; Laws & Gonzalez 2003; Mishenina et al. 2021).

Detailed quantitative estimates of the enhancements in the surface abundance of solar-type stars for both lithium isotopes were performed by Montalbán & Rebolo (2002). They considered the dissolution of between 1 M_{J} and 10 M_{J} of planetary material in the convective envelope of main-sequence stars with masses between 1.3 and 0.7 $M_{\odot}$, and the same metallicity as the Sun or larger by a factor of two. Long-lived enhancements were found to be viable for main-sequence stars of 1.3–1.1 $M_{\odot}$ for ^{6}Li, and extending down to 0.9 $M_{\odot}$ for ^{7}Li, which is destroyed at slightly higher temperatures.

In particular, for the super-lithium-rich F-type star in the NGC 6633 open cluster strong evidence for planetary accretion has been found by a detailed analysis of the chemical abundances of many elements including beryllium (Ashwell et al. 2005). Assuming a chemical composition of the planetary material around this star similar to solar system chondritic meteorites, the total amount of accreted material that would be necessary to account for the photospheric chemical enrichment adds up to just 1.1% of the Earth's mass.

Accretion of giant planets or brown dwarfs has been invoked to explain the unusual lithium enrichment and fast rotation of some red giant stars (Carlberg et al. 2013).

A comprehensive analysis combining GALAH survey lithium abundances and MESA evolutionary models to study the scenario of lithium enrichment by accretion of hot Jupiters has been presented by Soares-Furtado et al. (2021), who found that the strongest lithium enrichment from accretion of a hot Jupiter happens at early

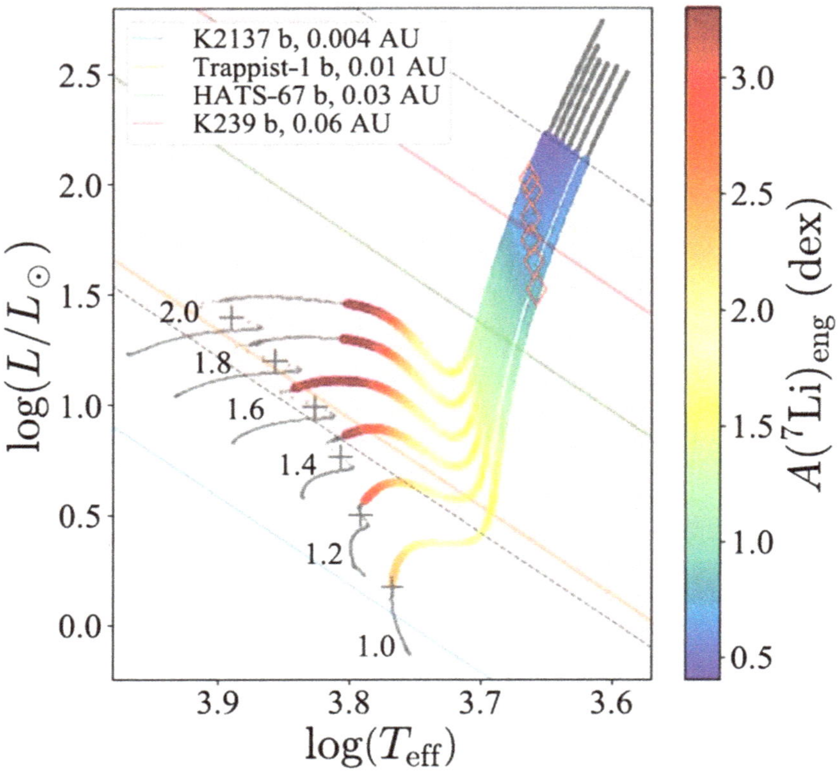

Figure 3.10. Theoretical evolutionary tracks of post-main-sequence evolution in the HRD. In the right y-axis, color-coded lithium abundances indicate the surface enrichment obtained by accretion of a hot Jupiter planet. The highest lithium abundances reach up to almost the cosmic value and occur for stars more massive than 1.4 $M_\odot$ at the beginning for post-main-sequence evolution. The pink region represents orbital separations between 0.1 and 0.01 au. Some examples of known exoplanets are indicated. Figure credit: Soares-Furtado, M. et al. 2021 "Lithium enrichment signatures of planetary engulfment events in evolved stars" AJ 162 273.

phases of post-main-sequence evolution in stars between 1.4 and 2.0 $M_\odot$ (Figure 3.10). Note that these calculations can explain high lithium abundance in post-main-sequence stars, but not higher than the cosmic value.

Bibliography

Amard, L., Palacios, A., Charbonnel, C., Gallet, F., & Bouvier, J. 2016, A&A, 587, A105
Amard, L., Palacios, A., Charbonnel, C., et al. 2019, A&A, 631, A77
Ashwell, J. F., Jeffries, R. D., Smalley, B., et al. 2005, MNRAS, 363, L81
Baraffe, I., Pratt, J., Goffrey, T., et al. 2017, ApJ, 845, L6
Biermann, L. 1948, ZA, 25, 135
Boesgaard, A. M., & Tripicco, M. J. 1986, ApJL, 302, L49
Boesgaard, A. M., & Budge, K. G. 1989, ApJ, 338, 875
Böhm, K. H. 1963, ApJ, 138, 297
Boothroyd, A. I., Sackmann, I. -J., & Fowler, W. A. 1991, ApJ, 377, 318

Brown, T. M., & Morrow, C. A. 1987, ApJ, 314, L21
Buldgen, G., Eggenberger, P., Noels, A., et al. 2023, A&A, 669, L9
Canuto, V. M., & Mazzitelli, I. 1991, ApJ, 370, 295
Carlberg, J. K., Cunha, K., Smith, V. V., & Majewski, S. R. 2013, AN, 334, 120
Cayrel, R., Cayrel de Strobel, G., Campbell, B., & Dappen, W. 1984, ApJ, 283, 205
Chaboyer, B. 1994, PASP, 106, 200
Charbonnel, C., & Talon, S. 2005, Sci., 309, 2189
Constantino, T., Baraffe, I., Goffrey, T., et al. 2021, A&A, 654, A146
D'Antona, F., & Mazzitelli, I. 1994, ApJS, 90, 467
Deliyannis, C. P., Demarque, P., & Pinsonneault, M. H. 1989, ApJ, 347, L73
Denissenkov, P. A. 2010, ApJ, 723, 563
Dumont, T., Palacios, A., Charbonnel, C., et al. 2021a, A&A, 646, A48
Dumont, T., Charbonnel, C., Palacios, A., & Borisov, S. 2021b, A&A, 654, A46
Duncan, D. K., & Jones, B. F. 1983, ApJ, 271, 663
Eddington, A. S. 1933, Natur, 132, 639
Endal, A. S., & Sofia, S. 1976, ApJ, 210, 184
Endal, A. S., & Sofia, S. 1981, ApJ, 243, 625
Feiden, G. A. 2016, A&A, 593, A99
García López, R. J., & Spruit, H. C. 1991, ApJ, 377, 268
Gonzalez, G. 1997, MNRAS, 285, 403
Graedel, T. E., Sackmann, I. -J., & Boothroyd, A. I. 1991, GeoRL, 18, 1881
Guzik, J. A., & Cox, A. N. 1995, ApJ, 448, 905
Hayashi, C. 1961, PASJ, 13, 450
Herwig, F., Bloecker, T., Schoenberner, D., & El Eid, M. 1997, A&A, 324, L81
Israelian, G., Santos, N. C., Mayor, M., & Rebolo, R. 2001, Natur, 411, 163
Jeffries, R. D., Jackson, R. J., Franciosini, E., et al. 2017, MNRAS, 464, 1456
Kawaler, S. D. 1987, PASP, 99, 1322
Kawaler, S. D. 1988, ApJ, 333, 236
Kippenhahn, R., Meyer-Hofmeister, E., & Thomas, H. C. 1970, A&A, 5, 155
Korn, A. J., Grundahl, F., Richard, O., et al. 2006, Natur, 442, 7103
Korn, A. J., Grundahl, F., Richard, O., et al. 2007, ApJ, 671, 402
Laughlin, G., & Adams, F. C. 1997, ApJ, 491, L51
Laws, C. H., & Gonzalez, G. 2003, ApJ, 595, 1148
Lydon, T. J., Fox, P. A., & Sofia, S. 1992, ApJ, 397, 701
Martín, E. L., & Claret, A. 1996, A&A, 306, 408
Michaud, G., Montmerle, T., Cox, A. N., et al. 1979, ApJ, 234, 206
Michaud, G., Tarasick, D., Charland, Y., & Pelletier, C. 1983, ApJ, 269, 239
Michaud, G. 1986, ApJ, 302, 650
Mishenina, T., Basak, N., Adibekyan, V., Soubiran, C., & Kovtyukh, V. 2021, MNRAS, 504, 4252
Montalbán, J. 1994, A&A, 281, 421
Montalbán, J., & Schatzman, E. 2000, A&A, 354, 943
Montalbán, J., & Rebolo, R. 2002, A&A, 386, 1039
Nelson, L. A., Rappaport, S., & Chiang, E. 1993, ApJ, 413, 364
Piau, L., & Turk-Chièze, S. 2002, ApJ, 566, 419
Pinsonneault, M. H., Kawaler, S. D., Sofia, S., & Demarque, P. 1989, ApJ, 338, 424

Pinsonneault, M. H., Walker, T. P., Steigman, G., & Vijay K. Narayanan 1999, ApJ, 527, 180
Prandtl, L. 1952, Essentials of Fluid Dynamics: With Applications to Hydraulics, Aeronautics, Meteorology, and other Subjects (New York: Hafner Publishing Company)
Pratt, J., Baraffe, I., Goffrey, T., et al. 2020, A&A, 638, A15
Rieutord, M., & Zahn, J. P. 1995, A&A, 296, 127
Schatzman, E. 1993, A&A, 279, 431
Schwarzschild, M. 1958, Structure and Evolution of Stars (Princeton, NJ: Princeton Univ. Press)
Sestito, P., & Randich, S. 2005, A&A, 442, 615
Shu, F., Najita, J., Ostriker, E., et al. 1994, ApJ, 429, 781
Siess, L., Dufour, E., & Forestini, M. 2000, A&A, 358, 593
Singh, H. P., Roxburgh, I. W., & Chan, K. L. 1998, A&A, 340, 178
Sills, A., & Deliyannis, C. P. 2000, ApJ, 544, 944
Soares-Furtado, M., Cantiello, M., MacLeod, M., & Ness, M. K. 2021, AJ, 162, 273
Somers, G., & Pinsonneault, M. H. 2016, ApJ, 829, 32
Spina, L., Randich, S., Palla, F., et al. 2014, A&A, 567, A55
Stephens, A., Boesgaard, A. M., King, J. R., & Deliyannis, C. P. 1997, ApJ, 491, 339
Swenson, F. J., & Faulkner, J. 1992, ApJ, 395, 654
Taylor, G. I. 1920, MNRAS, 80, 308
Tognelli, E., Prada Moroni, P. G., & Degl'Innocenti, S. 2016, MNRAS, 460, 3888
Varenne, O., & Monier, R. 1999, A&A, 351, 247
Wallerstein, G., Herbig, G. H., & Conti, P. S. 1965, ApJ, 141, 610
Weymann, R., & Sears, R. L. 1965, ApJ, 142, 174
Xiong, D. R. 1989, A&A, 213, 176
Zahn, J. P. 1974, Stellar Instability and Evolution IAU Symp. 59 D, *(Dordrecht: Reidel) 185*

Chapter 4

Lithium in the ISM and New-formed Stars

4.1 Lithium in the ISM

At the wavelength of interest for detection of lithium (670.8 nm), the ISM can be probed by high spectral resolution spectroscopy in the line of sight toward hot stars.

The first detection of lithium in the ISM was reported by Traub & Carleton (1973) and is shown in Figure 4.1. They used 38 hours of integration time with the Pepsios scanning spectrometer attached to the Smithsonian Astrophysical Observatory 1.52 m telescope located on Mount Hopkins (Arizona, USA).

The EW of the Li I resonance line in the ISM is very small because most of the lithium is ionized. A value of EW = 0.68 mÅ ± 0.09 mÅ was reported. The ionization correction for lithium was determined from the ionization balance for calcium. Using a model of Herbig (1968), an ionization ratio $n(\text{Li II})/n(\text{Li I}) = 156 \pm 30$ was derived.

The ratio of lithium to sodium in this interstellar cloud was found to be consistent with meteorite values in the solar system.

The absolute abundances of both elements are depleted by about a factor of six with respect to the solar system. It is assumed that this effect is caused by the depletion of both elements from the gas phase due to the formation of dust grains and molecules.

Determination of the absolute lithium abundance in the ISM is hampered by uncertainties in the ionization fraction and the depletion into lithium-bearing molecules and dust grains (Morton 1974).

An analysis of lithium and potassium for 10 lines of sight in the Galaxy was made by Steigman (1996), who found that the ratio of lithium to potassium is consistent with that for the solar system (Figure 4.2). The scatter around the mean value is consistent within the error bars.

On 1987 February 23 the brightest stellar explosion in over 400 years was discovered in the Large Magellanic Cloud (LMC). It was the naked-eye Type II supernova 1987A (SN 1987A), the result of the final collapse of a massive star.

doi:10.1088/2514-3433/acd8bbch4

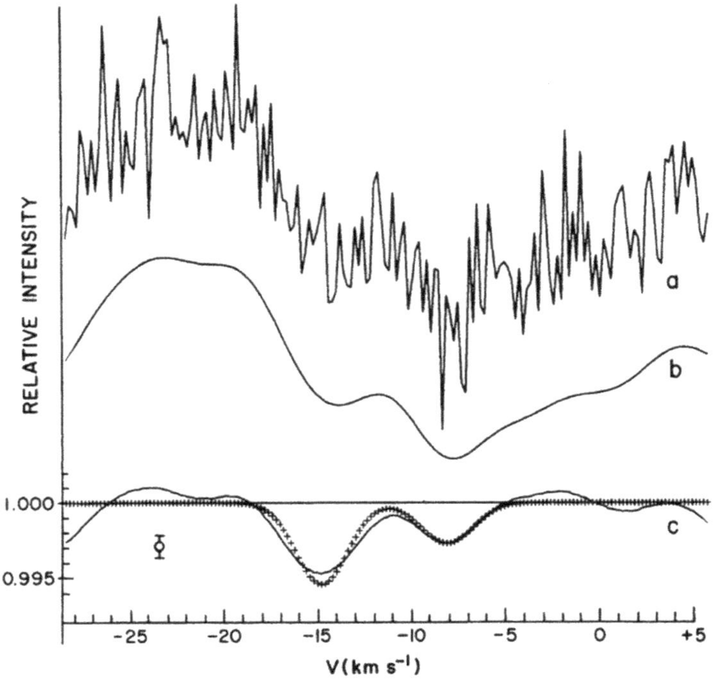

Figure 4.1. First detection of lithium in the ISM toward a very hot star. The upper spectrum shows the original data obtained with Pepsios at the Smithsonian Astrophysical Observatory. The middle spectrum is a smoothed version of the former. The lower spectrum has been corrected for instrumental effects. The main component of the Li I doublet (670.78 nm) is located at a radial velocity of -14.9 km s^{-1}, which is consistent with the known radial velocity of the main interstellar cloud in this line of sight. Figure credit: Traub, W. A., & Carleton, N. P. 1973 "Detection of interstellar lithium" ApJ 184 L11.

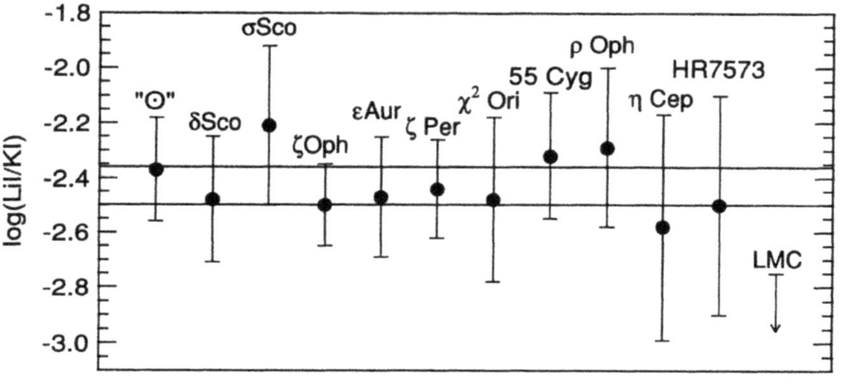

Figure 4.2. Lithium abundance relative to potassium for 10 lines of sight toward bright hot stars in the Galaxy. The meteoritic value for the solar system is shown with a solar symbol. An upper limit of the Large Magellanic Cloud is indicated with an arrow. Figure credit: Steigman, G. 1996 "Cosmic lithium: Going up or coming down" ApJ 457 737.

A compact relativistic object—a black hole or neutron star—may have remained at the center of the explosion.

Modeling of the multi-epoch multi-wavelength light curve and neutrino emission strongly suggests that the remnant is a heavily embedded protopulsar (Greco et al. 2022) that has so far eluded a firm confirmation via direct imaging (Matsuura et al. 2022).

Several observing teams took advantage of this unique opportunity to obtain very high-resolution spectra in an attempt to secure the detection of lithium in the ISM toward the LMC, but none of them were successful (e.g. Molaro & Vladilo 1989).

All the spectra collected at the ESO 1.4 m coudé Auxiliary Telescope were combined by Baade et al. (1991) to obtain an extremely sensitive 3σ upper limit of 0.073 mÅ to the equivalent width of the Li I resonance line in the ISM of the LMC in the line of sight toward SN 1987A. The corresponding upper limit on the lithium abundance relative to potassium appears to be slightly lower than the mean value of the Galaxy (Figure 4.2).

Interstellar lithium hydride (LiH) absorption from molecular clouds along the line of sight to the lensed quasar B0218+357 has been reported at mm wavelengths using the 30 m radio telescope at IRAM on the Pico Veleta in the Spanish Sierra Nevada (Combes & Wiklind 1998) and confirmed with the Combined Array for Research in Millimeter-wave Astronomy at Cedar Flat in California (Friedel et al. 2011), and with ALMA (Muller et al. 2020).

Attempts to measure the Li isotope ratio and the primordial Li abundance from interstellar LiH observations in the line of sight toward dark clouds in the Milky Way and intermediate redshift systems have not been successful yet (Neufeld et al. 2017; Muller et al. 2020).

A detection of Li I in the extragalactic ISM was achieved using UVES at the ESO VLT in Paranal (Howk et al. 2012). SK 143, a hot supergiant star (spectral type O9.5Ib) in the Small Magellanic Cloud (SMC) was observed at a spectral resolution of $R = 70{,}000$. The iron abundance of the SMC (from hot stars) is about a factor of four below that of the Milky Way, indicating that the ISM material in the SMC has been less recycled by successive generations of stars than the ISM material of the Milky Way.

The line of sight was carefully chosen for the presence of a weak radiation field and significant neutral ISM absorption. A very high signal-to-noise ratio (S/N = 275) allowed this team of observers to obtain a clear ($16\,\sigma$) detection of lithium (Figure 4.3). They estimated an ISM lithium abundance in the SMC of log $N(\mathrm{Li}) = 2.68 \pm 0.16$ and a lithium isotopic ratio of $R_{\mathrm{I}} = 0.125$, although they placed a significance of less than 3σ on the later value. This ISM lithium abundance in the SMC is about a factor of four below that of the solar system (from meteorites).

The lithium ISM abundance is free from the problems that may affect stars, and if there has not been significant Li enrichment in the SMC it is representative of the primordial value and is in good agreement with SBBN predictions. The possibility of Li enrichment in the SMC is discussed in Izzo et al. (2022).

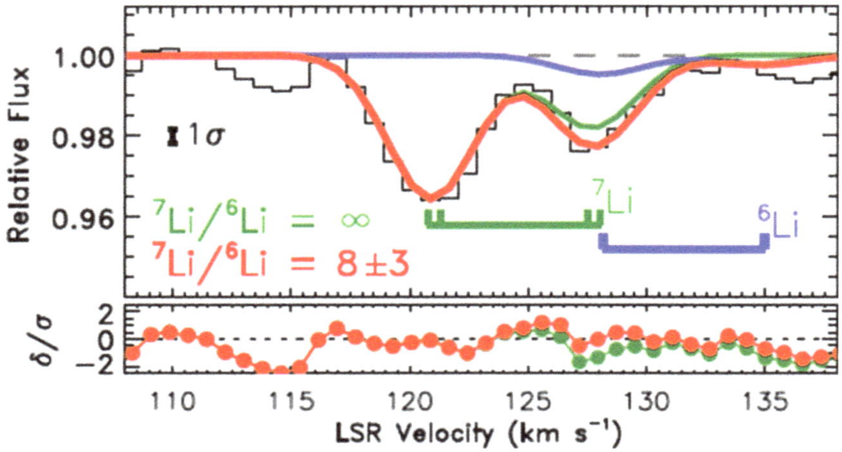

Figure 4.3. Detection of lithium in the ISM of the Small Magellanic Cloud. The velocity of the lithium-bearing cloud is centered at 121 km s^{-1}. The UVES/VLT data (black solid line) are compared with a best-fit model that includes a small contribution of ^{6}Li (red line) and no ^{6}Li at all (green line). The fit residuals for both cases are represented in the bottom panel. Figure credit: Howk, C. J. et al. 2012 "Observations of interstellar lithium in the low-metallicity Small Magellanic Cloud" *Natur* 489 121.

The ISM lithium isotopic ratio in the SMC is consistent to within the error bars with several determinations having been made for the Galaxy (Kawanomoto et al. 2009) and with the solar system (carbonaceous chondrite meteorites).

Nevertheless, significant enhancements of ^{6}Li have been reported in a few Galactic clouds (Lemoine et al. 1995; Knauth et al. 2003) and have been interpreted as possible evidence for local ISM enrichment with fresh lithium due to fusion of α particles. An observation of ^{6}Li (after subtraction of the dominant ^{7}Li component) in two ISM clouds is shown in Figure 4.4.

4.2 Lithium in New-formed Stars

The earliest optical manifestation of new-born solar-type stars are the so-called T Tauri stars (TTSs). These have been known for a long time to display strong lithium lines, which are indicative of high lithium abundance.

Sanford (1947) noticed the presence of the Li I resonance doublet among many other atomic absorption lines and some emission lines in a spectrogram of the star T Tauri, the prototype of the TTS.

The presence of Li I was included by Herbig (1962) as one of the spectroscopic criteria for the definition of classical T Tauri stars (CTTSs).

Quantitative criteria using the dependence of the Li I equivalent width with spectral type were proposed by Martín (1997) to distinguish between TTSs and post-TTSs, a more evolved class of pre-main-sequence stars.

An empty region in the diagram was dubbed the post-T Tauri gap. The difference in the locus of Li I equivalent width of the TTS and the members in young open

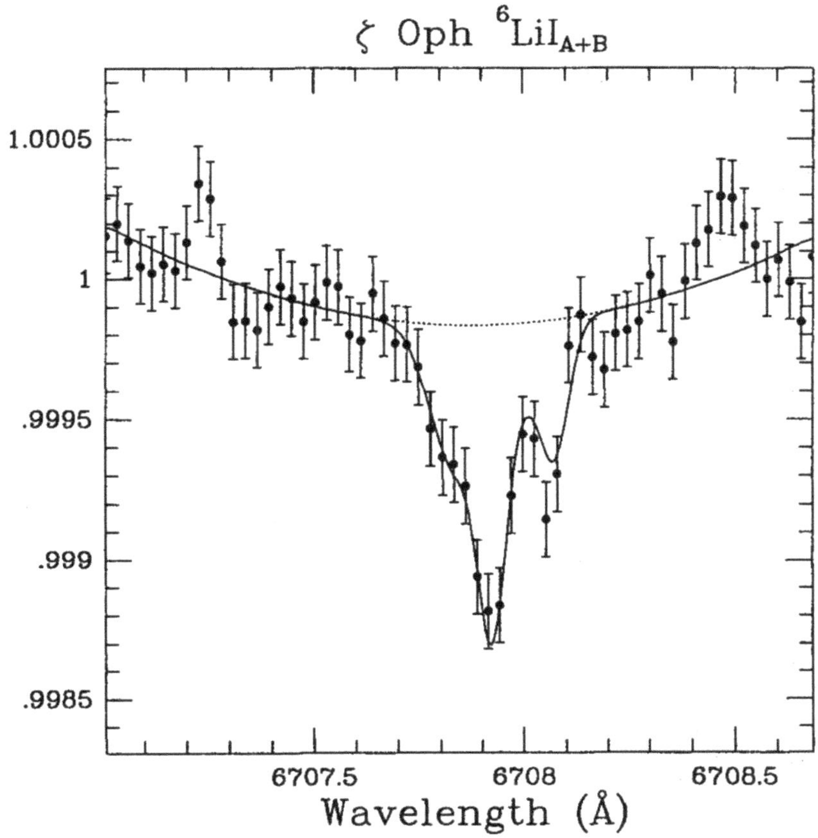

Figure 4.4. Blend of two ^{6}Li resonance doublets corresponding to two different ISM clouds. The contribution of ^{7}Li has been subtracted out. Datapoints obtained with the ESO 3.6 m telescope at La Silla fed with the Coudé Echelle Spectrometer (CES) are denoted by black circles with error bars. The resolving power of CES is $R = 10^4$. The best-fit model is depicted by the black solid line. The continuum level is fitted by the dotted line. Figure credit: Lemoine, M., Ferlet, R., & Vidal-Madjar, A. 1995 "The interstellar ^{7}Li/^{6}Li ratio. The line of sight to ζ Ophiuchi" A&A 298 879, reproduced with permission © ESO.

clusters indicated that lithium is a clear age indicator for K-type (orange) and M-type (red) dwarfs (Figure 4.5).

Deriving lithium abundances in T Tauri stars is a complicated task because the Li I resonance doublet at 670.8 nm is saturated, and because these very young stars often exhibit non-photospheric continuum flux called veiling, which is thought to be due to a hot boundary layer located between an accreting circumstellar disk and the stellar magnetosphere (Basri et al. 1991).

The effects of variable veiling of the Li I resonance doublet of the star AA Tau are shown in Figure 4.6.

Corrections for veiling can be estimated sometimes from a few absorption lines that are almost as deep as the Li I resonance feature, but the error bars in the lithium

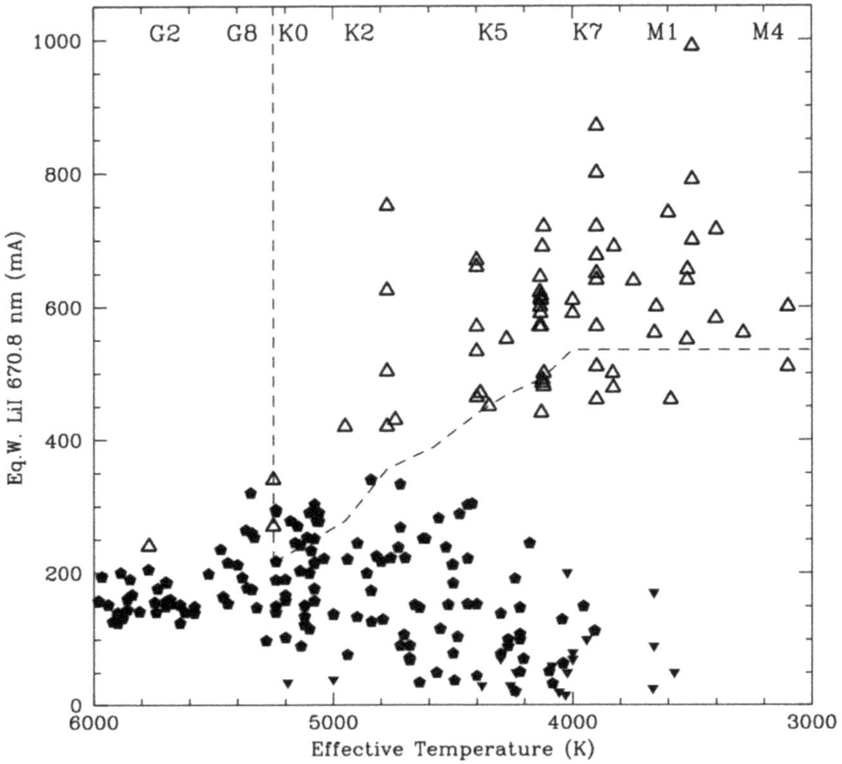

Figure 4.5. Dependence of the Li I resonance doublet equivalent width with spectral type (top abscissa label) or effective temperature (bottom abscissa label). Empty triangles denote Li I measurements in T Tauri stars while solid pentagons and inverted solid triangles denote Li I values or upper limits for stars in the open cluster IC2602, IC2391, IC4665, and the Pleiades, respectively. The vertical dashed line is located at 5250 K, which is considered to be the temperature limit for TTS classification. The dashed curve is a lithium isoabundance line for log N(Li) = 2.8. Figure credit: Martín, E. L. 1997 "Quantitative spectroscopic criteria for the classification of pre-main sequence low-mass stars" A&A 321 492, reproduced with permission © ESO.

abundance determination are considerable. Other times, for high mass accretion rates, the photospheric lines may be washed out by the veiling, and only the Li I feature remains detectable.

A way around the veiling problem has been the study of very young stars without active accretion, the so-called naked T Tauri stars (NTTSs) or weak-line T Tauri stars (WTTSs).

It has been found in WTTSs, after correcting for NLTE effects, which can be quite significant, that the locus of maximum Li abundances is A(Li) = 3.1 (Magazzù et al. 1992; Martín et al. 1994). This is slightly lower than the Li abundance found in the meteorites—A(Li) = 3.3—contrary to the expectation that the lithium abundance in the Galaxy has increased over the last 5 Gyrs.

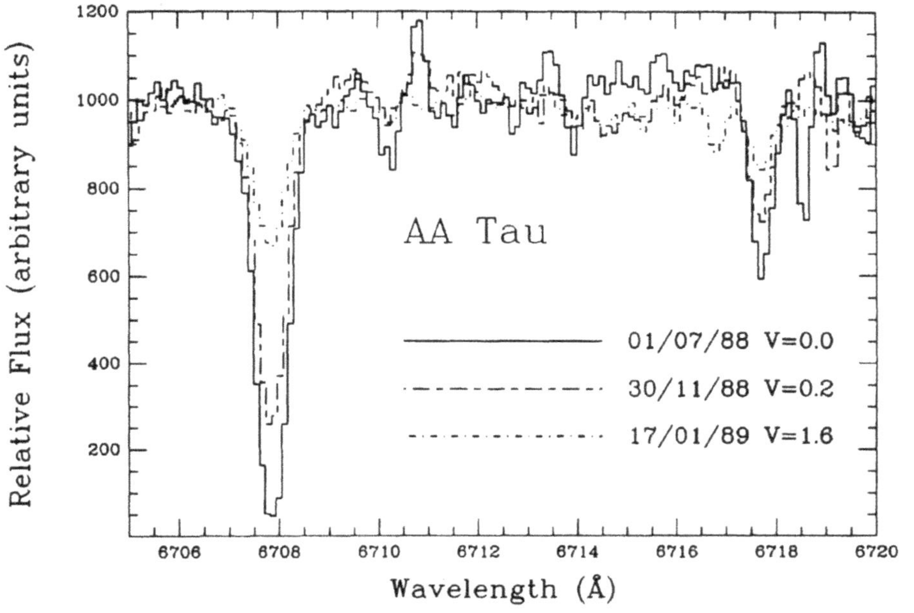

Figure 4.6. Dependence with veiling of the absorption of photospheric lines including the Li I resonance doublet in AA Tau, a TTS observed at high spectral resolution with the Echelle spectrograph of the Lick Shane 3 m telescope. Figure credit: Basri, G., Martín, E. L., & Bertout, C. 1991 "The lithium resonance line in T Tauri stars" A&A 252 625, reproduced with permission © ESO.

A possible way out of this paradox has been suggested by Molaro et al. (2012) who proposed that there is prompt PMS Li depletion due to enhanced overshooting. However, such a large amount of overshooting does not seem consistent with the understanding of Li depletion in solar-type stars (Baraffe et al. 2017a).

Another mechanism of very early Li depletion has been brought up by Baraffe et al. (2017b) who showed that burst accretion can induce significant PMS depletion during the first 10 Myr. These exotic ideas about very early PMS Li depletion remain speculative for the time being because they lack clear observational support.

In Figure 4.7 the spectrum of the Sun around the spectral region of the Li I resonance line at 6708 Å is shown together with two very young stars that are located near the Taurus–Auriga star-forming region. Both have ages estimated in the range 10 Myr to 20 Myr according to their positions in the HRD and the comparison with theoretical isochrones.

One of them (V1298 Tau) harbors a planetary system (David et al. 2019; Suárez Mascareño et al. 2021), and the other one is a double-lined spectroscopic binary, which is also a common proper-motion pair with V 1298 Tau.

4.3 Pre-main-sequence Lithium Burning

Before reaching the main sequence, where they spend most of their lifetimes burning hydrogen, stars must contract from very large initial sizes and increase their core temperature to about 3×10^6 K.

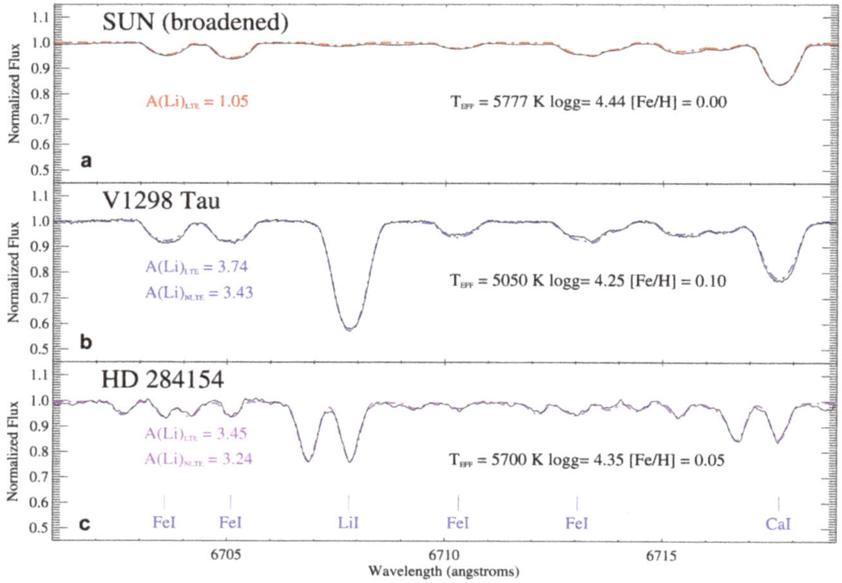

Figure 4.7. Spectral region of high-resolution spectra obtained with HARPS-North on the TNG around the Li I resonance doublet for the Sun, one very young solar-type star (V1298 Tau) and one double-lined young binary (HD284154). Optimum synthetic spectra that are able to fit the observed ones are shown. The Li abundances used in each theoretical spectrum are labeled. Note the large Li abundances, and the significant NLTE corrections in the young stars. Figure credit: Suárez Mascareño, A. et al. 2021 "Rapid contraction of giant planets orbiting the 20-million-year-old star V1298 Tau" NatAs 6232.

Low-mass stars, including the Sun, are thought to begin their lives with fully convective structures (Hayashi 1961), thus ensuring efficient chemical mixing, and contracting gravitationally at roughly constant surface temperature.

The energy released by gravitational contraction heats up the stellar core, which becomes radiative, and the radiative region expands toward the surface.

Several calculations of the lithium depletion during PMS contraction have arrived at different results for the case of the Sun (Ezer & Cameron 1963; Hayashi & Nakano 1963; Weymann & Moore 1963; Bodenheimer 1965; D'Antona & Mazzitelli 1994). For solar metallicity and one solar mass the models of Siess et al. (2000) predict that 37% of the initial lithium abundance still remains in the photosphere when the star reaches the zero-age main sequence (ZAMS).

It is still unclear whether or not the Sun has depleted any lithium during its PMS evolution, but it is now beyond dispute that lower-mass stars certainly do.

Theoretical PMS evolutionary models of Li depletion are shown in Figure 4.8. The models begin a zero age with constant temperature Hayashi tracks that run vertically in the HRD as stars contract gravitationally with fully convective internal structure. Lithium depletion is predicted to start on these tracks, first at 1.2 $M_{\odot}$, and later at lower mass. Stars with masses 0.8 $M_{\odot}$ and lower are predicted to deplete their lithium completely during PMS evolution.

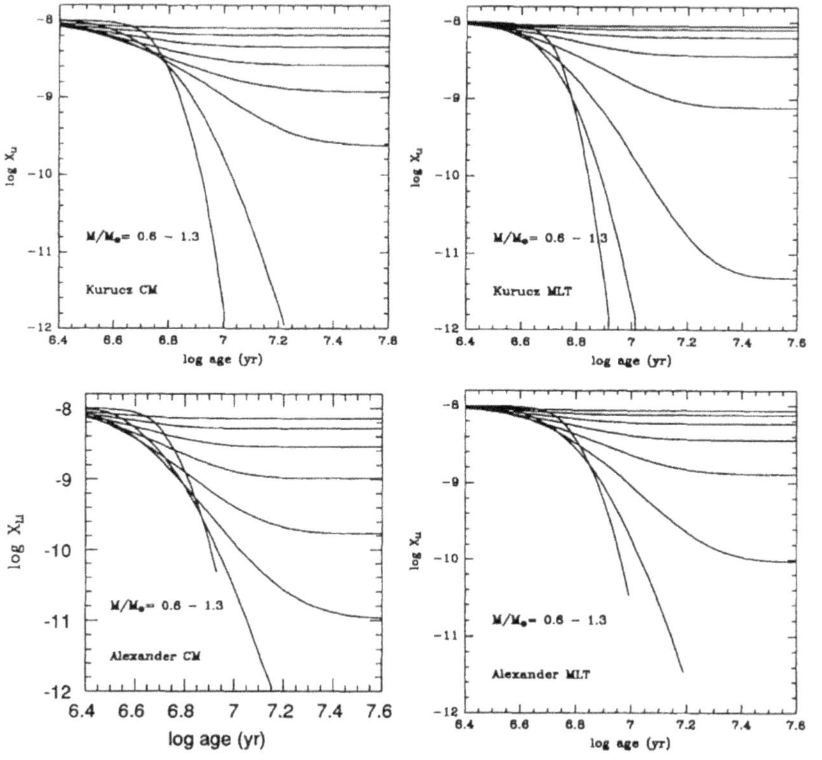

Figure 4.8. Theoretical PMS evolutionary depletion of Li for stars with masses between 1.30 and 0.60 $M_{\odot}$. Note that the starting level of undepleted lithium abundance is set at log $X_{\rm Li} = -8$, which is equivalent to log N (Li) = 3.0. The models obtained using different opacities (Alexander or Kurucz) and different prescriptions of convection—Canuto and Mazzitelli (CM) or Mixing Length Theory (MLT)—are shown in different panels. Figure credit: D'Antona, F., & Mazzitelli, I. 1994 "New pre-main-sequence tracks for $M < 2.5$ $M_{\odot}$ as tests of opacities and convection model" ApJS 90 467.

Higher mass stars still preserve some of their initial lithium content when they settle on the main sequence because they have developed large radiative zones in their interiors where chemical mixing is not efficient (Pinsonneault et al. 1990).

Within the first 10 Myr of PMS stellar evolution, the evolutionary models of D'Antona & Mazzitelli (1994) predict that Li depletion has halted in stars of mass >1.2 $M_{\odot}$, but that it is starting to happen in stars of mass < 0.60 $M_{\odot}$ (Figure 4.8). Lower mass stars take longer to reach the core temperatures needed for thermonuclear Li burning.

Bildsten et al. (1997) derived analytical relations between lithium abundance and the age, luminosity, and radius for contracting PMS stars. They suggested that lithium is a reliable age indicator for very young PMS low-mass stars. This idea was in tune with the suggestions made by previous researchers, such as Martín et al. (1994) and Basri et al. (1996).

The work of Lars Bildsten and his students provided a relatively simple analytical framework for this hypothesis. Subsequently, lithium has been exploited as a chronometer for dating different binary systems, young associations, and open clusters.

Detailed theoretical explorations of the dependence of convective PMS lithium burning on a variety of parameters such as chemical abundances, rotation, and magnetic effects on convection and starspots in the stellar surfaces have been made by Martín & Claret (1996), Ventura et al. (1998), Burke et al. (2004), Jackson & Jeffries (2014), Fu et al. (2016), and Mullan & MacDonald (2023).

Jackson & Jeffries (2014) studied the effects of magnetic starspots on the ages derived from the comparison of lithium depletion observations and model predictions.

A review of PMS Li depletion can be found in Jeffries (2006).

The so-called lithium depletion boundary (LDB) has been defined by several authors (e.g., Basri et al. 1996; Rebolo et al. 1996; Martín et al. 1998; Stauffer et al. 1998; Jeffries & Oliveira 2005; Juarez et al. 2014) as the location in the HRD in terms of effective temperature (or its proxies color or spectral type) or bolometric luminosity (or its proxy absolute magnitude) that separates the region where all objects are Li-depleted from the region where all objects have preserved their lithium.

The LDB is sharply defined if all members in an association or open cluster are coeval, and gets blurred if there is a significant spread in age.

Using data from the Gaia second data release (DR2), the membership around the location of LDB in 12 associations and open clusters was assessed and their LDB ages were revisited using several evolutionary tracks (Galindo-Guil et al. 2022). The results are summarized in Figure 4.9.

The LDB ages span over an order of magnitude, from about 20 Myr to over 600 Myr. In the next section, a closer look will be taken to the youngest part of this diagram where the onset of PMS lithium depletion takes place. The mid part of the diagram will be examined in the next chapter.

While the main focus of many studies of lithium among the members of young associations and clusters has been to use it as a chronometer to determine a single value for the age, some attention has also been devoted to investigate the dispersion of lithium abundance at constant effective temperature (as a proxy for mass).

The following effects have been considered:

(1) Effects from accretion of fresh material from circumstellar disks
(2) Starspots covering a significant fraction of the stellar surface
(3) Effects of magnetic fields and rotation on the lithium depletion rate
(4) Unresolved binaries where both components contribute to the optical spectrum
(5) Contamination by non-members of the association or cluster
(6) An age spread among the members in the association or cluster

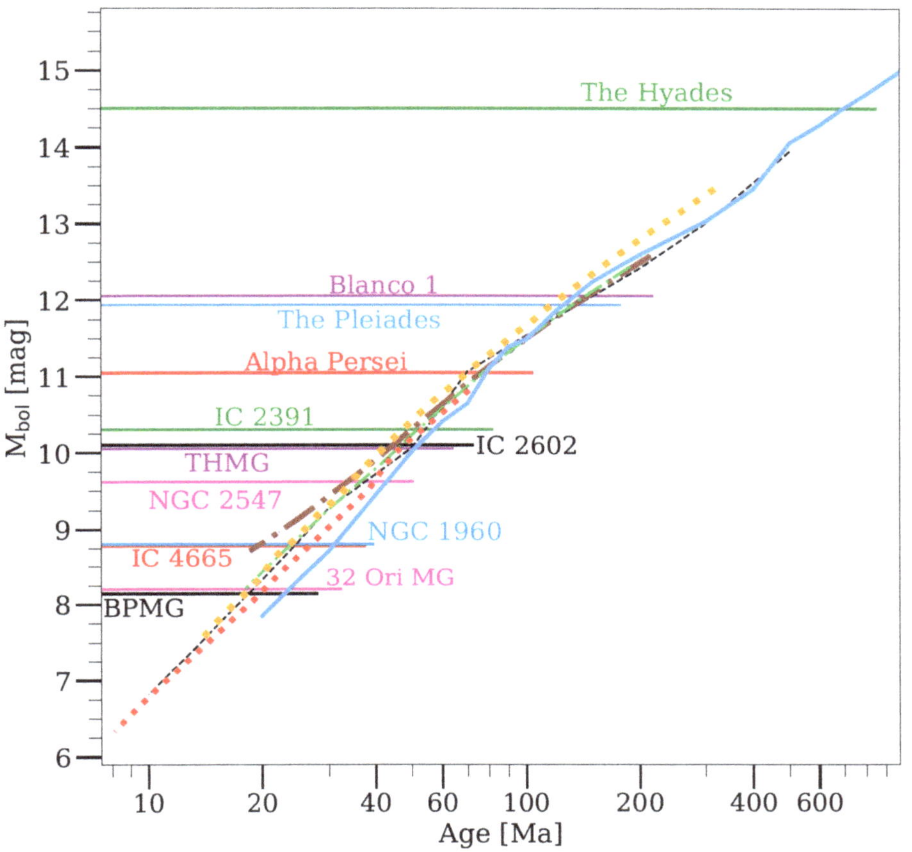

Figure 4.9. An absolute bolometric magnitude versus age diagram showing the location of the LDB for a number of associations and open clusters based on Gaia data. The older a coeval group is, the higher the magnitude (fainter luminosity) of the LDB. Theoretical PMS evolutionary tracks are denoted by the following line types: dashed black (D'Antona & Mazzitelli 1994), thick dash dotted brown (Burrows et al. 1997), dotted red (Siess et al. 2000), dash dotted green (Burke et al. 2004), solid blue (BT-Settl, Allard et al. 2012), dotted thick red (Tognelli et al. 2015). Figure credit: Galindo-Guil, F. J. et al. 2022 "Lithium depletion boundary, stellar associations and Gaia" A&A 664 A70, reproduced with permission © ESO.

4.4 Lithium as a Chronometer in Very Young Stars

Lithium is expected to be efficiently depleted during the PMS evolution of low-mass stars because they have deep convective envelopes. This theoretical prediction has motivated numerous studies of Li among the very young low-mass populations (age < 30 Myr) of associations and open clusters. In this section we take a close look at some of the most important ones spanning the range of ages between 1 Myr and 30 Myr.

The onset of Li depletion is observed at these ages as it rapidly carves a very distinctive pattern among the M-type members of coeval groups of stars.

For such very young stars, the studies of photospheric lithium abundance are complicated by the presence of active accretion disks, variable reddening, fast rotation, non-LTE effects and strong magnetic activity.

4.4.1 The Taurus–Auriga T-Association

The star T Tauri (distance = 145.14 pc) is the prototype of the T Tauri class of very young stars. It is a member of the very young Taurus–Auriga stellar association and star-forming region, and it is one the first stars in which the Li I resonance doublet was noticed (Sanford 1947).

Taurus–Auriga (hereafter TA) is so young that most of the mass in this region is in the form of dense interstellar molecular gas and star formation is still ongoing. Consequently, it is expected that the new-formed stars in the TA association are too young to show any lithium depletion.

Nevertheless, the presence of lithium-depleted stars in TA could be interesting from the point of view of estimating for how long the active processes of star formation may persist in a given molecular cloud complex. It has been proposed that star-formation may start as a low-efficiency long-timescale process before building up to higher-efficiency shorter-timescale outburst (Palla & Stahler 2000).

The census of stellar and substellar members of TA has been increasing steadily for decades.

Esplin & Luhman (2019) reported 79 new members using data from CFHT, Gaia, and UKIRT, and compiled a list of 519 objects that are likely to belong to the association.

They set a Li I EW > 0.4 Å as part of their positive membership criteria, because, they claimed, that most of the confirmed TA members meet such requirement (Basri et al. 1991; Magazzù et al. 1992; Martín et al. 1994).

Lithium was included as a TA membership flag in the work of Kraus et al. (2017) who considered those objects with Li I EW greater than the upper envelope of the Tuc–Hor association (which is about 40 Myr old) as most probably TA members.

Noteworthy patterns that can be mentioned from looking at Figure 4.10 reproduced from the work by Kraus et al. (2017) include the following:

1. The upper envelope of Li I EW values increases toward later spectral types as expected for cosmic lithium abundance as a function of temperature.
2. The dispersion in EW is about a factor of two for early-K and late-M types, and a factor of three for late-K and early-M types.
3. There could be a spread in ages among the members in TA which may account for the dispersion in lithium line strengths. Some evidence for two populations in the association, one younger and more concentrated toward the higher density regions in the molecular clouds, and the other (older) with a wider spatial distribution.

Krolikowski et al. (2021) compiled a list of 658 possible members of TA from multiple sources including Gaia. Their work has revealed 17 subgroups with mean ages ranging from 1.3 Myr to 13 Myr. The possibility that an age spread in the TA

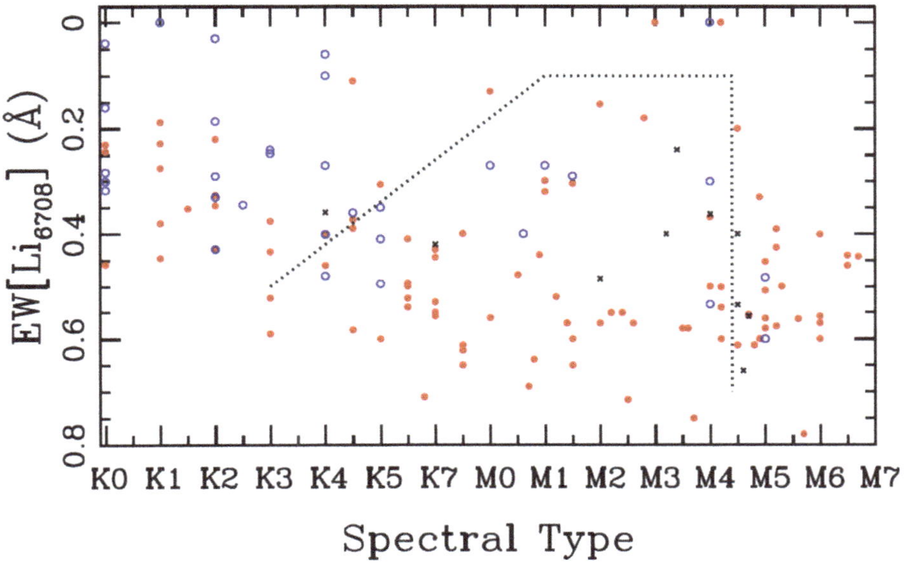

Figure 4.10. Li I resonance doublet equivalent widths measured in spectra of candidate members in the TA association with respect to spectral type. Filled red points denote sources that are most probably members based on criteria other than lithium. Empty blue points indicate sources that are probably non-members. Black crosses are objects that lack enough information to assess their membership status. The dashed line shows the upper envelope of Li I equivalent widths in the older (40 Myr old) Tuc–Hor association. Figure credit: Kraus, A. L. et al. 2017 "The greater Taurus–Auriga ecosystem. I. There is a distributed older population" ApJ 838 150.

population is contributing to the dispersion in lithium therefore seems plausible and deserves further scrutiny.

4.4.2 The Lupus T Association

The Lupus T association has similar properties to the TA association, but it is located in the southern hemisphere and is close to the Scorpius–Centaurus OB association.

Biazzo et al. (2017) studied lithium in Lupus using data collected with the visual arm of X-Shooter at the VLT. Intermediate-resolution ($R = 8800$) were obtained for 85 sources, and higher resolution (R = 17,400) for four sources in Lupus.

Equivalent widths of the Li I resonance doublet were measured by Gaussian fitting to the line profile using the IRAF task splot. Corrections for the blending of the Fe I line at 670.74 nm were made using the empirical prescription provided by Soderblom et al. (1993).

Many stars in Lupus show signs of active accretion from circumstellar disks. Biazzo et al. (2017) reported a careful study of the lithium abundances with appropriate corrections for veiling effects. Corrections for non-photospheric continuum (veiling) were needed for 35% of the stars in the sample. The effects of veiling corrections are illustrated in Figure 4.11.

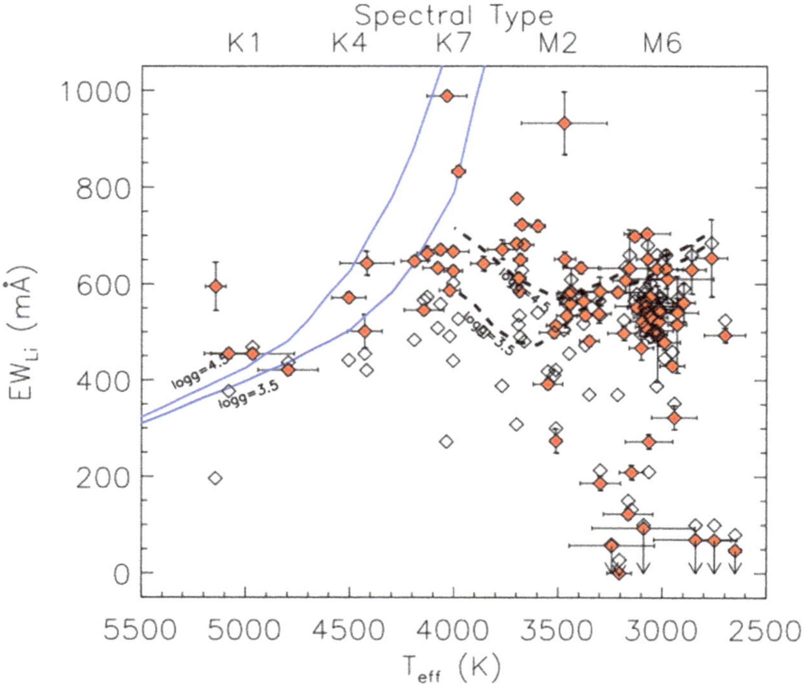

Figure 4.11. The effects of veiling corrections on the Li I resonance doublet EW as a function of spectral type (upper abscissa label) or effective temperature (T_{eff}, lower abscissa label) for stars in the Lupus T association. The EW values directly measured from the X-Shooter spectra are denoted with open symbols, and filled red symbols represent the EWs after correcting for veiling. The blue curves represent lithium isoabundance from the CoGs published by Pavlenko & Magazzù (1996) and Palla et al. (2007). Figure credit: Biazzo, K. et al. 2017 "X-Shooter spectroscopy of young stellar objects in Lupus. Lithium, iron and barium elemental abundances" A&A 605 A66, reproduced with permission © ESO.

The photospheric lithium abundance of the stars in Lupus were determined using the CoGs published by Pavlenko & Magazzù (1996) for temperatures hotter than 4000 K, and by Palla et al. (2007) for cooler stars.

The results can be seen in Figure 4.12. The pattern of observed lithium abundance for Lupus was compared with theoretical models of PMS lithium depletion for different ages (lithium isochrones) computed by Baraffe et al. (2015).

Most of the stars in Figure 4.12 have lithium abundance around log N(Li) = A (Li) = 3.1, which closely corresponds to the generally recognized cosmic value log N (Li) = A(Li) = 3.3.

Stars earlier than K4 (hotter than 4500 K) tend to have higher lithium abundances, probably owing to NLTE effects that were not accounted for in the analysis.

Stars later than M2 (cooler than 3500 K) show a very wide spread in lithium abundances.

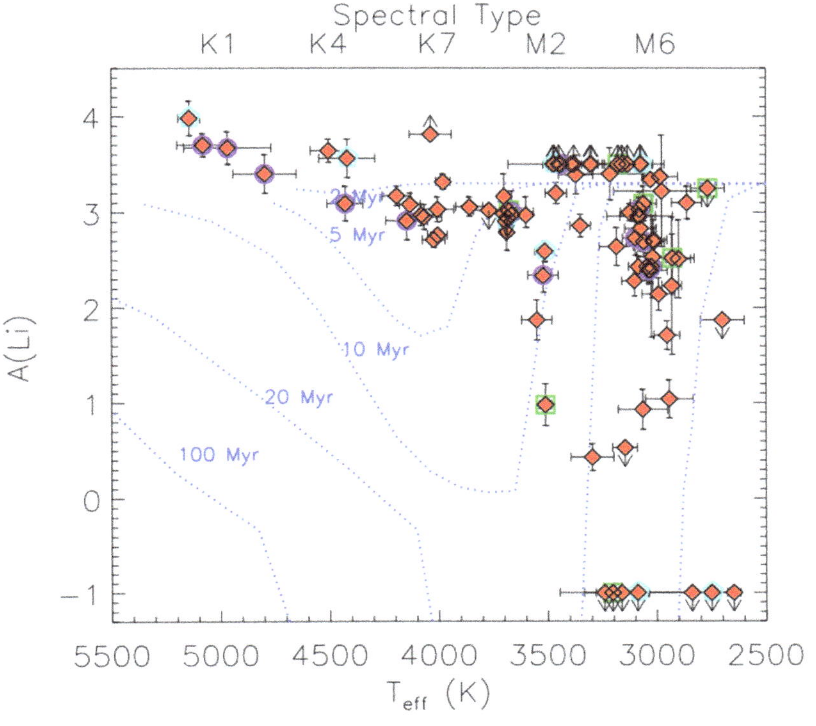

Figure 4.12. Lithium abundance as a function of spectral type (upper abscissa) or effective temperature (lower abscissa) for stars in the Lupus T association. Blue dotted curves represent lithium isochrones from Baraffe et al. (2015). Figure credit: Biazzo, K. et al. 2017 "X-Shooter spectroscopy of young stellar objects in Lupus. Lithium, iron and barium elemental abundances" A&A 605 A66, reproduced with permission © ESO.

Biazzo et al. (2017) considered three main possibilities to explain this spread:

1) A connection between lithium and rotation, such as predicted by some evolutionary models (e.g. Ventura et al. 1998). They searched for a correlation between lithium and the rotational broadening due to rotation ($V \sin i$) but did not find it.
2) An age spread in the association. They compared the ages and masses derived from the isochrones of Baraffe et al. (2015) with the lithium burning calculations of Bildsten et al. (1997).

They did not find a good agreement between the two sets of predictions. While the ages and masses of the lithium-depleted stars were in the range 1–7 Myr, and 0.2–0.3 $M_{\odot}$, respectively, from the Baraffe et al. models, the ages and masses from the Bildsten et al. models turned out to be older than 15 Myr and >0.5 $M_{\odot}$, respectively.

The disagreement between the Baraffe et al. (2015) predictions and the pattern of lithium abundances in Lupus is quite clear in Figure 4.12. It takes longer than 10 Myr to produce lithium depletion factors greater than a factor of 1000, and the models expect that the most lithium-depleted stars should be hotter than actually observed.

Bursts of accretion onto the newly born star with rates $>10^{-4}$ $M_{\odot}$ per year could compress the stellar core and initiate early lithium burning (Baraffe et al. 2017a, 2017b). However, Biazzo et al. estimated that not even those bursts could produce the very large lithium depletion factors observed among the M-type stars in Lupus.

After detailed discussion of all the scenarios mentioned above, Biazzo et al. (2017) concluded that there was no satisfactory explanation of the wide spread in lithium abundances observed among the M-type stars in Lupus. This problem is also present among the TA association M-type members and remains open for future investigations.

Searches for new members of the Lupus star-formation region keep adding more objects that are confirmed via spectroscopic analysis including the detection of the Li I resonance doublet (Majidi et al. 2023).

4.4.3 The Perseus Star-formation Complex and the Associated IC 348 Cluster

At a distance of around one third of a kiloparsec, the Perseus region hosts a number of star-forming clumps, dark clouds and a very young open cluster known as IC 348.

The presence of the Li I resonance doublet among 33 low-mass members of IC 348 was examined by Scott Dahm using the High Resolution Echelle Spectrometer (HIRES) on the 10 m Keck I telescope at MKO.

The spectral resolution of the data was $R = 33{,}000$. Typical signal-to-noise ratios in the lithium spectral region were 100. The spectral types estimated for the objects ranged from G0 to M0, and the ages and masses estimated using the theoretical isochrones of Siess et al. (2000) were between 1 and 10 Myr and between 2 and 0.5 $M_{\odot}$, respectively.

The Li I resonance doublet EWs of the IC 348 members observed by Dahm (2008) were generally higher than those of the Pleiades cluster, as illustrated in Figure 4.13, and as expected from the much lower age of IC 348 (around 6 Myr) compared to that of the Pleiades (around 110 Myr).

Despite the lack of correction for veiling effects, the IC 348 lithium data do not show a wide spread in lithium EW at any given temperature.

There is no appreciable Li depletion in the IC 348 data, except for one star which is noted to be a fast rotator with disk accretion (Dahm 2008). It would be interesting to investigate lithium among the M-type members in IC 348 where the spread in lithium is greater in the T associations.

If the spread in lithium at a given mass is taken to be an indicator of a spread in age, the comparison between IC 348 and the T associations in Lupus and TA suggests that the clustered mode of star formation tends to produce a tighter distribution of stellar age than the distributed star formation mode.

4.4.4 The Orion Giant Star-formation Complex and Associated Clusters

At a distance of around half a kiloparsec, the Orion region is the closest giant star-formation region to us. The most populous and densest cluster formed in this region is named the Orion Nebula Cluster (ONC) and is located at a distance of 414 pc (Menten et al. 2007).

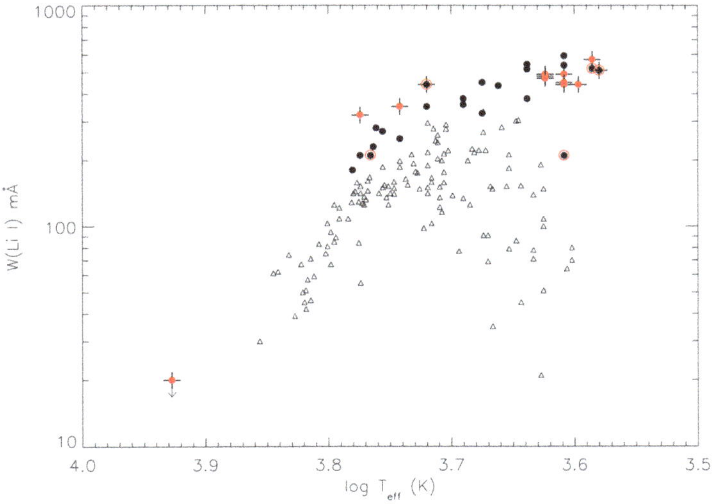

Figure 4.13. Measurements of Li I resonance doublet equivalent widths as a function of effective temperature for IC 348 and Pleiades low-mass stars. Solid back circles represent IC 348 stars without signs of accretion disks. Red symbols denote IC 348 stars with signs of accretion-disk activity. Open triangles represent Pleiades data from Soderblom et al. (1993). Figure credit: Dahm, S. E. 2008 "A spectroscopic examination of accretion diagnostics for near solar mass stars in IC 348" AJ 136 521.

The late Francesco Palla and collaborators drew attention to the pattern of lithium abundances among the low-mass members in the ONC as a way to test theories of cluster formation and evolution (Palla et al. 2005, 2007). They proposed that lithium-depleted stars associated with sites of ongoing star formation could be tracers of a low-efficiency long-lived phase of star production lasting for 10 Myr or more. They noted that such young lithium-poor stars had also been identified in the Upper Scorpius OB association (Martín 1998) and the σ Orionis open cluster (Sacco et al. 2007).

The observations analyzed by Palla et al. (2007) were obtained with the fiber-fed FLAMES spectrograph at the ESO VLT. Mid-resolution spectra (R = 19,300) were obtained for 51 stars of spectral types later than M1 in two regions of the ONC. Pseudo-equivalent widths (pEW) of the Li I resonance doublet were measured using direct integration of the line profile with respect to the local pseudo-continuum in the MIDAS environment. The majority of the stars needed corrections for veiling (see Figure 4.6).

Lithium abundances were derived by Palla et al. (2007) using new CoGs computed with NextGen model atmospheres (Hauschildt et al. 1999) and the WITA 6 code (Pavlenko 2001). This code includes the contribution of the TiO molecular bands to the local opacity. As a result, they inferred that four stars in their sample have lithium abundances significantly lower than the cosmic value.

Palla et al. used the analytical calculations of lithium depletion by Bildsten et al. (1997) to estimate lithium depletion ages between 14 Myr and 33 Myr for those four stars, considerably older than the bulk of the low-mass ONC population (1–2 Myr). They considered lithium depletion to be a useful probe for the history of star formation in the ONC.

An age spread is just one of the factors that may account for the large spread in Li I pEW values of M-type members in young stellar associations. The careful work of Palla and collaborators incorporated many of those effects, and suggests that an age spread plays a major role in the dispersion of lithium abundances among the M-type members of very young associations and clusters.

Evidence in support of an age spread in the ONC was provided by the OmegaCAM survey where three distinct pre-main sequences were identified in the CM diagram (Beccari et al. 2017), suggesting that there were three discrete bursts of star-formation with mean ages 1.24, 1.88, and 2.87 Myr. The age spread within each population is small (<1 Myr), consistent with the characteristic dynamical timescale of the ONC (Tan et al. 2006). This scenario of sequential star-formation implies an age spread within the ONC, but it is intrinsically different from the model of accelerated activity proposed by Palla & Stahler (2000).

Besides the ONC, other well-studied open clusters in the Orion complex are σ Ori and λ Ori. They are slightly closer (distance 350–400 pc) and older (age 1–10 Myr) than the ONC (Walter et al. 2008; Kounkel et al. 2014). It has been suggested that there are two stellar populations with different ages in the σ Ori cluster (Groups 1 and 2) and that the age of the λ Ori cluster is similar to that of the older group in σ Ori (Maxted et al. 2008).

Zapatero Osorio et al. (2002) derived Li abundances for a modest sample of K- and M-type members of the σ Ori cluster. A dispersion in Li abundances could be seen in their data, but the sample was too small. A possible spread in the EW values of the Li I resonance doublet at 670.8 nm was discussed with a larger sample in the study of Kenyon et al. (2005) but no clear conclusion was drawn as to what could be the main cause of it.

Sacco et al. (2008) observed 98 and 49 K6-M5 stars in the σ Ori and λ Ori clusters, respectively, using FLAMES at the VLT with a resolving power $R = 17{,}000$. They measured the Li I equivalent width by direct integration of the line profile in the interval 670.58–670.95 nm using the IRAF task splot. Their results are illustrated in Figure 4.14.

While most of the stars in their sample have strong Li I features with pEW > 400 mÅ, there are a handful with weaker lines. Sacco and coworkers considered them as possibly affected by veiling, or else as non-members of the clusters, although they also mentioned the possibility that some of them could be lithium-depleted members with ages greater than the mean value for the clusters.

Spectroscopic detection of the Li I feature at 670.8 nm has been included as a criterion for membership and pEW measurements have been reported in the most recent works on the census of the σ Ori cluster that consider several hundreds of stars (Caballero et al. 2012, 2019; Hernández et al. 2014), but the question of

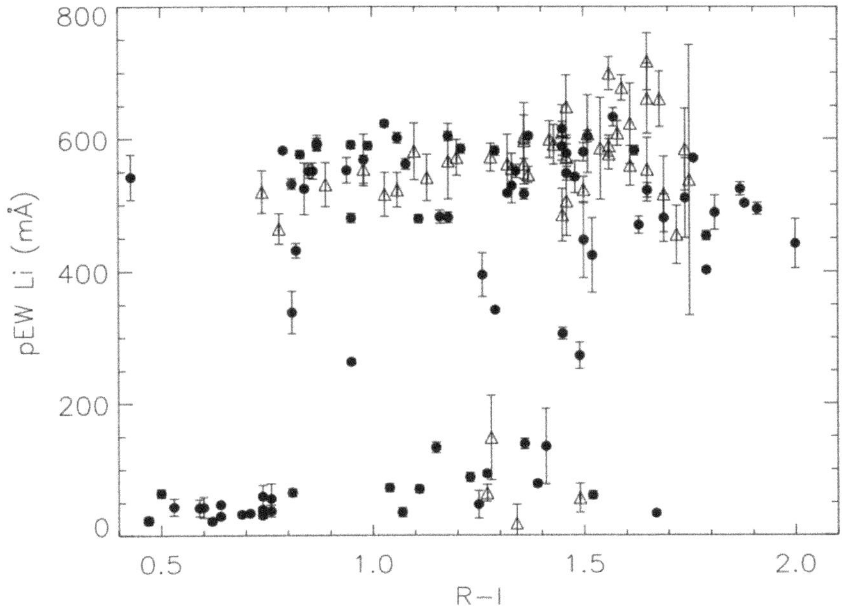

Figure 4.14. Pseudo-equivalent widths of the Li I resonance doublet with respect to the *R–I* color for stars in the σ Ori (filled circles) and λ Ori (open triangles) clusters. Figure credit: Sacco, G. G. et al. 2008 "FLAMES spectroscopy of low-mass stars in the young clusters σ Ori and λ Ori" A&A 488 167, reproduced with permission © ESO.

whether or not there is a genuine spread in lithium indicative of an age spread remains open.

The jury is still out in deciding whether or not there is evidence for an age spread among the members of the σ Ori and λ Ori open clusters.

4.4.5 The NGC 2264 Cluster

NGC 2264 is a rich (around 2000 members) very young (age 1–5 Myr; Rebull et al. 2002; Dahm 2008) open cluster located at a distance of about 760 pc (Park et al. 2000). It still has embedded stars and molecular clouds, indicating that star formation is still going on (Teixeira et al. 2012).

The Gaia ESO survey (GES) obtained precise Li I EW measurements of the low-mass population in NGC 2264 using FLAMES at the VLT (Lanzafame et al. 2015).

A combined study of the lithium data with rotation periods from Cody et al. (2014) was made by Bouvier et al. (2016) who reported that fast rotators are more lithium rich than slow rotators over the range of temperatures between 4400 K and 3800 K (see Figure 4.15).

As possible explanations for the lithium-rotation correlation, Bouvier et al. discussed magnetic activity, steady disk accretion, and planet engulfment, but they did not find any of those scenarios to be completely satisfactory.

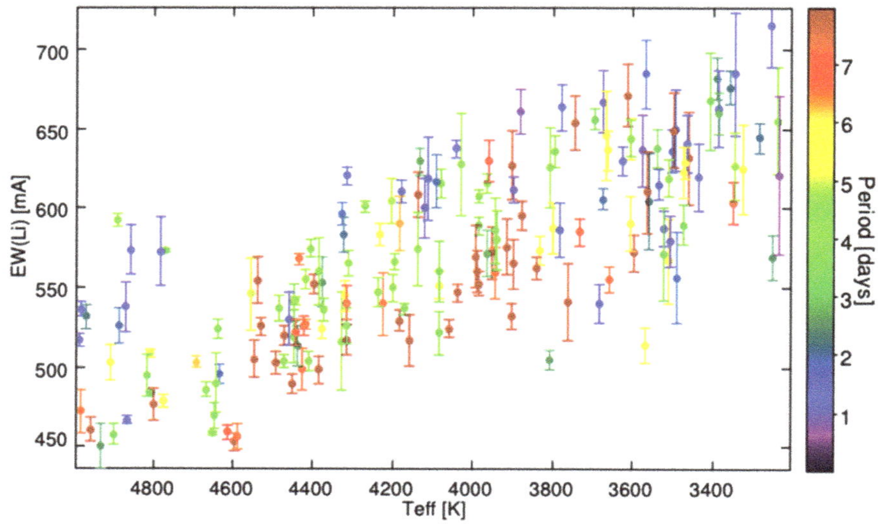

Figure 4.15. Pseudo-equivalent widths (pEW) of the Li I resonance doublet with respect to effective temperature for stars in the NGC2264 cluster. The color scales reflect the different rotation periods. Note that the blue symbols tend to have higher values of EW, particularly between 4400 K and 3800 K. Figure credit: Bouvier, J. et al. 2016 "The GAIA-ESO survey: A lithium-rotation connection at 5 Myr?" A&A 590 A78, reproduced with permission © ESO.

Another possibility that might be considered is the possible effect of rotational broadening on the measurement of the pEW because of line blending and continuum determination.

4.4.6 The TW Hydrae Association

The star TW Hydrae (TW Hya for short) is a mid-K type classical T Tauri star located at a distance of 60.13 pc from us.

A group of T Tauri stars and brown dwarfs associated with TW Hya has been revealed by several teams (de la Reza et al. 1989; Gregorio-Hetem et al. 1992; Kastner et al. 1997; Webb et al. 1999; Murphy et al. 2015).

A high-spectral resolution study of lithium among 32 members in the TW Hya association was undertaken by Mentuch et al. (2008). The study was based on data collected with the MIKE Echelle spectrograph at the 6.5 m Magellan–Clay Telescope at Las Campanas Observatory (Chile). Mentuch et al. used it to obtain high-resolution (R = 60,000) spectra.

Lithium abundances were derived from the Li I resonance doublet at 670.8 nm and the subordinate line at 610.4 nm by comparing the observed spectra with a grid of 240 LTE synthetic spectra computed with the PHOENIX code (Hauschildt & Baron 1999). Their results are illustrated in Figure 4.16.

Mentuch et al. (2008) estimated an age of 12 ± 8 Myr from the comparison of their lithium data with the predictions of evolutionary models and they noted the consistency with other methods such as dynamical expansion (de la Reza et al. 2006).

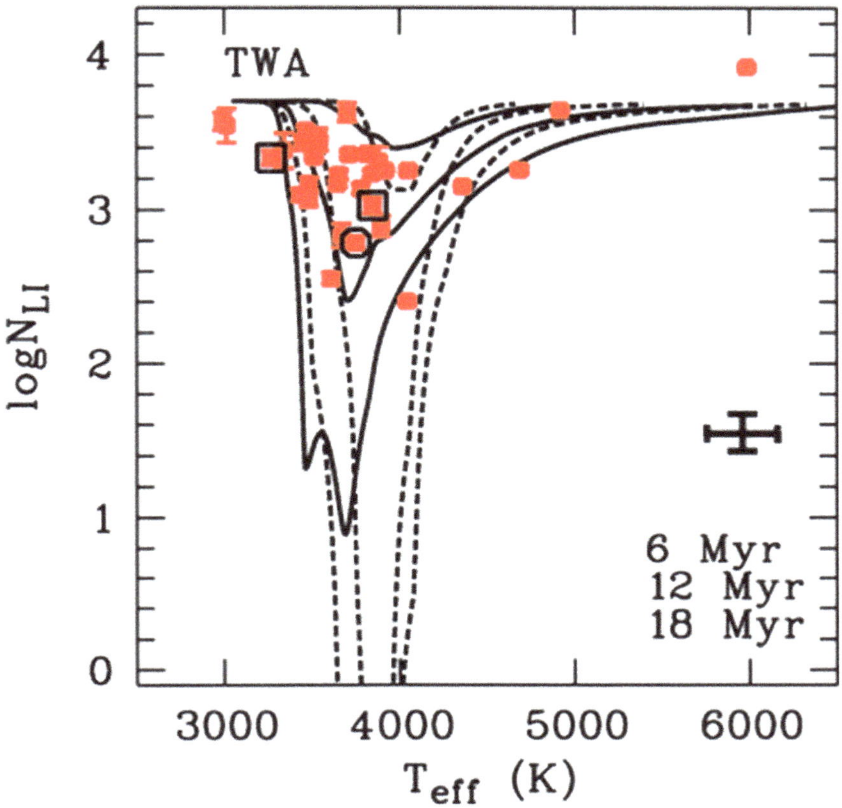

Figure 4.16. Lithium depletion curves versus effective temperature for three different ages (labeled) and two different models (Baraffe et al. 1998, solid line; Siess et al. 2000, dashed line). Red points denote observed abundances of TW Hya members. Encircled points indicate fast rotators. The black cross indicates the typical empirical error bar in abundance and temperature. Figure credit: Mentuch, E. et al. 2008 "Lithium depletion of nearby young stellar associations" ApJ 689 1127.

As in the case of the T Tauri associations, there is a spread of more than an order of magnitude in lithium abundance at constant effective temperature among the late-K and M-type members that deserves further scrutiny in order to try to disentangle the various factors that contribute to it.

4.4.7 The β Pictoris Moving Group

Beta Pictoris (β Pic for brevity) is a young nearby (distance = 19.44 pc) relatively hot (A6V) main-sequence star that harbors an extensive dusty circumstellar disk (Janson et al. 2021) and two superJupiter planets (Lacour et al. 2021).

A group of 24 coeval objects with spectral types ranging from A0 to M7 moving in space together with β Pic was first identified by Barrado y Navascués et al. (1999) and confirmed by Zuckerman et al. (2001).

It is known as the β Pic moving group (hereafter BPMG). Subsequent explorations increased the number of confirmed members to more than a hundred (Torres et al. 2006; Lepine & Simon 2009; Kiss et al. 2011; Schlieder et al. 2012; Shkolnik et al. 2012, 2017; Malo et al. 2014).

The distances of the BPMG members range from 9 pc to 73 pc, making it the nearest known young stellar association, although with high spatial dispersion. Its youth, large spatial extent, high velocity dispersion and the presence of wide binaries—semimajor axis > 1000 astronomical units (AU)—indicate that this association was not formed as part of a dense open cluster.

The metallicity of the BPMG has been estimated to be very similar to that of the Sun (Viana-Almeida et al. 2009). There have been about 20 different studies of lithium in BPMG members. Here just a handful of representative studies are selected.

Mentuch et al. (2008) used the MIKE Echelle spectrograph at the 6.5 m Magellan–Clay Telescope to obtain high-resolution (R = 60,000) spectra of 23 objects in the BPMG.

Lithium abundances were derived from the Li I resonance doublet at 670.8 nm and the subordinate line at 610.4 nm by comparing the observational spectra with a grid of 240 LTE synthetic spectra computed with the PHOENIX code (Hauschildt & Baron 1999). The adopted lithium abundances for the members in the BPMG are shown in Figure 4.17 together with two different sets of evolutionary models.

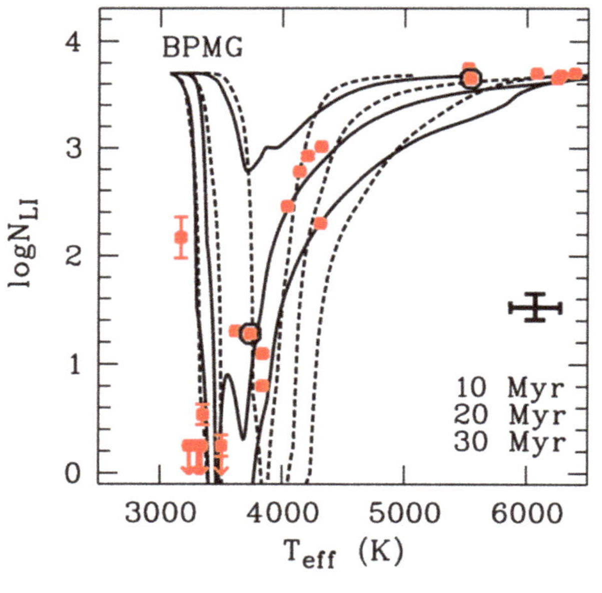

Figure 4.17. Lithium depletion curves versus effective temperature for three different ages (labeled) and two different models (Baraffe et al. 1998, solid line; Siess et al. 2000, dashed line). Red points denote observational abundances among BPMG members. Encircled points indicate fast rotators. The black cross denotes the typical empirical error bar. Figure credit: Mentuch, E. et al. 2008 "Lithium depletion of nearby young stellar associations" ApJ 689 1127.

Mentuch et al. (2008) noted that none of the evolutionary tracks provide a perfect fit to the data, but they favor a BPMG age of 21 ± 9 Myr based on the Baraffe et al. (1998) models.

They used an initial lithium abundance of log *N*(Li) = 3.7, which is higher than the lithium abundance in the solar system meteoritic material and in the T Tauri stars. Taking an initial lithium abundance of log *N*(Li) = 3.3 consistent with meteorites and T Tauri stars decreases the age by 5 Myr, i.e. 16 ± 9 Myr.

Yee & Jensen (2010) obtained lithium observations for additional members in BPMG using the 4 m telescope at Cerro Tololo, and they derived lithium abundances following the method of Sestito et al. (2008).

They calculated ages for each of the 10 individual stars in their sample using three different sets of models (D'Antona & Mazzitelli 1997, 1998; Baraffe et al. 1998; Siess et al. 2000). They reported that the ages obtained from the isochrone fitting of the locations of the stars in the HRD are systematically lower than the ages derived from the PMS lithium depletion rates.

Additional estimates of the BPMG age based on lithium depletion and on isochrone fitting of the HRD give values in the range from 17 Myr to 29 Myr (Binks & Jeffries 2014; Malo et al. 2014; Mamajek & Bell 2014; Bell et al. 2015; Messina et al. 2016; Galindo-Guil et al. 2022).

The improved agreement between the isochrone and lithium ages comes from the introduction into the PMS evolutionary models of effects that account for the magnetic activity in the stars (Feiden 2016).

An independent calculation of the BPMG age, based on dynamical expansion using Gaia Data Release 2 (DR2) data and radial velocities (age = 18.5 ± 2.4 Myr; Miret-Roig et al. 2020) makes the association somewhat younger than the age derived from the updated study of the lithium depletion also using the Gaia DR2 (age = 24.3 ± 0.3 Myr; Galindo-Guil et al. 2022) and the age of 25 ± 1 Myr given by the empirical modeling of Jeffries et al. (2023) using a training set of 52 clusters observed by the GAIA-ESO survey.

The role of rotation was studied by Messina et al. (2016) who compiled Li I EWs and rotation periods for 66 members in the BPMG. They found that fast rotators show less lithium depletion than slow rotators; they used evolutionary models that incorporate the effects of magnetic fields to compare with the data (Figure 4.18). They derived a linear correlation between Li I equivalent width and rotation period, and used it to make corrections for the lithium strength in the slow rotators.

The Dartmouth evolutionary models for three different ages are shown. These models incorporate rotation-induced magnetic fields that inhibit convective energy transport, slowing down the contraction rate and making the stellar surfaces cooler than in standard models.

The corrections for the lithium-rotation trend and the utilization of the Dartmouth evolutionary models (Feiden & Chaboyer 2013; Feiden 2016), which incorporate the effects of rotation-induced magnetic fields on the efficiency of convective mixing, improve the agreement between the data and the theory. Nevertheless, significant lithium dispersion still remains for stars that appear to be coeval and have the same mass and rotation rate.

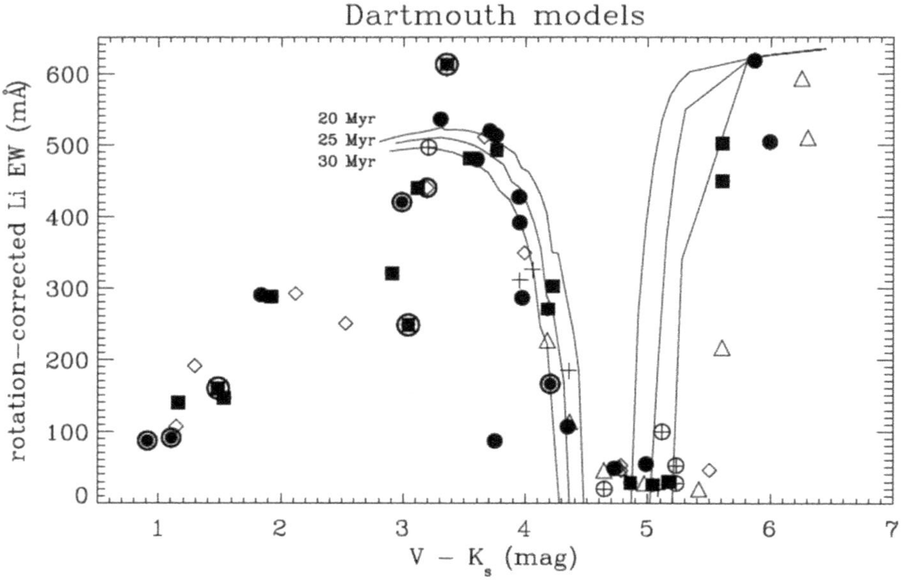

Figure 4.18. Li I resonance doublet EW, corrected for rotation (see text), versus effective temperature for 66 members of the BPMG. The different symbols used for the datapoints have the following meaning: Single stars (black circles), close binaries (<60 AU, triangles), close triples (<30 AU, encircled), wide binaries (black squares), wide triples (+ sign) and spectroscopic binaries (empty diamonds). Figure credit: Messina, S. et al. 2016 "The rotation-lithium depletion correlation in the β Pictoris association and the LDB age determination" A&A 596 A29, reproduced with permission © ESO.

The possibility of an age spread playing a role in shaping the pattern of lithium abundances in the BPMG cannot be dismissed completely, although it does not seem so necessary to invoke it as in the case of the T associations.

4.4.8 The 32 Ori Association

The 32 Ori group is named after the B-type binary 32 Orionis. A group of young stars comoving with the massive binary was discovered by Mamajek (2007). The mean distance to the group is 93 pc, making it one of the nearest to the Sun. Lithium observations have been used as a membership criterion to identify low-mass stars belonging to the group (Alcalá et al. 2000).

A census of the 32 Ori group was carried out by Bell et al. (2017). The group includes 46 members, with spectral types ranging from B5 to L1. This team observed 32 Ori members with the Wide Field Spectrograph at the ANU 2.3 m telescope in Siding Springs Observatory (Australia). The instrumental configuration they used provided a spectral resolving power $R = 7000$.

The EW of the Li I resonance doublet was measured by fitting a Gaussian to the line profile. The EW values obtained are shown in Figure 4.19 as a function of spectral type and color.

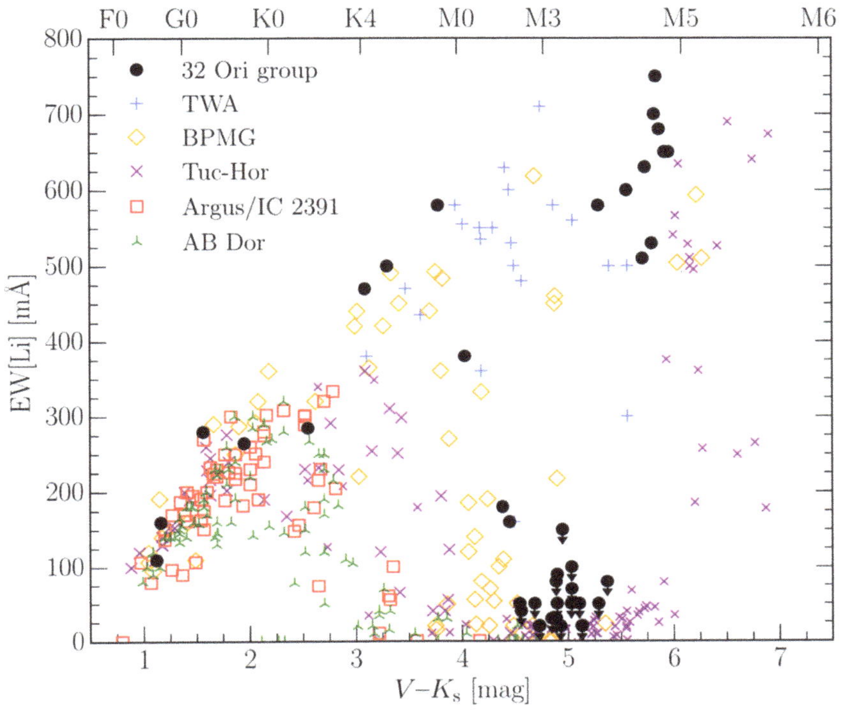

Figure 4.19. Filled black dots denote Li I resonance doublet equivalent widths measured in spectra of members of the 32 Ori association with respect to spectral type (upper abscissa) or $V - K_s$ color (lower abscissa). Other symbols are used for different groups, as shown in the upper right corner. Figure credit: Bell, C. P. M., Murphy, S. J., & Mamajek, E. E. 2017 "A stellar census of the nearby, young 32 Orionis group" MNRAS 468 1198.

Using astrometric and photometric data from the Gaia second data release, spectroscopic data obtained Keck HIRES, as well as light curves from the Kepler extended mission (K2), Stauffer et al. (2020) provided evidence for membership in 32 Ori of two peculiar lithium-rich M-type stars that were found toward the Pleiades cluster by Oppenheimer et al. (1997). They are not included in Figure 4.18, but their Li I resonance doublet EW values are entirely consistent with those of 32 Ori members for their spectral class (M5), strongly supporting membership of 32 Ori rather than in the Pleiades.

The pattern of Li I EWs with respect to color in 32 Ori is quite similar to that in the BPMG. The updated analysis of Galindo-Guil et al. (2022) provides very similar Li-based ages for both associations (25.0 ± 0.8 Myr for 32 Ori, 24.3 ± 0.3 Myr for BPMG), but Luhman (2022a) finds a value younger by 3 Myr for 32 Ori with respect to the BPMG based on color–magnitude diagrams using the third data release of Gaia.

The Li I EW values in 32 Ori for M0 and M5 members tend to be higher than those in the BPMG (see Figure 4.19), which lends support the conclusion reached by Luhman (2022a).

The age obtained in the empirical best-fit analysis of lithium depletion by Jeffries et al. (2023) is 24 ± 1 Myr, in agreement with that by Galindo-Guil et al. (2022).

4.4.9 The Scorpius–Centaurus Complex

The hottest stars are called O-type and B-type. They are massive stars that have short lives owing to rapid consumption of their internal fuel (H, He, and the rest of the chemical elements up to Fe) and end up as Type II supernovae. They trace the spiral arms of the Milky Way because they are found near the stellar nurseries where they are born.

Groups of hot stars were recognized and classified by Pannekoek (1929) and Ruprecht (1966). The term "OB association" was coined by Ambartsumian (1947) to recognize them. Thanks to the Hipparcos satellite, a census of the nearest OB associations was carried out (de Zeeuw et al. 1999).

The nearest OB association to the Sun is called Scorpius–Centaurus (Sco–Cen). It is subdivided into three subgroups named Lower Centaurus–Crux, Upper Centaurus–Crux, and Upper Scorpius. Compact stellar groups and star-forming regions are possibly associated with this large complex of young stars. In fact, Preibisch & Mamajek (2008) consider the Chamaeleon, Lupus, and Rho Ophiuci star-forming regions, and the Corona Australis, TW Hya (see section 4.4.2) and V1062 Sco groups to be part of the same sequence of star formation events that gave birth to the Sco–Cen association.

The presence of the Li I resonance doublet has been used as a membership criterion to classify stars and brown dwarfs as potentially belonging to the Sco–Cen association (Walter et al. 1994; Martín 1998; Preibisch et al. 2001; Zerjal et al. 2019, 2023; Luhman 2022a).

Luhman (2022b) has identified 10,509 candidate members using the early installment of the Gaia third data release. For those objects with available measurements of the Li I EW, he produced the plot shown in Figure 4.20. The main trends that are apparent from this figure are the following:

1. The upper envelope of Li I resonance doublet EWs increases steadily from F-type to mid-M type stars as expected from the larger fraction of lithium atoms in the neutral state in the atmospheres of cooler stars. This envelope is representative of the initial (undepleted) lithium abundance in new-born stars.
2. A large spread in Li I EWs at any spectral type, which increases for M-type stars. Possible effects that may account for this spread were discussed in the context of the TA association (section 4.1).
3. A dip in the upper envelope of Li I EWs in the M-type members in the Upper Centaurus–Lupus (UCL) and Lower Centaurus Crux (LCC) subgroups that is not present among the Upper Scorpius subgroup. This dip indicates the onset of lithium depletion in the bulk of the M-type population of the older LCC subgroup (20 Myr), whereas it is not yet present either the younger Upper Scorpius (11 Myr) or in the Lupus and Ophiucus T Tauri aggregates.

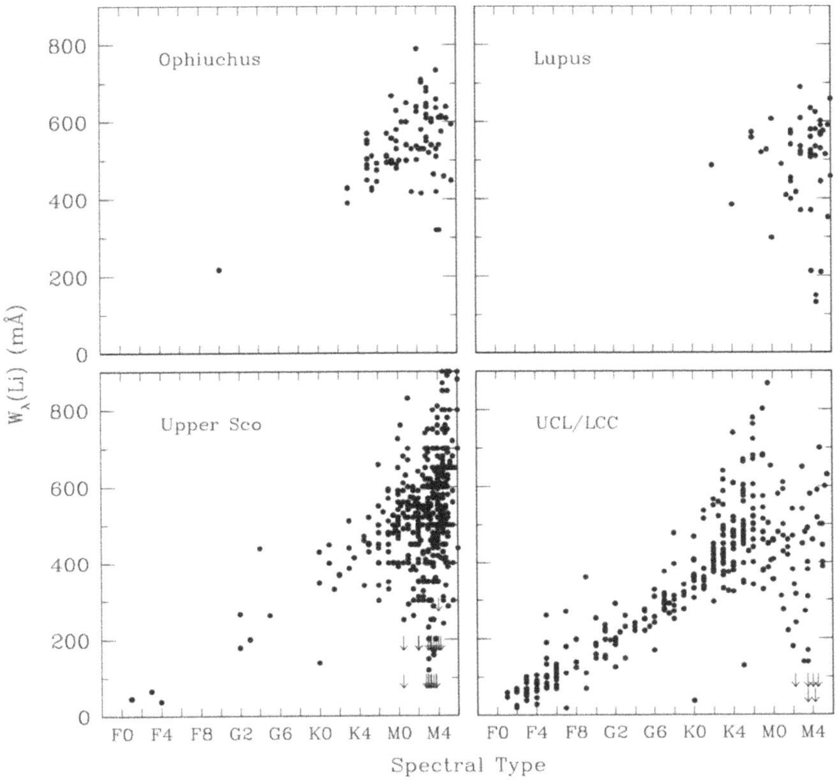

Figure 4.20. Compilation of Li I resonance doublet EWs from different sources as a function of spectral types for various subgroups and star-formation regions that have been proposed to belong to the Sco–Cen complex. Figure credit: Luhman, K. L. 2022b "A census of the stellar populations in the Sco–Cen complex" AJ 163 24.

Since the number of members of the different groups of Sco–Cen is quite large, a detailed statistical study of lithium depletion among their low-mass population is worthwhile. A recent reassessment of the ages of different groups in Sco–Cen using the CHRONOSTAR Bayesian method for kinematic age determination confirms that some younger groups show on average less lithium depletion than older groups (Zerjal et al. 2023).

4.4.10 The Gamma Velorum Cluster

The Wolf–Rayet/O-type star binary system Gamma Velorum is the hottest and most massive member in a young open cluster discovered by Pozzo et al. (2000). It is located at a distance of 336 pc (Maíz Apellániz et al. 2022).

The Gaia-ESO spectroscopic survey (GES) obtained lithium data for 198 low-mass members in the Gamma Velorum cluster (Jeffries et al. 2017) confirmed using radial velocities.

The Li-depletion factors predicted by the evolutionary models were converted into the observational plane using the CoGs provided by Soderblom et al. (1993) for $T_{\rm eff} > 4200$K and by Zapatero Osorio et al. (2002) for $T_{\rm eff} < 4200$K.

Jeffries et al. stated that the evolutionary tracks could not fit the lithium data at all. To circumvent this problem, they proposed an increase of around 10% in the radii of PMS stars due to the effects of magnetic starspots (Jackson & Jeffries 2014). Spots make the stars cooler and less luminous, and they affect the relationships between color and temperature. By applying a constant fraction of 25% of flux blocked at the stellar surface to the models of Dotter et al. (2008) and Baraffe et al. (2015), the agreement between the Li-depletion isochrones and the lithium data in Gamma Velorum was greatly improved, as illustrated in Figure 4.21.

Jeffries et al. (2017) showed that radius inflation leads to a delay in the onset of lithium depletion and to a determination of LDB ages greater than those derived

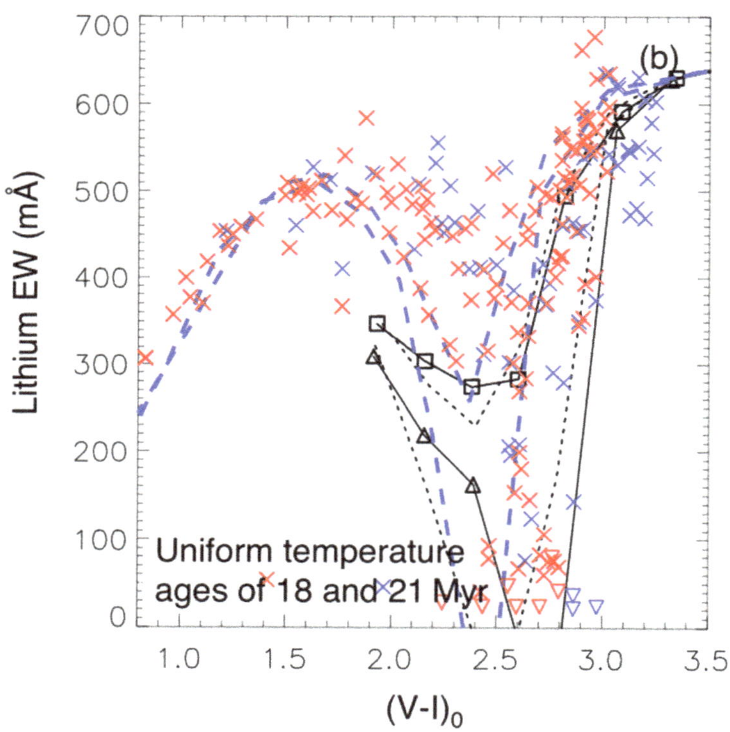

Figure 4.21. Measurements of Li I resonance doublet EWs as a function of dereddened $V - I$ color for Gamma Velorum low-mass cluster members. Blue crosses denote suspected binaries and red crosses represent single stars. No difference in the Li I EW is noticed between these two populations. Two Li-depletion isochrones are plotted corresponding to ages 18 and 21 Myr. Open symbols on each isochrone indicate the location of masses from 0.8 to 0.2 $M_\odot$ in steps of 0.1 $M_\odot$. Dotted lines represent models by Dotter et al. (2008), and solid curves come from the models by Baraffe et al. (2015). The models have been shifted with a constant radius inflation owing to the effects of magnetic starspots. Figure credit: Jeffries, R. et al. 2017 "The GAIA-ESO survey: Lithium depletion in the Gamma Velorum cluster and inflated radii in low-mass pre-main-sequence stars" MNRAS 464 1456.

from standard models. They noted that those results were in good agreement with previous work by Somers & Pinsonneault (2014). Their conclusion is that the LDB age of Gamma Velorum is in the range 18–21 Myr, significantly older than the one obtained from isochrone fitting of the CMD.

4.4.11 The Tucana–Horologium Association

The Tucana–Horologium association (Tuc–Hor for brevity) is a group of nearby (distance around 40 pc) young (20–40 Myr) comoving stars that was discovered independently by Torres et al. (2000) and Zuckerman & Webb (2000). It is similar in size to the BPMG and much larger than the TW Hya association.

Kraus et al. (2014) built upon previous searches for members in Tuc–Hor by Song et al. (2003); Malo et al. (2013) and Rodríguez et al. (2013) and increased the census of members to almost 400. They reported high-resolution ($R = 35000$) spectroscopic observations with MIKE for 13 known members and 191 candidate late-type members (spectral types in the range K3 to M6). Their measurements of Li I EWs with respect to spectral type are shown in Figure 4.22.

The presence of lithium was used by Kraus et al. (2014) as a criterion for membership in the association, rejecting 19 stars with spectral types K4 and earlier being rejected because of the lack of Li I detection (EW < 0.1 Å). For late-K and early-M spectral types lithium cannot be used as a membership criterion because it is largely depleted by the age of this association.

Kraus et al. (2014) reported 3σ confidence levels for Li I detections of 0.05 Å for K-type stars and 0.07 Å for early M-type stars. This precision is more than enough

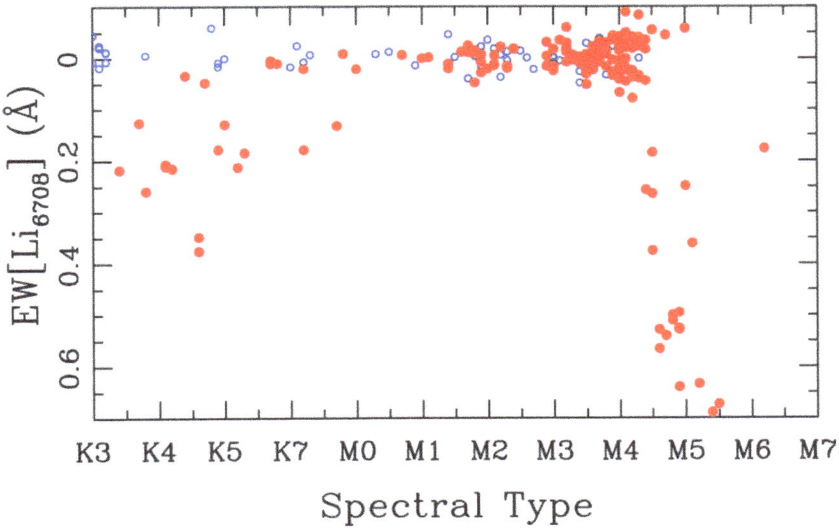

Figure 4.22. Li I resonance doublet EWs measured in the spectra of candidate members of the Tuc–Hor association with respect to spectral type. Filled red points denote most likely members based on criteria other than lithium. Empty blue points indicate probable non-members. Figure credit: Kraus, A. L. et al. 2014 "A stellar census of the Tucana–Horologium moving group" AJ 147 146.

to detect the Li I resonance doublet in these stars for cosmic lithium abundance or for significant lithium depletion.

From the comparison of the lithium and spectral type data with the models of D'Antona & Mazzitelli (1997) and Baraffe et al. (1998), Kraus et al. (2014) obtained a mean age of 40 Myr for the association, with an uncertainty of about 10%.

The lithium age of Tuc–Hor has been found to be consistent with the dynamical age obtained from extensive traceback analysis of members of the association included in the Gaia third data release and with radial velocity measurements from high-resolution spectroscopy (Galli et al. 2023).

4.4.12 The Octans Association

The Octans association began to be recognized as a group of X-ray emitting solar-type stars around the south celestial pole (Torres et al. 2003). A mean distance of 141 ± 34 pc and age of 10 Myr were estimated by Torres et al. (2008).

A study of Octans carried out by Murphy & Lawson (2015) using the Wide Field Spectrograph on the ANU 2.3 m telescope in Siding Spring Observatory (Australia) obtained $R = 7000$ spectra for 29 new members with spectral types between K5 and M4. The EWs of the Li I resonance doublet were measured by Gaussian fitting to the line profile. Multiple fits with different choices of the integration limits and the pseudo-continuum level provided an assessment of the EW uncertainty for each star. The results are shown in Figure 4.23.

The LDB is not completely sampled in Octans owing to the lack of data for colors redder than $V - K = 5$, corresponding to spectral type M4 and later. Nevertheless, the drop in Li I EWs for early-M spectral types indicates an age in the range 30–40 Myr. Murphy & Lawson (2015) noted that there were members with no detection of Li I that could be significantly older.

4.4.13 Very Young Binaries

Observations of lithium have been used to constrain the age of very young binary systems.

Stauffer et al. (1995) obtained a high-resolution (R = 43,733) spectrum of HR 4796B using the Echelle spectrograph at the CTIO 4 m telescope. HR 4796 A and B is a visual binary system (angular separation 7.7″, Jura et al. 1993) where the primary is an early-type (A0) star with a dusty disk and the secondary is a late-type (M2.5) star.

The Li I EW value of HR 4796B measured by Stauffer et al. using Gaussian fitting to the resonance feature at 670.8 nm was 0.55 ± 0.03 Å, which is similar to those found in T Tauri stars (Martín et al. 1994). Using isochrone fitting and the strength of the Li I feature, Stauffer et al. estimated an age of 8 ± 2 Myr for this binary system.

Similar studies of lithium and isochrone age for the young cool companions of bright early-type stars with disks have led to age constraints for Fomalhaut (Barrado y Navascués et al. 1997) and Vega (Barrado y Navascués 1998).

More recently, comparative studies between lithium depletion and isochrone fitting have been started for young multiple systems with brown dwarfs and stars

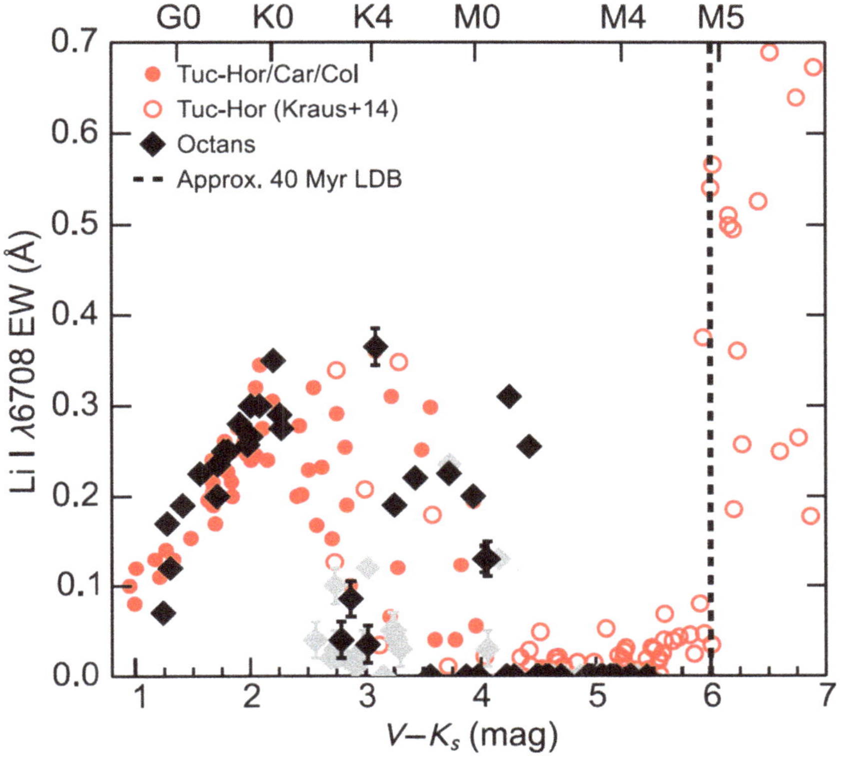

Figure 4.23. Li I resonance doublet EWs measured in spectra of members of the Octans association (filled black diamonds) with respect to spectral type and color. Stars not confirmed as members on the basis of kinematic criteria are shown by gray symbols. Filled red points denote the EW measurements in the Tuc–Hor, Carina and Columba complex and empty red circles represent the EW values provided by Kraus et al. (2014) in Tuc–Hor. Figure credit: Murphy, S. J., & Lawson, W. A. 2015 "New low-mass members of the Octans stellar association and an updated 30–40 Myr lithium age" MNRAS 447 1267.

(Martín et al. 2022). This approach offers a promising avenue for testing the calibration of age for young stars and deserves further exploration.

Bibliography

Ambartsumian, V. A. 1947, J Évolyutsiya zvezd i astrofizika

Alcalá, J. M., Covino, E., Torres, G., et al. 2000, A&A, 353, 186

Allard, F., Homeier, D., & Freytag, B. 2012, RSPTA, 370, 2765

Baade, D., Cristiani, S., Lanz, T., et al. 1991, A&A, 251, 253

Baraffe, I., Chabrier, G., Allard, F., & Hauschildt, P. H. 1998, A&A, 337, 403

Baraffe, I., Homeier, D., Allard, F., & Chabrier, G. 2015, A&A, 577, A42

Baraffe, I., Pratt, J., Goffrey, T., et al. 2017a, ApJ, 845, L6

Baraffe, I., Elbakyan, V. G., Vorobyov, E. I., & Chabrier, G. 2017b, A&A, 597, A19

Barrado y Navascués, D., Stauffer, J. R., Hartmann, L., & Balachandran, S. C. 1997, ApJ, 475, 313

Barrado y Navascués, D. 1998, A&A, 339, 831
Barrado y Navascués, D., Stauffer, J. R., Song, I., & Caillault, J. -P. 1999, ApJ, 520, L123
Basri, G., Martin, E. L., & Bertout, C. 1991, A&A, 252, 625
Basri, G., Marcy, G. W., & Graham, J. R. 1996, ApJ, 458, 600
Beccari, G., Petr-Gotzens, M. G., Boffin, H. M. J., et al. 2017, A&A, 604, A22
Bell, Cameron P. M., Mamajek, Eric E., & Naylor, T. 2015, MNRAS, 454, 593
Bell, C. P. M., Murphy, S. J., & Mamajek, E. E. 2017, MNRAS, 468, 1198
Biazzo, K., Frasca, A., Alcalá, J. M., et al. 2017, A&A, 605, 66
Bildsten, L., Brown, E. F., Matzner, C. D., & Ushomirsky, G. 1997, ApJ, 482, 442
Binks, A. S., & Jeffries, R. D. 2014, MNRAS, 438, L11
Bodenheimer, P. 1965, ApJ, 142, 451
Bouvier, J., Lanzafame, A. C., Venuti, L., et al. 2016, A&A, 590, A78
Burke, C. J., Pinsonneault, M. H., & Sills, A. 2004, ApJ, 604, 272
Burrows, A., Marley, M., Hubbard, W. B., et al. 1997, ApJ, 491, 856
Caballero, J. A., Cabrera-Lavers, A., García-Álvarez, D., & Pascual, S. 2012, A&A, 546, A59
Caballero, J. A., de Burgos, A., Alonso-Floriano, F. J., et al. 2019, A&A, 629, A114
Cody, A. M., Stauffer, J., Baglin, A., et al. 2014, AJ, 147, 82
Combes, F., & Wiklind, T. 1998, A&A, 334, L81
D'Antona, F., & Mazzitelli, I. 1994, ApJS, 90, 467
D'Antona, F., & Mazzitelli, I. 1997, Mem. S. Astron. Italiana, 68, 807
D'Antona, F., & Mazzitelli, I. 1998, Brown Dwarfs and Extrasolar Planets ASP Conf. Series, 134, 442
Dahm, S. E. 2008, AJ, 136, 521
David, T. J., Cody, A. M., Hedges, C. L., et al. 2019, AJ, 158, 79
de la Reza, R., Torres Carlos, A. O., Quast, G., Castilho, B. V., & Vieira, G. L. 1989, AJ, 343, L61
de la Reza, R., Jilinski, E., & Ortega, V. G. 2006, AJ, 131, 2609
de Zeeuw, P. T., Hoogerwerf, R., de Bruijne, J. H. J., Brown, A. G. A., & Blaauw, A. 1999, AJ, 117, 354
Dotter, A., Chaboyer, B., Jevremović, D., et al. 2008, ApJS, 178, 89
Esplin, T. L., & Luhman, K. L. 2019, AJ, 158, 54
Ezer, D., & Cameron, G. W. 1963, Icar, 1, 422
Feiden, G. A., & Chaboyer, B. 2013, ApJ, 779, 183
Feiden, G. A. 2016, A&A, 593, A99
Friedel, D. N., Kemball, A., & Fields, B. D. 2011, ApJ, 738, 37
Fu, X., Bressan, A., Molaro, P., & Marigo, P. 2016, IAU Symp., 317, 300
Galindo-Guil, F. J., Barrado, D., Bouy, H., Olivares, J., et al. 2022, A&A, 664, A70
Galli, P. A. B., Miret-Roig, N., Bouy, H., Olivares, J., & Barrado, D. 2023, MNRAS, 520, 6245
Greco, E., Miceli, M., Orlando, S., et al. 2022, ApJ, 931, 132
Gregorio-Hetem, J., Lepine, J. R. D., Quast, G. R., Torres, C. A. O., & de La Reza, R. 1992, AJ, 103, 549
Hauschildt, P. H., & Baron, E. 1999, JCoAM, 109, 41
Hauschildt, P. H., Allard, F., Ferguson, J., Baron, E., & Alexander, D. R. 1999, ApJ, 525, 871
Hayashi, C. 1961, PASJ, 13, 450
Hayashi, C., & Nakano, T. 1963, PThPh, 30, 460
Herbig, G. H. 1962, AdA&A, 1, 47

Herbig, G. H. 1968, ZA, 68, 243
Hernández, J., Calvet, N., Perez, A., et al. 2014, ApJ, 794, 36
Howk, J. C., Lehner, N., Fields, B. D., & Mathews, G. J. 2012, Natur, 489, 121
Izzo, L., Molaro, P., Cescutti, G., et al. 2022, MNRAS, 510, 5302
Jackson, R. J., & Jeffries, R. D. 2014, MNRAS, 445, 4306
Janson, M., Brandeker, A., Olofsson, G., & Liseau, R. 2021, A&A, 646, A132
Jeffries, R. D., Randich, S., & Pasquini, L. 2006, Chemical Abundances and Mixing in Stars in the Milky Way and Its Satellites ESO Astrophysics Symposia (Berlin: Springer) 163
Jeffries, R. D., & Oliveira, J. M. 2005, MNRAS, 358, 13
Jeffries, R. D., Jackson, R. J., Franciosini, E., et al. 2017, MNRAS, 464, 1456
Jeffries, R. D., Jackson, R. J., Wright, N. J., et al. 2023, MNRAS, 523, 802
Juarez, A. J., Cargile, P. A., James, D. J., & Stassun, K. G. 2014, ApJ, 795, 143
Jura, M., Zuckerman, B., Becklin, E. E., & Smith, R. C. 1993, ApJ, 418, L37
Kastner, J. H., Zuckerman, B., Weintraub, D. A., & Forveille, T. 1997, Sci., 277, 67
Kawanomoto, S., Kajino, T., Aoki, W., et al. 2009, ApJ, 701, 1506
Kenyon, M. J., Jeffries, R. D., Naylor, T., Oliveira, J. M., & Maxted, P. F. L. 2005, MNRAS, 356, 89
Kiss, L. L., Moór, A., Szalai, T., et al. 2011, MNRAS, 411, 117
Knauth, D. C., Federman, S. R., & Lambert, D. L. 2003, ApJ, 586, 268
Kounkel, M., Hartmann, L., Loinard, L., et al. 2014, ApJ, 790, 49
Kraus, A. L., Shkolnik, E. L., Allers, Katelyn, N., & Liu, M. C. 2014, AJ, 147, 146
Kraus, A. L., Herczeg Gregory, J., Rizzuto Aaron, C., et al. 2017, ApJ, 838, 150
Krolikowski, D. M., Kraus, A. L., & Rizzuto, A. C. 2021, AJ, 162, 110
Lacour, S., Wang, J. J., Rodet, L., et al. 2021, A&A, 654, L2
Lanzafame, A. C., Frasca, A., Damiani, F., et al. 2015, A&A, 576, A80
Lemoine, M., Ferlet, R., & Vidal-Madjar, A. 1995, A&A, 298, 879
Lepine, S., & Simon, M. 2009, AJ, 137, 3632
Luhman, K. 2022a, AJ, 164, 151
Luhman, K. 2022b, AJ, 163, 24
Magazzù, A., Rebolo, R., & Pavlenko, Y. V. 1992, ApJ, 392, 159
Maíz Apellániz, J., Barbá, R. H., Fernández Aranda, R., et al. 2022, A&A, 657, A131
Majidi, F. Z., Alcalá, J. M., Frasca, A., et al. 2023, A&A, 671, A46
Malo, L., Doyon, R., Lafrenière, D., et al. 2013, ApJ, 762, 88
Malo, L., Artigau, É., Doyon, R., et al. 2014, ApJ, 788, 81
Mamajek, E. E., Elmegreen, B. G., & Palous, J. 2007, Triggered Star Formation in a Turbulent ISM. IAU Symp., 237, 442
Mamajek, E. E., & Bell, C. P. M. 2014, MNRAS, 445, 2169
Martín, E. L., Rebolo, R., Magazzù, A., & Pavlenko, Y. V. 1994, A&A, 282, 503
Martín, E. L., & Claret, A. 1996, A&A, 306, 408
Martín, E. L. 1997, A&A, 321, 492
Martín, E. L. 1998, AJ, 115, 351
Martín, E. L., Basri, G., Gallegos, J. E., et al. 1998, ApJ, 499, L61
Martín, E. L., Lodieu, N., & del Burgo, C. 2022, MNRAS, 510, 2841
Matsuura, M., Wesson, R., Arendt, R. G., et al. 2022, MNRAS, 517, 4327
Maxted, P. F. L., Jeffries, R. D., Oliveira, J. M., Naylor, T., & Jackson, R. J. 2008, MNRAS, 385, 2210

Menten, K. M., Reid, M. J., Forbrich, J., & Brunthaler, A. 2007, A&A, 474, 515
Mentuch, E., Brandeker, A., van Kerkwijk, M. H., Jayawardhana, R., & Hauschildt, P. H. 2008, ApJ, 689, 1127
Messina, S., Lanzafame, A. C., Feiden, G. A., et al. 2016, A&A, 596, A29
Miret-Roig, N., Galli, P. A. B., Brandner, W., et al. 2020, A&A, 642, A179
Molaro, P., & Vladilo, G. 1989, Ap&SS, 156, 107
Molaro, P., Bressan, A., Barbieri, M., Marigo, P., & Zaggia, S. 2012, MSAIS, 22, 233
Morton, D. C. 1974, ApJ, 193, L35
Mullan, D. J., & MacDonald, J. 2023, RNAAS, 7, 59
Muller, S., Roueff, E., Black, J. H., et al. 2020, A&A, 637, A7
Murphy, S. J., Lawson, W. A., & Bento, J. 2015, MNRAS, 453, 2220
Murphy, S. J., & Lawson, W. A. 2015, MNRAS, 447, 1267
Neufeld, D. A., Goldsmith, P. F., Comito, C., & Schmiedeke, A. 2017, ApJ, 837, 52
Oppenheimer, B. R., Basri, G., Nakajima, T., & Kulkarni, S. R. 1997, AJ, 113, 296
Palla, F., & Stahler, S. W. 2000, ApJ, 540, 255
Palla, F., Randich, S., Flaccomio, E., & Pallavicini, R. 2005, ApJ, 626, L49
Palla, F., Randich, S., Pavlenko, Y. V., Flaccomio, E., & Pallavicini, R. 2007, ApJ, 659, L41
Pannekoek, A. 1929, PuAms, 2, 1
Park, B.-G., Sung, H., Bessell, M. S., & Kang, Yong H. 2000, AJ, 120, 894
Pavlenko, Y. V., & Magazzù, A. 1996, A&A, 311, 961
Pavlenko, Y. V. 2001, ARep., 45, 144
Pinsonneault, M. H., Kawaler Steven, D., & Demarque, P. 1990, ApJS, 74, 501
Pozzo, M., Jeffries, R. D., Naylor, T., et al. 2000, MNRAS, 313, L23
Preibisch, T., Guenther, E., & Zinnecker, H. 2001, AJ, 121, 1040
Preibisch, T., & Mamajek, E. E. 2008, Handbook of Star Forming Regions, Volume II: The Southern Sky (San Francisco, CA: ASP Monograph Publ.) 5, 235
Rebolo, R., Martin, E. L., Basri, G., Marcy, G. W., & Zapatero-Osorio, M. R. 1996, ApJ, 469, L59
Rebull, L. M., Wolff, S. C., Strom, S. E., & Makidon, R. B. 2002, AJ, 124, 546
Rodríguez, David, R., Zuckerman, B., Kastner, J. H., et al. 2013, ApJ, 774, 101
Ruprecht, J. 1966, BAICz, 17, 33
Sacco, G. G., Randich, S., Franciosini, E., Pallavicini, R., & Palla, F. 2007, A&A, 462, L23
Sacco, G. G., Franciosini, E., Randich, S., & Pallavicini, R. 2008, A&A, 488, 167
Sanford, R. F. 1947, PASP, 59, 134
Schlieder, J. E., Lépine, S., & Simon, M. 2012, AJ, 144, 109
Sestito, P., Palla, F., & Randich, S. 2008, A&A, 487, 965
Shkolnik, E. L., Anglada-Escudé, G., Liu, M. C., et al. 2012, ApJ, 758, 56
Shkolnik, E. L., Allers, K. N., Kraus, A. L., Liu, M. C., & Flagg, L. 2017, AJ, 154, 69
Siess, L., Dufour, E., & Forestini, M. 2000, A&A, 358, 593
Soderblom, D. R., Jones, B. F., Balachandran, S., et al. 1993, ApJ, 106, 1059
Somers, G., & Pinsonneault, M. H. 2014, ApJ, 790, 72
Song, I., Zuckerman, B., & Bessell, M. S. 2003, ApJ, 599, 342
Stauffer, J. R., Hartmann, L. W., & Barrado y Navascues, D. 1995, ApJ, 454, 910
Stauffer, J. R., Schultz, G., & Kirkpatrick, J. D. 1998, ApJ, 499, L199
Stauffer, J., Barrado, D., David, T., et al. 2020, AJ, 160, 30
Steigman, G. 1996, ApJ, 457, 737

Suárez Mascareño, A., Damasso, M., Lodieu, N., et al. 2021, NatAs, 6, 232
Tan, J. C., Krumholz, M. R., & McKee, C. F. 2006, ApJ, 641, L121
Teixeira, P. S., Lada, C. J., Marengo, M., & Lada, E. A. 2012, A&A, 540, A83
Tognelli, E., Prada Moroni, P. G., & Degl'Innocenti, S. 2015, MNRAS, 449, 3741
Torres, C. A. O., et al. 2000, AJ, 120, 1410
Torres, C. A. O., Quast, G. R., de La Reza, R., da Silva, L., & Melo, C. H. F. 2003, Galactic Star Formation Across the Stellar Mass Spectrum ASP Conf. Ser. 287 (San Francisco, CA: ASP) 439
Torres, C. A. O., Quast, G. R., da Silva, L., et al. 2006, A&A, 460, 695
Torres, C. A. O., Quast, G. R., Melo, C. H. F., & Sterzik, M. F. 2008, Handbook of Star Forming Regions, Volume II: The Southern Sky (San Francisco, CA: ASP Monograph Publ.) 5, 757
Traub, W. A., & Carleton, N. P. 1973, ApJ, 184, L11
Ventura, P., Zeppieri, A., Mazzitelli, I., & D'Antona, F. 1998, A&A, 331, 1011
Viana Almeida, P., Santos, N. C., Melo, C., et al. 2009, A&A, 501, 965
Walter, F. M., Vrba, F. J., Mathieu, R. D., Brown, A., & Myers, P. C. 1994, AJ, 107, 692
Walter, F. M., Sherry, W. H., Wolk, S. J., & Adams, N. R. 2008, Handbook of Star Forming Regions 4, (San Francisco, CA: ASP) 732
Webb, R. A., Zuckerman, B., Platais, I., et al. 1999, ApJ, 512, L63
Weymann, R., & Moore, E. 1963, ApJ, 137, 552
Yee, J. C., & Jensen, E. L. N. 2010, ApJ, 711, 303
Zapatero Osorio, M. R., Béjar, V. J. S., Pavlenko, Y., et al. 2002, A&A, 384, 937
Žerjal, M., Ireland, M. J., Nordlander, T., et al. 2019, MNRAS, 484, 4591
Žerjal, M., Ireland, M. J., Crundall, T. D., Krumholz, M. R., & Rains, A. D. 2023, MNRAS, 519, 3992
Zuckerman, B., & Webb, R. A. 2000, ApJ, 535, 959
Zuckerman, B., Song, I., Bessell, M. S., & Webb, R. A. 2001, ApJ, 562, L87

Eduardo Martín

Chapter 5

Lithium in Main and Post-main-sequence Stars

Stars spend most of their lifetimes in a very stable configuration where the gravitational contraction is counterbalanced by the thermonuclear energy coming from hydrogen fusion in the core. This long phase is called the stellar main sequence.

The Sun is a main-sequence star. In the course of its evolution, the Sun has depleted its initial content of lithium by more than two orders of magnitude with respect to the rest of the solar system. This process may have taken place during the evolution of the Sun while it was gravitationally contracting in the PMS phase and while it has been steadily fusing H into He on the main-sequence.

It is likely that the loss of angular momentum of the Sun through magnetic braking and the depletion of Li are connected. However, while most solar twins, including the Sun itself, share similar magnetic activity and rotation rates, there is a wide spread in Li abundances, the Sun belonging to the class of Li-poor stars.

There is controversy in the literature concerning whether or not the solar lithium abundance is normal for its age and mass.

The Sun can be put in context by looking at other stars over a range of ages and masses. These can provide a broader picture of how the Sun may have evolved in the past and how it is likely to change in the future.

A fully consistent picture of how the Sun has evolved to reach its current status is still missing. Lithium depletion is clearly one of the key ingredients that remain to be properly accounted for in order to reach a proper understanding of the evolution of the Sun and the solar-type stars.

Before settling on the main-sequence (pre-main-sequence evolution) and after departing from the main-sequence (post-main-sequence evolution), stars evolve more rapidly and other mechanisms take effect that alter the lithium composition. Some of these mechanisms make lithium decay (convection and dilution) and others enhance it (stellar flares, He- and Li-flashes, and hot bottom burning in red giants).

Lithium is a complex chronometer in stars because the time evolution of its surface abundance depends on a variety of physical phenomena.

doi:10.1088/2514-3433/acd8bbch5

5.1 Lithium in Solar-type Stars

The history of how the lithium abundance of the Sun has dwindled by a factor of about 200 during its 4560 million years of existence, and how it can be expected to change in the future, can be traced by looking at stars that are very similar to the Sun in mass and overall chemical composition (so-called metallicity) but have different ages.

Stellar ages can be obtained using the following methods:

(1) Isochrone fitting to the location in the HRD of ensembles of coeval stars of different masses, such as those belonging to young associations, open clusters, and globular clusters (Sandage & Eggen 1969; Yi et al. 2001)
(2) Generalization of the previous method to billions of stars across the Milky Way using Gaia parallaxes (Howes et al. 2019)
(3) Calibrations of the spin-down rate of solar-type stars (so-called gyrochronology) using relationships such as the Skumanich law (Skumanich 1972), although more advanced to account for rotation data from the Kepler and Gaia space missions (Curtis et al. 2020; Gruner & Barnes 2020)
(4) Calibrations of the rate of Li depletion on the surface of stars and brown dwarfs (particularly the LDB method)
(5) Calibrations of the rate of decay of unstable isotopes (so-called nucleochronology; Reeves 1991; Plez et al. 2004; Barbuy et al. 2011)
(6) Cooling times of degenerate objects (brown dwarfs and white dwarfs)
(7) Asteroseismological modeling (Bazot et al. 2018)

For more details on the methods employed to obtain stellar ages, the reader can consult the reviews by Soderblom (2010) and Barrado (2016).

The determination of lithium abundances in solar-type stars is a complicated task because the strongest lithium line available for observation is usually rather weak, and is known to be blended with other atomic and molecular lines, as illustrated in Figure 5.1 where the Li I resonance doublet is shown for the Sun and two solar-type stars with Li abundance higher than solar.

The need for high spectral resolution makes lithium observations quite demanding in terms of telescope time, and this fact, in combination with the very low Li abundances in many stars due to surface Li depletion implies that the total numbers of stars for which Li abundances are available is relatively scarce with respect to other more abundant elements as illustrated in Figure 5.2. As a result, the studies of Li behavior are hampered by low number statistics that sometimes lead to controversial results in the literature.

Solar analogs (also called solar twins) are stars that have properties very similar to those of the Sun (Cayrel de Strobel et al. 1981). The Li abundance of solar analogs has been studied in many field stars and in open clusters.

A summary of the observational scene is illustrated in Figure 5.3, which is taken from the recent work of Carlos et al. (2020). The Li depletion in solar twin stars is correlated with age, although for a given age there is significant spread, which indicates that other factors may be at play.

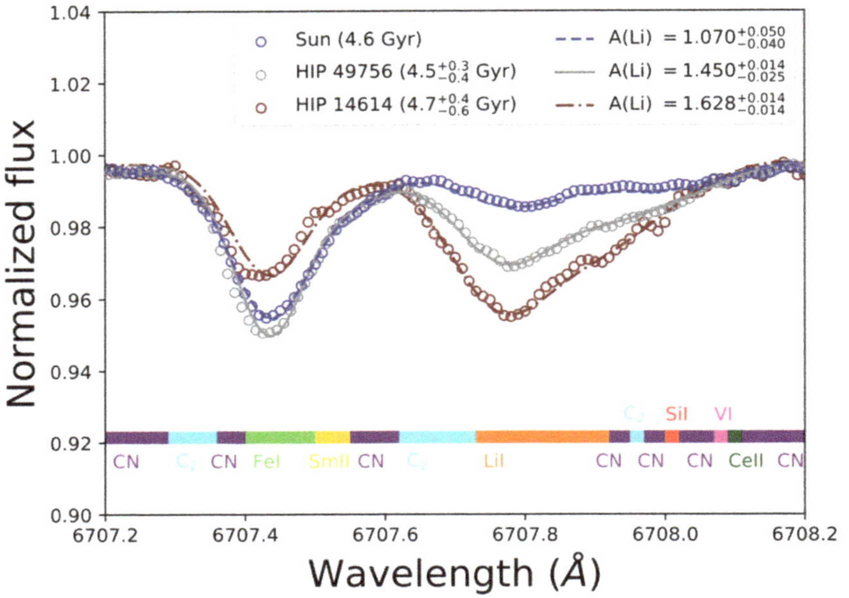

Figure 5.1. Spectral region around the Li I resonance doublet. High-resolution (R = 115,000) spectroscopic data from HARPS are denoted with open circles, and synthetic spectral fits are shown as curves of different colors for the Sun and two solar-type stars. The identification of the dominant absorbing species is provided in the color-coded bar at the bottom. Figure credit: Carlos, M. et al. 2019 "The Li age correlation: the Sun is unusually Li deficient for its age" MNRAS 485 4052.

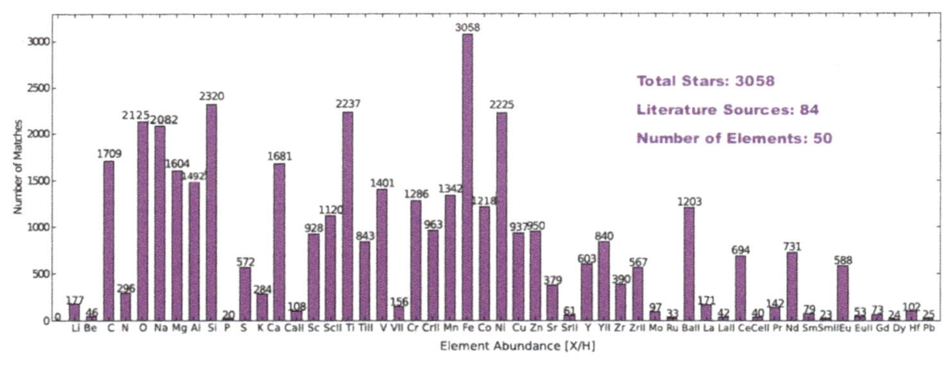

Figure 5.2. Histogram of star counts in the Hypatia catalog (FGKM stars within 150 pc of the Sun) with measured chemical abundances of 50 elements. Note that only 177 stars have lithium abundance determination, much less than the 3058 stars that have iron abundance. Figure credit: Hinkel, N. et al. (2014) "Stellar abundances in the solar neighborhood: The hypatia catalog" AJ 148 54

According to Carlos et al. (2019), a linear fit to the Li depletion versus age correlation can be described with the following equation:

$$A(\mathrm{Li}) = (-0.20 \pm 0.02)\ \mathrm{age} + 2.44\ (\pm 0.10). \qquad (5.1)$$

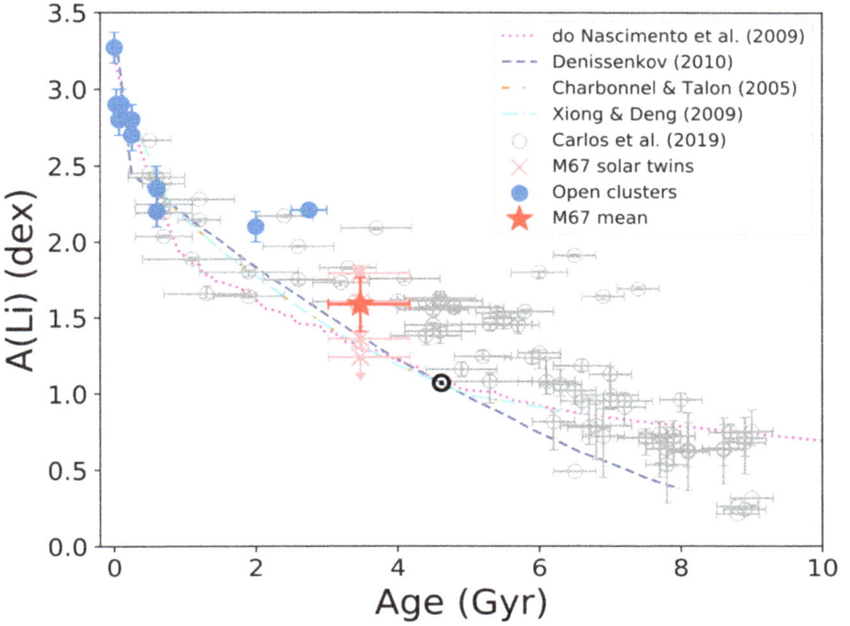

Figure 5.3. Lithium abundance versus age for solar analogs in open clusters (blue circles), the nearly solar-age cluster M67 (pink crosses and red star) and the scattered population of the Milky Way (gray circles). The location of the Sun is represented with the usual symbol of an open circle with a dot in the middle. Several models of lithium depletion are represented with color-coded curves. Figure credit: Carlos, M. et al. 2020 "Li abundances for solar twins in the open cluster M67" MNRAS 492 245.

After subtracting a fit with this equation, the residuals of lithium abundance are found to be slightly correlated with iron abundance, indicating that Li depletion tends to increase for higher stellar metallicity, in agreement with previous results from Delgado Mena et al. (2014).

The correlation of increased Li depletion with larger metallicity for solar twins has been corroborated by the study of Martos et al. (2023) using a sample of 118 stars with high resolution (R = 115,000) spectra obtained with HARPS.

Of particular interest is the open cluster M67 because it has age estimates from 3.4 to 5.4 billion years that span the solar age (Gao 2018). Carlos et al. (2020) observed three M67 solar twins using the Gemini remote access to the ESPaDOns Spectrograph at the CFHT in the MKO. This combination provided the high spectral resolution (R = 45,000), which is mandatory to resolve the rather weak Li I resonance doublet from adjacent Fe I lines in these stars.

Lithium abundances for the stars observed by Carlos et al. (2020) and for other stars from the literature were derived in a self-consistent way. The abundance analysis was carried out using a 1D LTE spectral synthesis code and model atmospheres, which are known to be good enough approximations for this case. Nevertheless, NLTE corrections were applied to the abundances.

Two M67 members have $T_{\rm eff}$ slightly greater than that of the Sun (5874 K and 5843 K) corresponding to masses 6% and 4% > 1.00 $M_\odot$. Both of them have Li abundances clearly greater than that of the Sun, namely A(Li) = 1.79 and A (Li) = 1.57, respectively.

The two M67 members with T_{eff} closest to that of the Sun are YBP 1194 and YBP 1303, with 5786 K and 5768 K, respectively. One of them has a Li abundance of A (Li) =1.36 (Liu et al. 2016), greater than that of the Sun, and the other one has an upper limit of A(Li) <1.24 according to Carlos et al. (2020), which suggests that it may be as lithium depleted as the Sun. The mean lithium abundance for the M67 solar twins, including upper limits, was determined to be A(Li) = 1.5 ± 0.2 dex, which is slightly higher than that of the Sun.

Clearly there is room for improvement because more solar twins in M67 should be studied at higher spectral resolution and S/N. The Li abundances in these four M67 members suggest that there is a steep increase in Li depletion with decreasing stellar mass at the solar age, but more data are needed. The usually adopted definition of solar twins as stars with T_{eff} within ±100 K of the Sun may be too broad for making a fine comparison of the Li abundances of these stars with that of the Sun. Stars with the same T_{eff} as that of the Sun but older have lower mass than the Sun, and the young Sun was cooler than at present.

Spectroscopic studies of solar-type stars belonging to two open clusters older than the Sun (Berkeley 32 and NGC 188) have found that they have much higher Li abundance than the Sun (Randich et al. 2003; Randich et al. 2009) and little dispersion. This result is intriguing and cannot be entirely ascribed to the role of metallicity because NGC 188 has a metallicity similar to that of the Sun.

An intriguing question is whether or not the Sun is an unusual star with respect to similar stars in the solar neighborhood. This problem has been addressed from several perspectives, including rotational broadening (dos Santos et al. 2016; Beck et al. 2017), metallicity and galactic orbit (Robles et al. 2008), magnetic activity (Lorenzo-Oliveira et al. 2018), Li depletion (Beck et al. 2017; Carlos et al. 2019) and the presence of planets (Delgado Mena et al. 2014; Martos et al. 2023).

While there is consensus that the Sun is a typical star in most of its properties, lithium stands out as a possible exception.

The various results reported in the literature are summarized below and lead to seemingly contradictory conclusions, feeding ongoing controversy about the lithium depletion in the Sun:

(1) Pasquini et al. (1994) found that the Sun belongs to a class of lithium-poor stars that comprises about half of the solar-type stars in the solar neighborhood.
(2) Delgado Mena et al. (2014) suggested that this class of Li-poor solar analogs could be associated with the presence of massive planets.
(3) The quest for solar siblings by Mahdi et al. (2016) did not find a dearth of solar twins with Li abundances similar to, or even lower than that of the Sun, although identification of a true monozygotic solar twin, meaning a solar twin formed in the same parental cloud as the Sun, remains elusive.
(4) Beck et al. (2017) claimed the solar Li depletion to be normal as compared with single solar analogs for which rotational periods have been obtained

from light curves derived from data obtained with the Kepler space telescope.

(5) In confrontation with the previous conclusions, Carlos et al. (2019) have stated that the solar Li abundance is anomalous in the sense that it is lower than 90% of the field stars of similar age, mass, and metallicity. However, this result does not have an overwhelming statistical significance (2 σ).

(6) In addition to lithium, Boesgaard et al. (2022a) also studied beryllium in a sample of solar twins. They confirm the wide spread in lithium abundances, but not in beryllium abundances, and they note that the Li/Be ratio is not consistent with the predictions of evolutionary models for solar-mass stars.

It seems as if the jury is still out concerning this crucial matter regarding our understanding of the lithium abundance in the Sun. It is likely that a much larger sample of solar analogs will be needed, together with additional theoretical insights.

5.2 Lithium in Stars More Massive Than the Sun

Solar-type type stars are also known as G-type stars in the Harvard spectroscopic classification system (Cannon 1917). The spectral class that is immediately earlier than G is called the F-type, and it corresponds to stars that, on the main sequence, are slightly hotter and more massive than the Sun. They have thinner convective regions than the Sun, and hence it was expected that lithium in their surface could remain undepleted during the main-sequence phase.

Surprisingly, some F-type stars deplete considerable amounts of lithium. In fact, there is a spectacular lithium dip among F-type stars in the Hyades open cluster (Boesgaard & Tripicco 1986; Boesgaard et al. 2016), which has also been seen in the older cluster NGC 752 (Pilachowski & Hobbs 1988; Sestito et al. 2004; Castro et al. 2016; Boesgaard et al. 2022a). The lithium dip in the F-type members of the Hyades and NGC 752 open clusters is shown in Figure 5.4.

NGC 752 (1.34 Gyr) is twice as old as the Hyades (0.67 Myr), which favors lithium depletion, but it is less metal-rich by about 70%, which is thought to reduce lithium depletion owing to lower opacities.

For stars with temperatures between 6500 K and 5700 K the surface lithium abundances of the members of NGC 752 are systematically lower by a factor of about two than their Hyades counterparts, suggesting that age plays a more important role than metallicity.

The lithium dip in F-type main-sequence stars prompted the development of evolutionary models that include microscopic diffusion (Michaud 1986; Michaud et al. 2015).

Boesgaard et al. (2022b) noted that there is also evidence for spin down and find the comparison between these two clusters to be well explained by the mechanism of rotationally induced mixing.

Stars on the hot side of the lithium dip—those with temperatures between 7000 K and 6500 K—show no evidence of increasing depletion with age; in fact, the contrary may be the case, with metallicity playing a more important role than age in

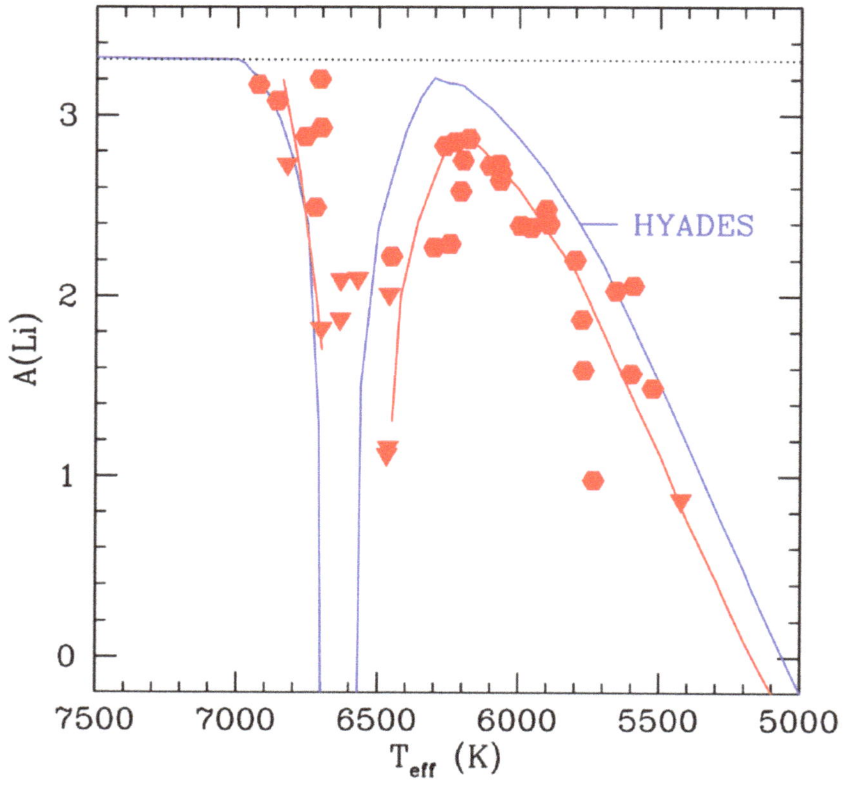

Figure 5.4. Lithium abundances versus effective temperature for F- and G-type members in the NGC 752 open cluster (red symbols). The blue curves denote the locus of lithium abundances for members in the Hyades cluster. Figure credit: Boesgaard, A. M. et al. 2022b "Lithium and beryllium in NGC 752-an open cluster twice the age of the Hyades" ApJ 927 118.

determining the lithium abundances for these early F-type stars. This property makes these stars particularly interesting to investigate the relationship between Li and Fe abundance in the Galaxy (Randich et al. 2020; Dantas et al. 2022; Nepal et al. 2023).

5.3 Lithium in Orange Dwarfs

Main-sequence stars with masses lower than that of the Sun are called orange dwarfs, and are classified as K-type. They are cooler and have deeper convective zones than the Sun. Consequently, chemical mixing is more efficient and the lithium-poor material from the stellar interior gets to the surface on a timescale that is shorter for lower-mass stars than for the Sun (Herbig 1965).

This mass-dependent effect on lithium depletion can be seen clearly in Figure 5.5, where a steep decline in the lithium abundances with cooler $T_{\rm eff}$ for open clusters of different ages is observed. The range of $T_{\rm eff}$ values (6300 K – 3000 K) spans from solar-type stars (G class), to orange dwarfs (K class) and reaches into the even cooler red dwarfs (M class).

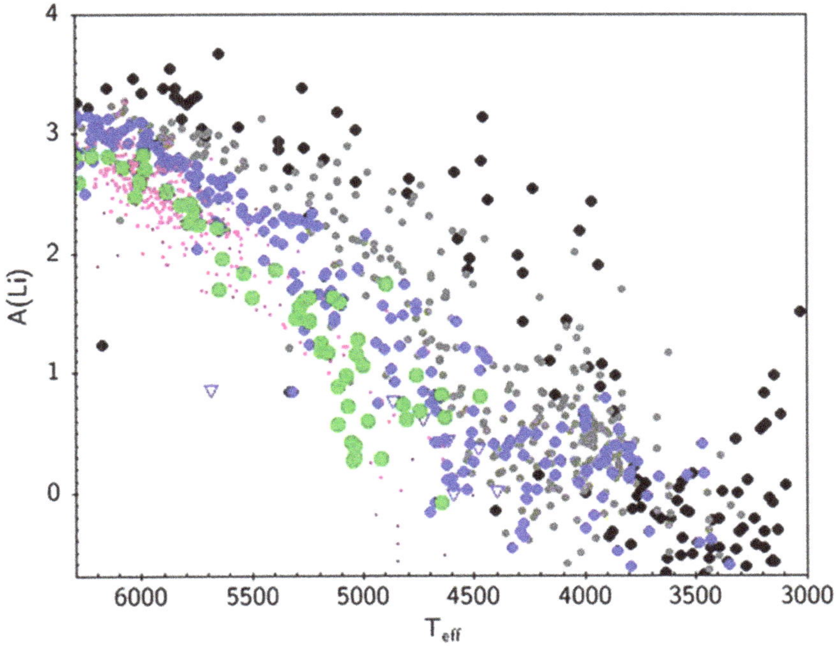

Figure 5.5. Observed decline of lithium abundances for decreasing T_{eff} in six open clusters of similar chemical composition to the Sun, but different ages. The data comes from the work of Cantat-Gaudin et al. (2020). The following different colors are used for identification: black for IC 2602 and NGC 2547 (ages between 30 and 50 Myr), gray for NGC 2516 (age = 100 Myr), blue for NGC 3532 (age = 400 Myr), green for NGC 2355 (age = 1000 Myr), and pink for Berkeley 39 (age = 6000 Myr). Inverted empty triangles denote upper limits. Figure credit: Randich, S., & Magrini, L. 2021 "Light elements in the universe" FrASS 8 6.

While the general trends of declining lithium abundances with decreasing temperature and increasing age are well understood within the framework of convective mixing in conventional stellar models, the observed wide spread in lithium abundances among stars that share similar mass and belong to the same open cluster suggests that there are additional factors at play.

Duncan & Jones (1983) first noticed a wide lithium abundance spread, which they attributed to a large age spread, among late G- and K-type stars in the Pleiades cluster. These results were confirmed and reinforced by Butler et al. (1987).

The issue of a wide age spread in the Pleiades was considered to be a red herring by Soderblom et al. (1993) who carried out an extensive study of lithium in over 100 cluster members with spectral types F, G and K. They considered it unlikely that there is an age spread greater than 20 Myr in the Pleiades, and favored instead the explanation that there is a direct connection between lithium and rotation.

The results of Soderblom et al. were confirmed with additional observations by García López et al. (1994) who showed that there is a strong correlation between lithium abundance and rotational broadening among Pleiades K-type members with masses estimated between 0.9 and 0.7 $M_{\odot}$. This effect is illustrated in Figure 5.6.

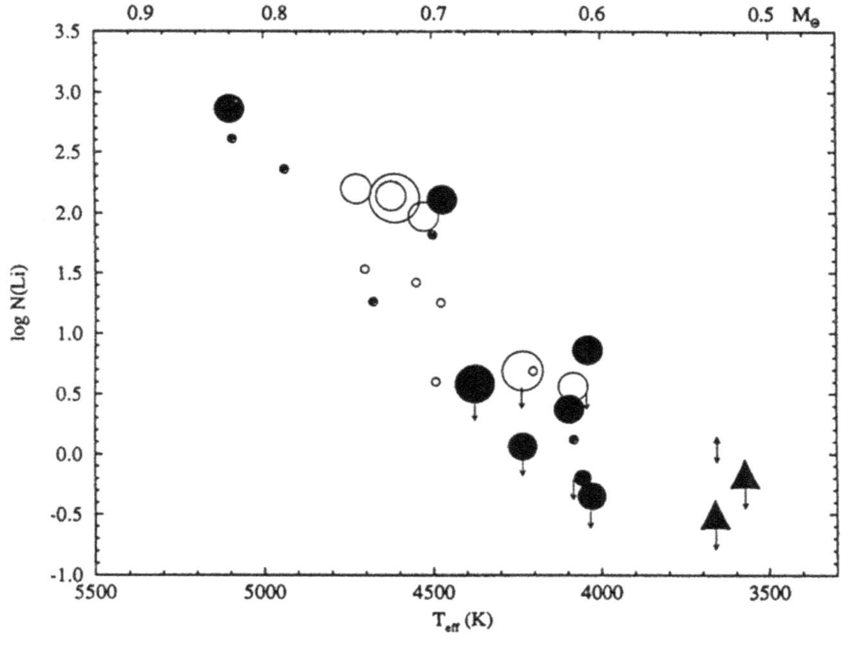

Figure 5.6. The temperature (or mass) and rotation dependence of lithium abundances for K- and early-M type stars in the Pleiades cluster. The size of the symbols scale with the projected rotational velocity ($V \sin i$). The color of the symbols corresponds to data obtained by different teams. Figure credit: García López, R., Rebolo, R., & Martín, E. L. 1994 "Pre-main-sequence lithium burning. II. Pleiades low mass stars" A&A 282 518, reproduced with permission © ESO.

They noted that in the mass range 0.9–0.7 $M_\odot$ all rapid rotators have high lithium abundance with respect to the slow rotators in the same mass domain. They also noted that for stars cooler than 4300 K (masses lower than 0.7 $M_\odot$), the correlation between lithium and rotation breaks down.

The existence of a physical connection between lithium abundance and rotation among Pleiades members with masses between 0.90 and 0.75 $M_\odot$ (spectral types from K0 to K4) has been confirmed with new spectroscopic and rotation data (from the Kepler space mission) in studies carried out by Barrado et al. (2016) and Bouvier et al. (2018) (figure 5.7).

The lithium–rotation connection has also been found among late-G and early-K type members in the AB Doradus association (Ortega et al. 2007) and the stellar stream Psc-Eri (Arancibia-Silva et al. 2020), both of which have ages approximately similar to that of the Pleiades open cluster.

Magnetic fields are considered by Somers & Stassun (2017) as the cause of the correlation between lithium and rotation because they are thought to inflate the stellar radii and diminish the central temperatures of stars. They show that there is a triple correlation between lithium, rotation, and radii among Pleiades K-type stars (Figure 5.8) in the sense that the fastest rotators have the highest lithium abundances and are the most inflated.

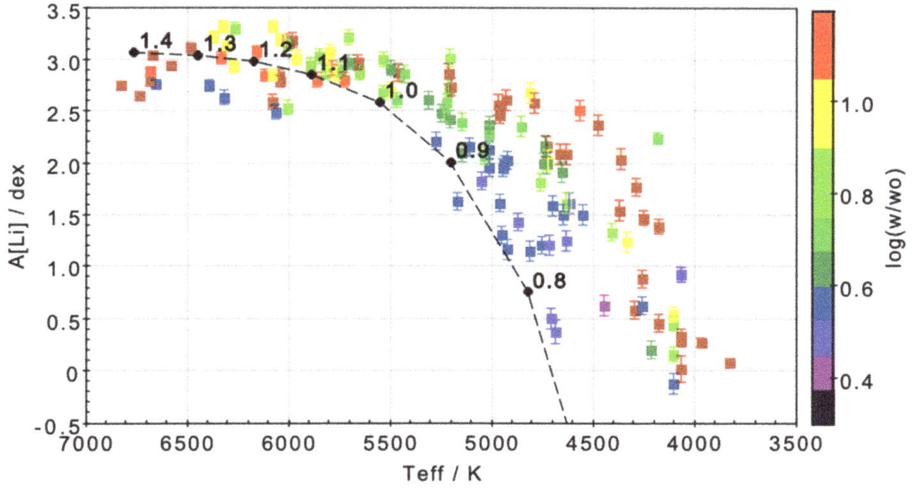

Figure 5.7. The temperature and rotation dependence of lithium abundances for G- and K-type stars in the Pleiades cluster. The color code of the symbols reflects the stellar angular velocity. A Li-depletion isochrone from Baraffe et al. (2015) for an age of 125 Myr is shown as a black dashed curve with masses in solar units ($M_\odot$) indicated on it. Figure credit: Bouvier, J. et al. 2018 "The lithium–rotation connection in the 125 Myr-old Pleiades cluster" A&A 613 A63, reproduced with permission © ESO.

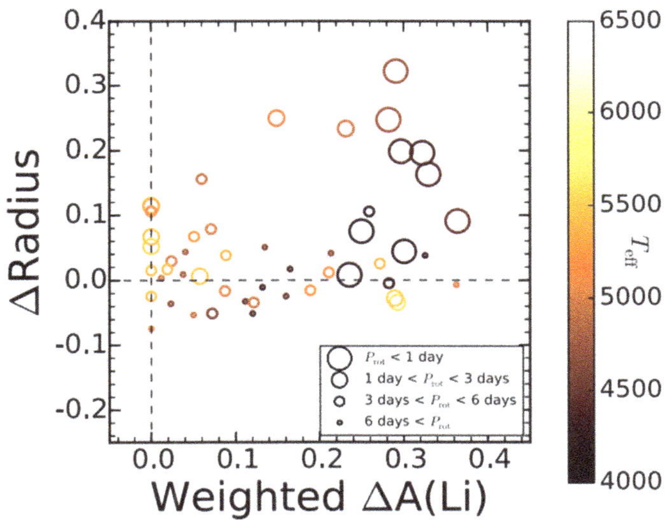

Figure 5.8. Tripartite correlation between radius inflation, lithium enhancement, and fast rotation among Pleiades low-mass stars. The Y-axis represents the departure in observed stellar radius with respect to the predicted value for its $T_{\rm eff}$ from the models by Somers & Pinsonneault (2015). The X-axis reflects the lithium enhancement, which is an analog of the radius departure but weighted by the width of the lithium dispersion of the corresponding $T_{\rm eff}$. The size of the symbols increases with faster rotation (shorter $P_{\rm rot}$). The color of the symbols denotes the $T_{\rm eff}$. The correlation is best seen among the K-type stars. Figure credit: Somers, G., & Stassun, K. G. 2017 "A measurement of radius inflation in the Pleiades and its relation to rotation and lithium depletion" AJ 153 101

The dispersion in lithium among K-type stars seems to reach a maximum in the Pleiades cluster, and diminishes for younger and older associations and open clusters. An example of a rich cluster slightly older than the Pleiades is M35 (NGC 2168, age = 150 Myr, Meibom et al. 2009).

Anthony-Twarog et al. (2018) obtained intermediate resolution (R = 17,000) spectra for 85 stars in M35 using the Hydra multi-object spectrograph on the 3.5 m WIYN telescope at KPNO. They measured the EW of the Li I resonance doublet using the splot package in the IRAF environment. The EW uncertainties were estimated by performing manual independent measurements by different people and calculating the standard deviations from the mean values.

The lithium abundances were derived following the interpolation computational scheme employed by Steinhauer & Deliyannis (2004). The results are illustrated in Figure 5.9, where M35 is compared with a younger cluster (Pleiades) and two older clusters (Hyades and Praesepe).

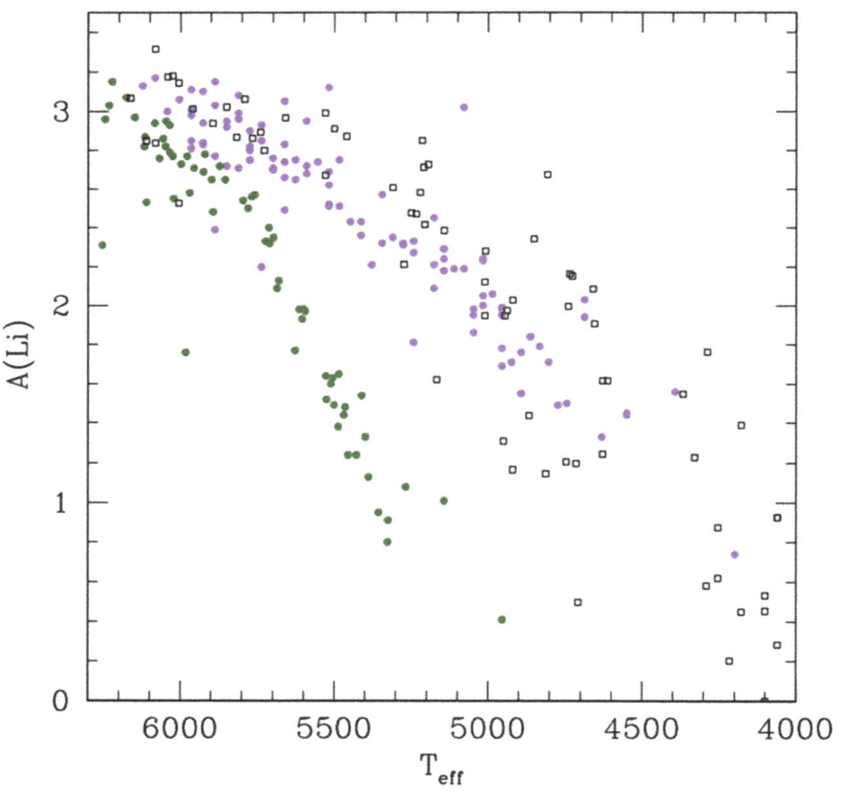

Figure 5.9. Lithium abundance versus effective temperature for members of M35 (filled purple circles), and for single star members of the Pleiades (black squares, data from Bouvier et al. 2018) and the Hyades and Praesepe open clusters (filled green circles, data from Cummings et al. 2017). Figure credit: Anthony-Twarog, B. et al. 2018 "Li in open clusters: Cool dwarfs in the young, subsolar-metallicity Cluster M35 (NGC 2168)" AJ 156 38.

The distribution of lithium abundances among the members of M35 and the Pleiades is quite similar, except for K-type stars with $T_{\rm eff}$ in the range 5000 K to 4000 K, where there is a clear discrepancy. The Pleiades K-type members show a wider spread in lithium abundance than the stars in M35. The discrepancy cannot be explained by an age effect because the lower boundary of lithium abundances in the Pleiades is located below that in M35, which is contrary to expectation given that the Pleiades cluster is younger.

Anthony-Twarog et al. (2018) discussed several possible ways to resolve this conundrum, including magnetic spots, interstellar reddening, and metallicity effects, but they were unable to reach firm conclusions.

By the age of 650 Myr, the lithium is gone in K-type stars, as can be seen in the Hyades/Praesepe sample shown in Figure 5.9. Main-sequence lithium depletion clearly takes place in late-G and K-type stars between the age of M35 and that of Hyades/Praesepe. It is interesting to look at clusters and groups of ages in the range 150–650 Myr. Two such systems are M34 and Latyshev 2 (also known as Group X, see Olivares et al. 2023 for a detailed discussion). The pattern of lithium observed in both systems is shown in Figure 5.10, extracted from the work of Newton et al. (2022).

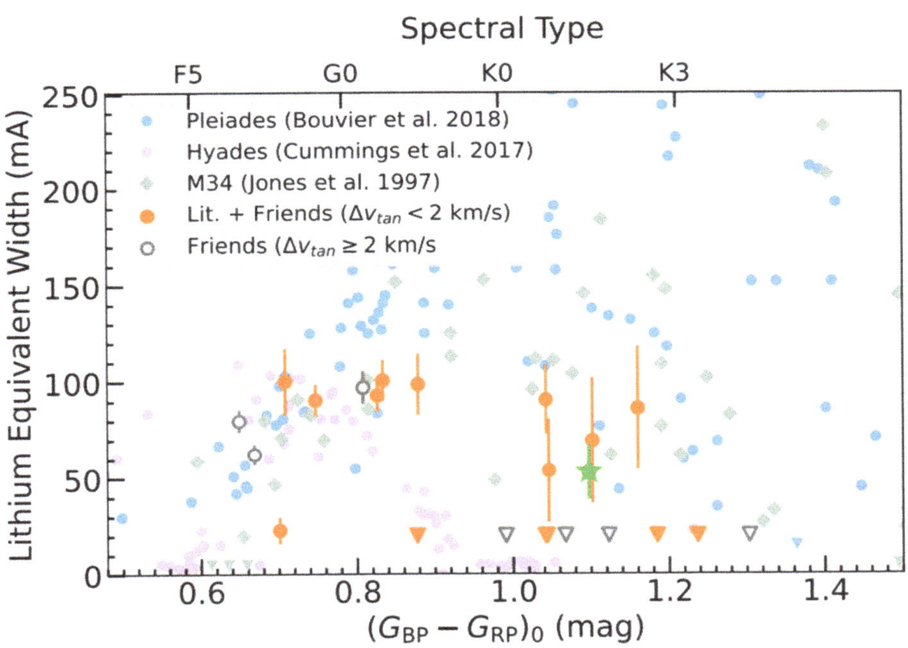

Figure 5.10. Lithium EW versus spectral type or dereddened Gaia color for members in several open clusters and associations with ages in the range from 110 Myr (Pleiades, filled blue circles) to 650 Myr (Hyades, filled gray circles). The lithium EW values for late-G and early-K members in the M34 open cluster (age = 200–300 Myr, green diamonds) and Latyshev 2 (also known as Group X, age = 250–300 Myr, orange symbols) tend to fall between the values of the younger (Pleiades) and older (Hyades) clusters. The location of a K-type star in Latyshev 2 (Group X) that hosts a transiting exoplanet is marked with a green star. Figure credit: Newton, E. R. et al. 2022 "TESS Hunt for Young and Maturing Exoplanets (THYME) VII: Membership, rotation and lithium in the young cluster Group-X and a new young exoplanet" AJ 164 115.

These authors used the Bayesian code of Stanford-Moore et al. (2020) to derive a lithium-based median age of 257 ± 27 Myr for the Latyshev 2 open cluster that is in good agreement with the age of 300 ± 50 Myr obtained from gyrochronology.

The mean Li I resonance doublet strength in M35 and Latyshev 2 that have ages in the range 200–300 Myr tends to be intermediate between those of the Pleiades and the Hyades, supporting the assumption that age is indeed a major factor in the pattern of lithium depletion among late-G and K-type main-sequence stars, but there is still a spread in lithium at a given color or spectral type that indicates that another property is also playing a role. The existence of a connection between lithium and rotation in the Latyshev 2 open cluster has not been studied yet.

If rapid rotation is sustained over long periods of time, lithium depletion can be inhibited in orange dwarfs. For example, the tidally locked binary V471 Tau in the Hyades cluster, composed of a K dwarf and a white dwarf, has a lithium abundance which is a factor of 100 higher than the level of lithium in the single K-type stars in the same cluster (Martín et al. 1997).

5.4 Lithium in Red Dwarfs

Stars with masses of about 0.6 $M_{\odot}$ and lower settle on the main sequence as red dwarfs. They belong to the M-type dwarf spectral class, cooler than 4000 K, and their optical spectra become dominated by the presence of conspicuous absorption bands from metallic molecules such as titanium oxide.

Red dwarfs have deep convective regions and below a mass of about 0.35 $M_{\odot}$, they are thought to be completely convective. As a consequence, they are expected to deplete lithium very efficiently even before they reach the main-sequence (see Figure 4.8).

Theoretically, main-sequence M-type stars with lithium should not exist. Any red dwarf that shows a detectable Li I resonance doublet is peculiar. It could be either a very young PMS star, a result of some kind of binary evolution, an evolved post-main-sequence star, a brown dwarf (for spectral type later than M6; Basri 2000), or a companion to a black hole or neutron star where Li production has taken place.

Spectroscopic surveys for lithium among field M dwarfs are useful for identifying such objects (Phan-Bao et al. 2017). The large area spectroscopic surveys carried out with the LAMOST telescope are starting to provide lithium data for thousands of stars, but they do not yet fully reach the red dwarf regime.

Figure 5.11 illustrates a recent unsupervised search for lithium in LAMOST DR5 carried out by Wheeler et al. (2021). The Li I resonance doublet was detected in 8428 stars out of a total of 430,000 stars included in LAMOST DR5 (about 2%). The threshold for detection was set at EW > 0.15 Å. The coolest temperature considered in this study corresponds to the boundary between the orange (K-type) and red (M-type) dwarfs.

A wide gap opens up between high-gravity (main-sequence dwarfs) and low-gravity (post-main-sequence subgiants and giants) toward the cooler extreme of the Kiel diagram shown in Figure 5.11. This is a spectroscopic version of the HRD, which has some key diagnostic advantages for the study of stellar structure and evolution (Langer & Kudritzki 2014). Lithium-rich cool stars exist on both sides of

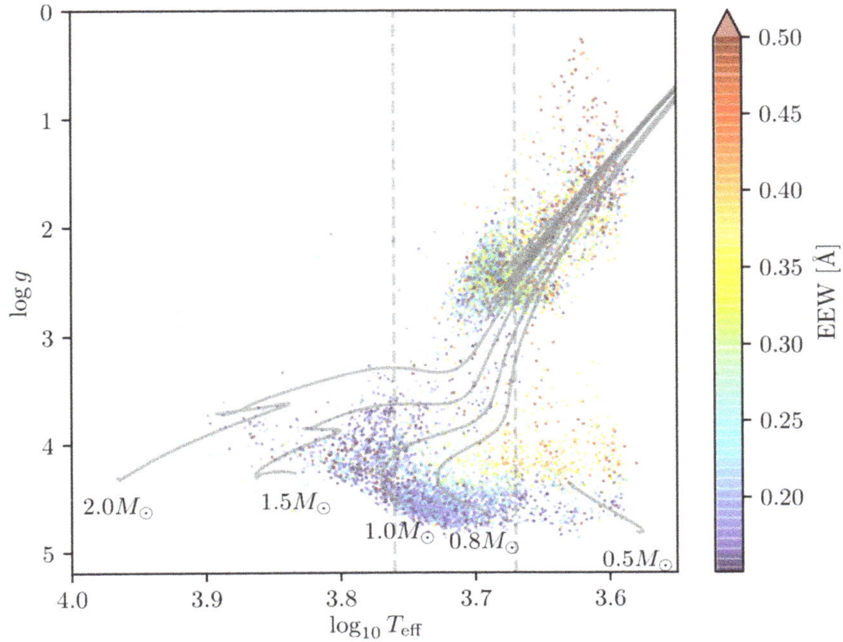

Figure 5.11. Kiel diagram for 8428 stars with Li I resonance doublet detected with LAMOST DR5 spectra. The color bar provides the key to the measured Li I EW in the LAMOST spectra. Isomass theoretical curves (solar metallicity evolutionary models) are plotted in gray, and the corresponding masses are indicated. Figure credit: Wheeler, A. J. et al. 2021 "An unsupervised method for identifying X-enriched stars directly from spectra: Li in LAMOST" ApJ 908 247.

the gap. The high-gravity ones are probably PMS stars and the low-gravity ones are post-main-sequence subgiants and giants. Their high lithium abundances are the result of different physical processes.

5.5 Lithium in Evolved Stars

When hydrogen gets exhausted in the stellar interiors, stars increase in size and evolve off the main sequence in their final journey to their death.

The first report of intense absorption identified as due to the Li I resonance doublet in a post-main-sequence star is due to McKellar (1940) who noted it in a spectrum of the asymptotic giant branch (AGB) star WZ Cassiopeae obtained at Dominion Astrophysical Observatory (DOA) in Victoria (British Coumbia, Canada). Analysis of spectra obtained at Okayama Astrophysical Observatory for this star by Hirai (1969) indicated that the lithium abundance must be enhanced by a factor of 1000 with respect to the solar system (cosmic value).

Conspicuous Li I doublet features (equivalent width between 0.5 Å and 0.2 Å) were reported in the S-type stars HD216672, R Cyg, R Lyn and T Cam (HD190629) from photographic spectrograms obtained at the Haute Provence Observatory (Bretz 1966). Surprising variations in the lithium line strength were noticed on

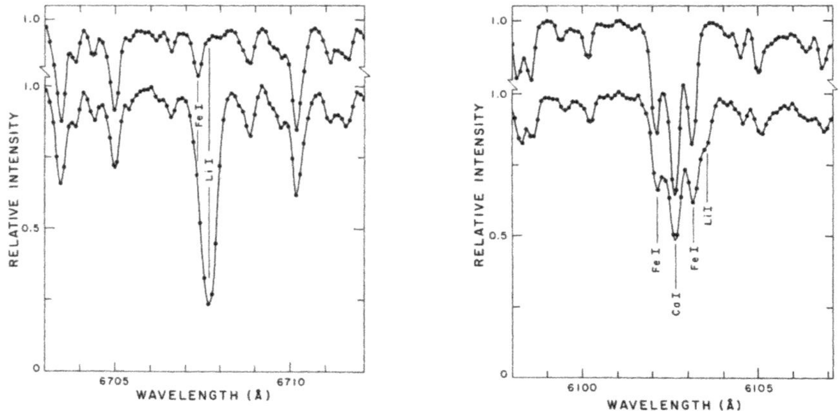

Figure 5.12. DOA spectra for two K-type giants, one with strong Li I lines (HD112127) and the other without. Figure credit: Wallerstein, G., & Sneden, C. 1982 "A K giant with unusually high abundance of lithium: HD 112127" ApJ 255 577.

timescales of days that were not accompanied by similar changes in other lines such as the Ca I at 671.768 nm. HR 8626 was also reported to have a very strong Li I feature (Baird et al. 1975), as well as some weak G-band giants (Hartoog 1978).

The presence of strong Li I features in stars evolved off the main sequence came as a surprise because it had been predicted that any lithium surviving the PMS and the main-sequence evolution would be diluted by mixing with deeper layers during the radius expansion of the star on its way to becoming a red giant.

Iben (1967) calculated dilution factors of between 3 and 60. For example, the red giant descendant of the Sun was predicted to have a lithium abundance about a factor of 2000 below cosmic owing to the combined effects of main-sequence depletion and post-main-sequence dilution. Using the Iben's maximum dilution factor, a lithium abundance $A(\text{Li}) > 1.5$ is usually considered as the threshold for the lithium-rich phenomenon in red giants.

A noteworthy case of enhanced lithium absorption in a red giant was reported by Wallerstein & Sneden (1982) from spectra obtained at the DOA 1.2 m telescope. Their spectra for two K-type giant stars are shown in Figure 5.12. The two spectra are very similar, except for the lithium features.

Additional examples of lithium-rich post-main-sequence stars were registered by Luck (1982), Hänni (1984, 1988), Lambert & Sawyer (1984), Luck & Sneden (1986), and Gratton & D'Antona (1988).

Brown et al. (1989) undertook the first large spectroscopic survey to assess the frequency of anomalously strong Li I features among apparently normal G- and K-type giants.

They observed 644 stars over two years using the coudé spectrograph at the 2.1 m telescope in McDonald Observatory (Texas, USA). The spectral resolution was about 26,000 and the S/N > 150. They compared the observational spectra with LTE synthetic spectra calculated with the code written by Sneden (1983) to derive surface lithium abundances. Examples of those comparison are illustrated in Figure 5.13.

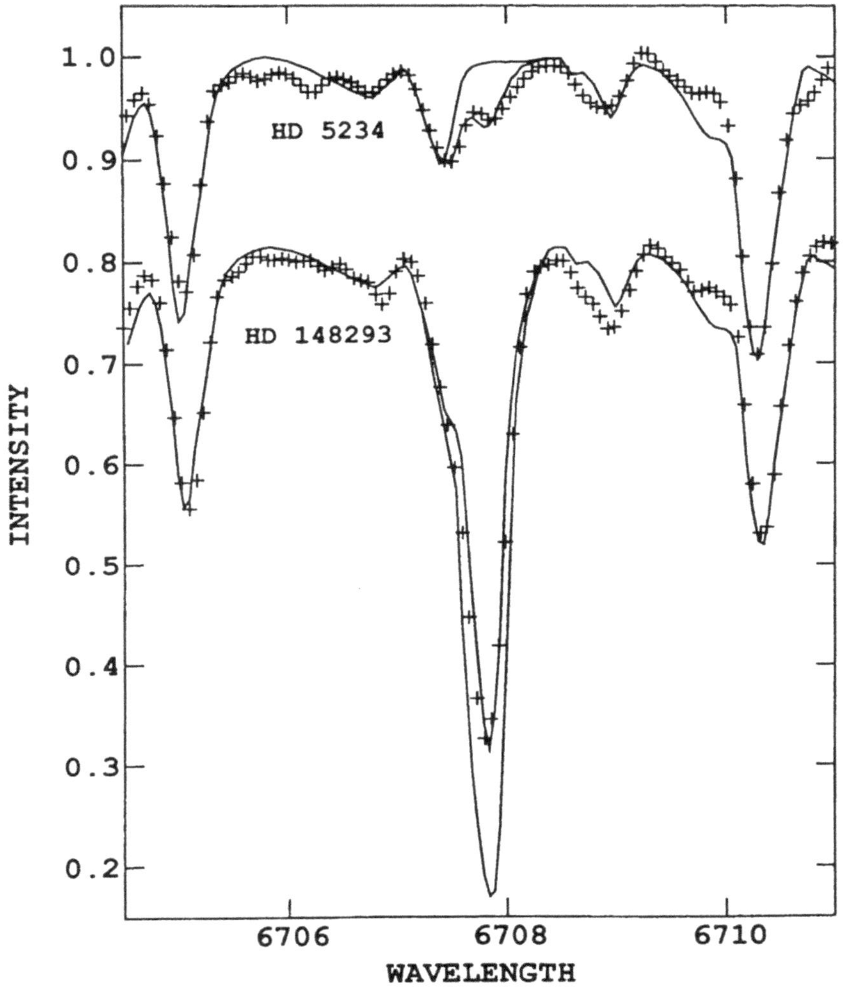

Figure 5.13. McDonald spectra around the Li I resonance doublet for two normal red giants (plus signs) compared with theoretical spectra for different lithium abundances (solid lines). Figure credit: Brown, J. A. et al. 1989 "A search for lithium rich giant stars" ApJS 71 293.

The McDonald lithium survey by Brown et al. yielded the statistical distribution of lithium abundances shown in Figure 5.14. The vast majority of the giant stars have lithium abundances below the maximum value predicted by stellar evolution theory (log N(Li) = 1.5; Iben 1967). In fact, only 12 stars (1.9%) have anomalously high lithium abundances.

Several scenarios were invoked to explain the unusual lithium excess in some red giants; namely, the Cameron–Fowler mechanism during He burning, inhibition of convective mixing, and accretion of lithium-rich planets. Brown et al. found that the majority (60%) of the lithium-rich stars could be explained by the He-burning effect, but they could not rule out that other mechanisms were at work for some stars.

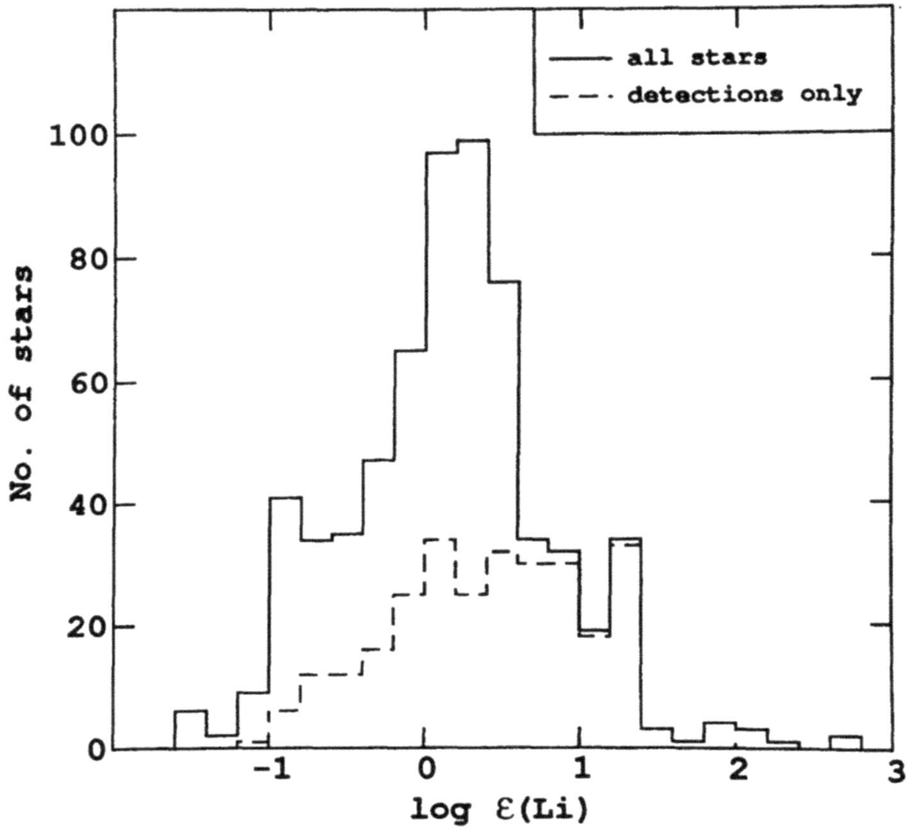

Figure 5.14. Distribution of lithium abundances among normal G and K-type giants observed by the McDonald lithium survey. Below a lithium abundance of $A(\text{Li}) = 1.5$ dex most of the values are upper limits because the Li I resonance doublet was not detected. Figure credit: Brown, J. A. et al. 1989 "A search for lithium rich giant stars" ApJS 71 293.

The high luminosity of red giant stars made it possible to detect lithium even in very distant stars beyond the Milky Way. For example, Smith & Lambert (1990) reported observations of 5 and 27 red giants in the Large Magellanic Cloud (LMC) and the Small Magellanic Cloud (SMC), respectively. They used the Cassegrain cross-dispersed Echelle spectrometer on the CTIO 4 m telescope to obtain spectra with a resolving power of 20,000 and S/N between 50 and 100. Representative spectra can be seen in Figure 5.15.

Smith & Lambert (1990) noted that the Li-rich phenomenon among red giants appears to be restricted to oxygen-rich stars (C/O ratio < 1) in a well-defined region of the evolutionary diagram. They interpreted their results as evidence in favor of the Cameron–Fowler transport mechanism (Cameron & Fowler 1971) operating in the envelope-burning phase of intermediate-mass stars (see also Scalo & Ulrich 1973), and suggested that those red giants could be of prime importance for enriching the ISM with freshly synthesized lithium.

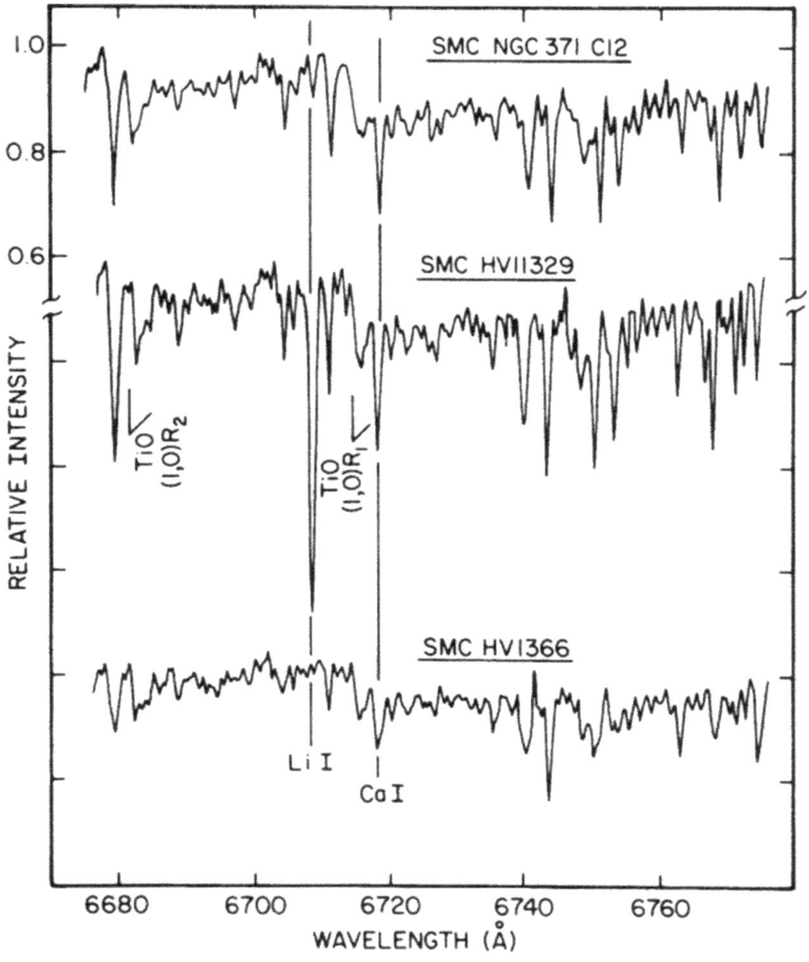

Figure 5.15. CTIO spectra of three red giants in the SMC. Only one of them, namely HV 11329, displays a very strong Li I feature. Figure credit: Smith, V. V. & Lambert, D. L. 1990 "On the occurrence of enhanced lithium in magellanic cloud red giants" ApJ 361 L69.

Sackmann & Boothroyd (1992) calculated that the mass range of the Li-rich red giants in the Magellanic Clouds should be from 4 to 6 times the mass of the Sun; and that they could reach peak surface lithium abundance of log N(Li) = 4.5, i.e. about a factor 10 larger than the cosmic value. These stars could shed about 10% of their mass in a dozen He flashes lasting 10^3–10^5 years, which could be a significant source of lithium enrichment in the ISM.

Forestini & Charbonnel (1997) and Lattanzio & Forestini (1999) performed additional computations of lithium production by hot bottom burning during the asymptotic giant branch (AGB) evolution.

Travaglio et al. (2001) confirmed the previous theoretical results and considered the effects of superwind episodes which increase the impact on galactic lithium enrichment.

Palacios et al. (2001) considered the case of shell lithium burning during the red giant bump and found that it could feed thermal instability and mixing. They dubbed this phase the "lithium flash". While the "lithium flash" could be relatively short lived, lasting about 2×10^4 yr, it could produce a lithium enrichment at the stellar phase for about 6×10^6 yr.

Some carbon-rich (C/O ratio > 1) red giants in the Milky Way have been found to contain a lithium overabundance with respect to the cosmic value.

Abia et al. (1991) discovered one super-lithium-rich carbon-rich red giant (C-giant in the following) among a sample of 250 stars that they observed with the ESO 1.4 m CAT telescope at La Silla observatory (Chile) and the INT in the ORM.

The spectrum of this C-giant, namely IY Hya, is shown in Figure 5.16. They derived the surface lithium abundance of this C-giant and of three more C-giants from the literature using a spectral synthesis code that included the presence of molecular bands of CN and TiO produced by the low surface gravity and cool effective temperature of the stars. The best fit between the observational and theoretical spectra can be seen in Figure 5.16.

The surface lithium abundance that Abia et al. (1991) obtained for IY Hya was log N(Li) = 5.4, which is about a factor of 100 higher than the cosmic value, and strongly favors a mechanism of lithium synthesis during the evolutionary history of

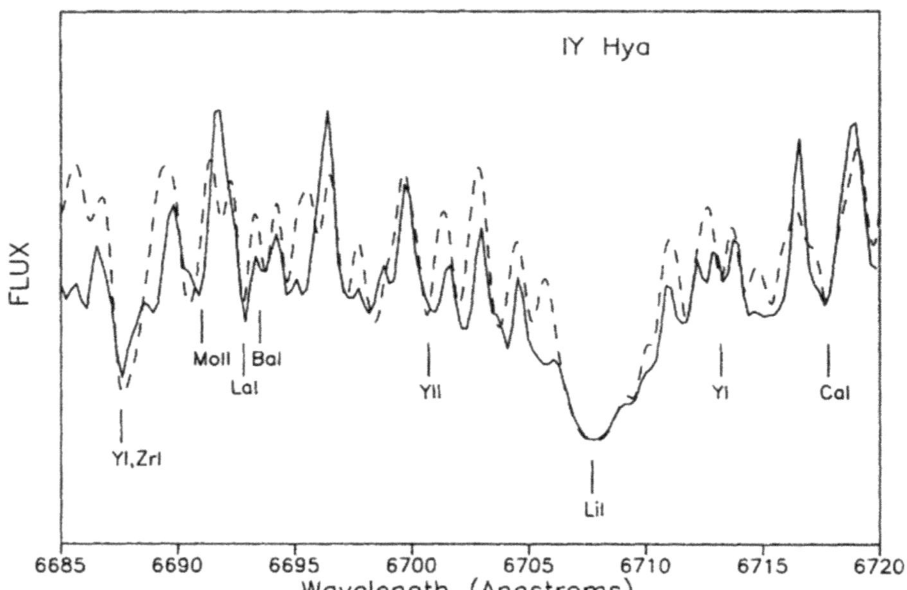

Figure 5.16. INT spectrum of the super-lithium-rich C-giant IY Hya (solid line). The best-fit synthetic spectrum is superimposed with a dashed line. Identified atomic lines identified are indicated. The spectral synthesis also includes a forest of CN and TiO molecular features. From their analysis, the authors infer for this C-giant an effective temperature of 3000 K and a surface lithium abundance of log N(Li) = 5.4. Figure credit: Abia, C. et al. 1991 "IY Hya: A new super-Li-rich carbon star" A&A 245 L1, reproduced with permission © ESO.

this C-giant. Nevertheless, they noted that the positions of the Li-rich C-giants in the evolutionary diagram fell outside the region where the Li-rich red giants in the Magellanic Clouds had been found, thus indicating an added layer of complexity in the mechanism of lithium production in red giants.

A large spectroscopic study of 614 C-giants in the LMC carried out by Hatzidimitriou et al. (2003) with the 2-dF multi-fiber spectrograph at the AAT found no super-Li-rich and only six Li-rich stars (about 1%), indicates that the incidence of lithium production by those stars in the Magellanic Clouds is about an order of magnitude lower than in the Milky Way, where 12.5% of the C-stars are Li-rich and 1.5% are super-Li-rich (Boffin et al. 1993).

No satisfactory agreement between the observations of Li-enriched red giants and the models of hot bottom burning could be reached in the analysis of Hatzidimitrou et al. This problem has motivated the development of additional theoretical work on the post-main-sequence evolution and lithium production of intermediate-mass single stars by Palmerini et al. (2011), and of binary evolution involving nova-polluted mergers by Sengupta et al. (2013).

The study of lithium in red giants has been extended to galaxies beyond the Milky Way and the Magellanic Clouds. Domínguez et al. (2004) reported the discovery of a Li-rich C-giant, called D461, in the Draco dwarf spheroidal galaxy, which was observed as part of study of red giants in nearby galaxies using the ISIS spectrograph on the WHT.

This dwarf galaxy has a metallicity about a factor of 100 below that of the Milky Way (e.g. Aparicio et al. 2001). The Li I resonance feature in D461 is shown in Figure 5.17. A surface lithium abundance of log $N(\mathrm{Li}) = 3.5$ was derived from spectral synthesis fitting, which is slightly higher than the cosmic value in the Milky Way.

The mass and age estimates for D461 were in the range of 1.2–2.0 $M_{\odot}$, and between 1 and 3 Gyr, and the favored mechanism for lithium enrichment was thermal pulses bringing up to the stellar surface the ashes of He burning (Domínguez et al. 2004).

These results have been confirmed and put into a wider context by Kirby (2015) with a study of lithium abundances of almost 3000 red giants located in eight dwarf spheroidal satellite galaxies of the Milky Way, which was carried out using data obtained with the DEIMOS multi-object spectrograph on the 10 m Keck II telescope in MKO.

They found Li-rich red giants in the Canes Venatici I, Fornax, Leo I and II, and Sculptor dwarf galaxies. These stars appear to be rare (about 1%), but they appear to represent a universal phase of post-main-sequence evolution where lithium is produced, rather than destroyed. Kirby concluded that the Li-rich giants in dwarf galaxies did not have any properties other than lithium that distinguished them from the bulk of normal giants. An additional Li-rich red giant has been reported by Menzies et al. (2015) in the dwarf galaxy IC 1613.

An important clue to explaining the Li-rich phenomenon in red giants comes from the spectroscopic study of the chromospheric activity in nearly 300 stars carried out using the high-resolution Habitable Planet Finder spectrograph at the HET (Sneden et al. 2022). These authors used the code SPECTRE (available at http://www.as.utexas.edu/~chris/spectre.html) to measure the EWs of the chromospheric He I spectral lines in the HET spectra ($R = 55{,}000$).

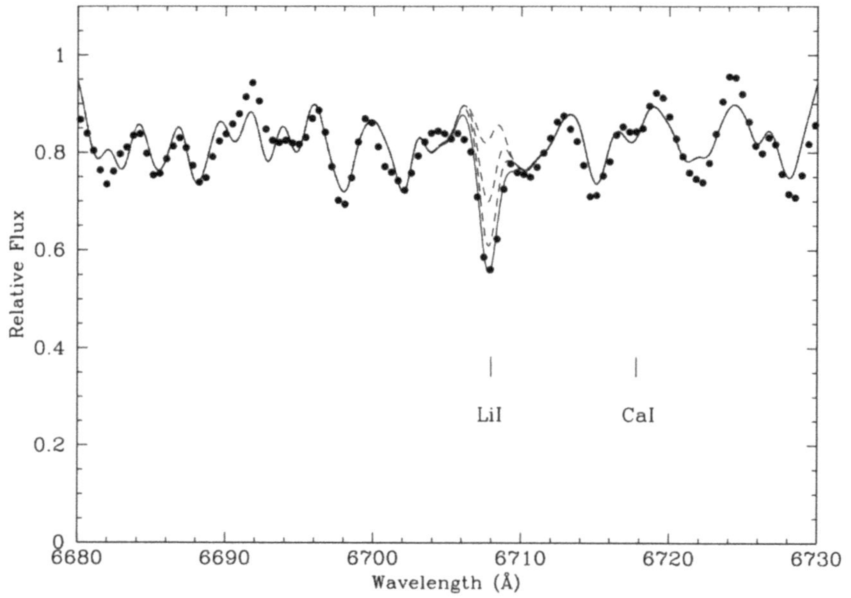

Figure 5.17. WHT spectrum with resolving power $R = 6000$ of the lithium-rich C-giant D461 (solid dots) in the Draco dwarf spheroidal galaxy. The best-fit synthetic spectrum is superimposed with a solid line. Dashed lines denote theoretical spectra with lower values of lithium abundance. From their analysis, the authors infer an effective temperature of 3600 K, an iron abundance of [Fe/H] = −2.0, a C/O ratio = 5, and a surface lithium abundance of log N(Li) = 3.5 for this C-giant. Figure credit: Domínguez, I. et al. 2004 "First detection of a lithium-rich carbon star in the Draco dwarf galaxy: Evidence for a young stellar population" A&A 422 1045, reproduced with permission © ESO.

A summary of the lithium abundances with respect to chromospheric emission in the He I line at 10830 nm is shown in Figure 5.18. From this study, Sneden et al. conclude that Li-rich red giants tend to be chromospherically active and they are strongly concentrated toward the thin disk of the Milky Way. They interpret these results as evidence in support of a connection between lithium enrichment and post-main-sequence binary evolution.

5.6 Lithium as a Stellar Chronometer

Main-sequence stellar evolution is a very slow process for most stars; consequently, it is difficult to derive ages for the majority of individual stars dispersed across the Milky Way using isochrone fitting in the H–R diagram. Orthogonal and independent dating methods are useful, such as asterosismology, gyrochronology, magnetochronology, nucleochronology (using radioactive nuclei such as uranium), or the study of space motions. In this context, lithium depletion provides an alternative chronometer of stellar evolution.

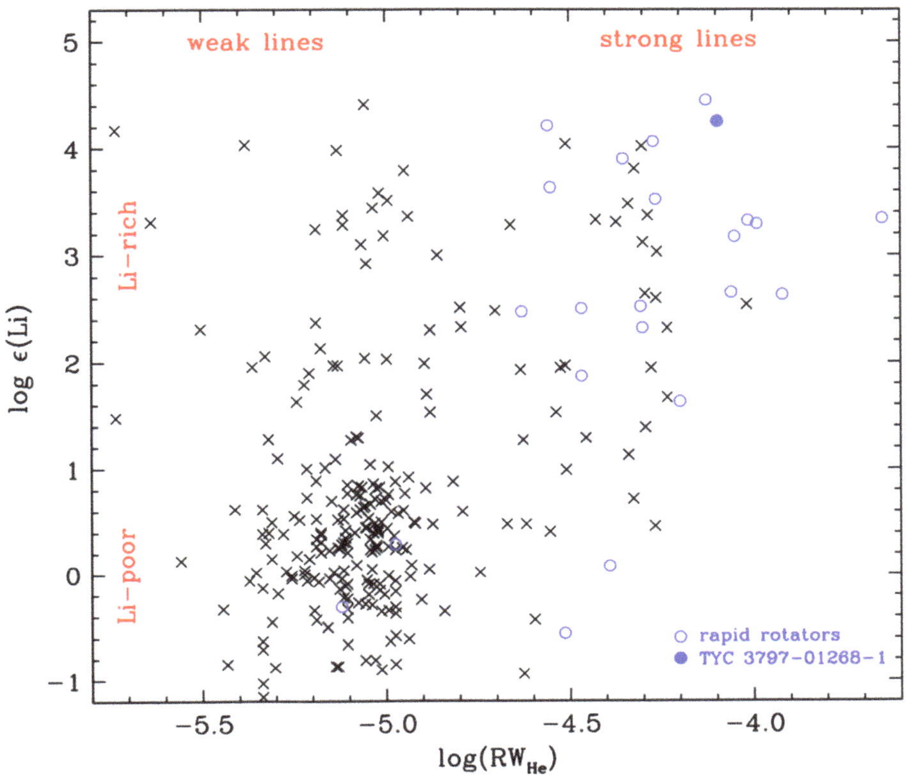

Figure 5.18. Lithium abundances versus reduced EWs of the He I 10830 nm line (chromospheric indicator) for almost 300 red giants in the Milky Way. The weakly shaded stripes separate the sample into four subsets: Li-rich weakly active stars (upper left), Li-poor weakly active stars (lower left), Li-rich strongly active stars (upper right) and Li-poor strongly active stars (lower right). Blue circles denote stars for which rotational broadening could be detected in the HPF spectra. The remarkable super-Li-rich star TYC 3797-01268-1 (HD 233517) star is marked with a filled blue circle. Figure credit: Sneden, C. et al. 2022 "The active chromospheres of lithium-rich red giants" ApJ 940 12.

The overall decrease of the Li I resonance doublet EW as a function of age has been illustrated in the so-called "lithium plot" by Erik Mamajek, available at the following link: http://www.pas.rochester.edu/~emamajek/images/li.jpg

Mamajek's lithium plot represents the Li I EW data for 11 stellar associations and open clusters with ages between a few Myr and a few Gyr as a function of effective temperature. It provides a quick-look summary of the dependence of the strength of the Li I resonance doublet with respect to age and effective temperature or spectral type. The age effects are calibrated using stellar associations and open clusters ranging from less than 5 Myr to about 4 Gyr (three orders of magnitude).

It is clear from the "lithium plot" that despite the wide spread in lithium at any given age and temperature, the upper envelope of Li I EW decreases for older stellar populations, thus giving support to the idea that lithium depletion can be used as a chronometer. Inspired by this plot, a Bayesian code called BAFFLES was developed to estimate ages using Li I EW values and $B - V$ colors for F, G and K-type stars (Stanford-Moore et al. 2020). The precision of lithium as a chronometer depends on the type of stars.

Building up on the idea of using lithium as a stellar chronometer, an extensive empirical model of the pattern of the EW of the Li I resonance doublet at 670.8 nm for ages between 2 and 6000 Myr, T_{eff} between 6500 and 3000 K and metallicity around solar has been developed by Jeffries et al. (2023) using the GAIA-ESO spectroscopic survey for 6200 stars in 52 open clusters.

Obtaining ages of individual main-sequence stars that do not belong to any known association or open cluster remains a challenging task as it is quite common that the ages obtained from different methods do not provide an agreement to better than about an order of magnitude. Jeffries et al. (2023) applied their model to estimate ages of stellar associations and stars hosting exoplanets.

An overall glance at the evolution of lithium in single stars is represented in Figure 5.19. Lithium is a complex and versatile chronometer because its evolution with time depends on different physical mechanisms that include those mentioned in the figure.

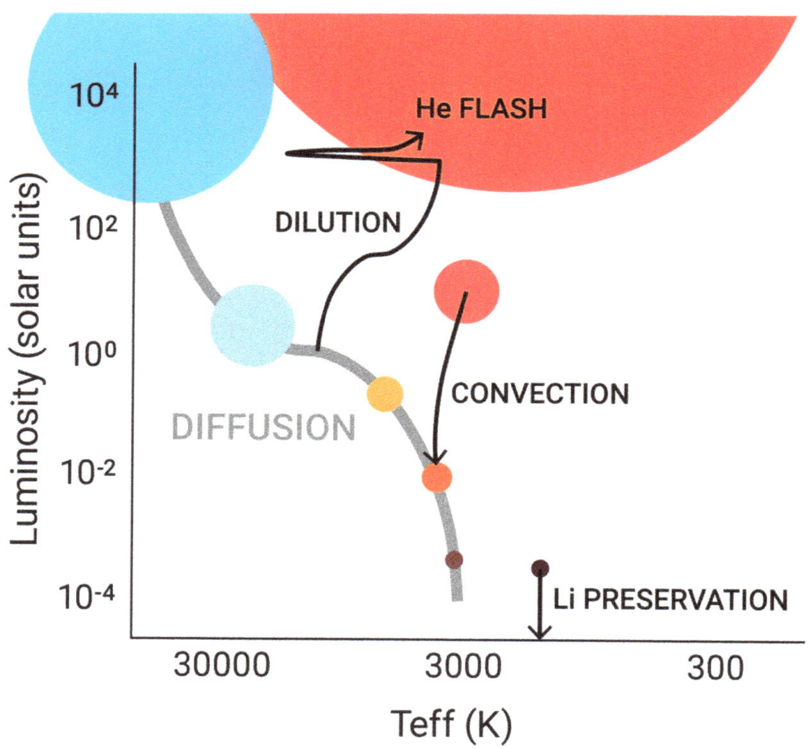

Figure 5.19. Schematic representation of the main features that dominate the behavior of the surface lithium abundance throughout the evolution of low-mass stars. Figure credit: the author with assistance from Gaby Pérez.

Bibliography

Abia, C., Boffin, H. M. J., Isern, J., & Rebolo, R. 1991, A&A, 245, L1
Aparicio, A., Carrera, R., & Martínez-Delgado, D. 2001, AJ, 122, 2524
Arancibia-Silva, J., Bouvier, J., Bayo, A., et al. 2020, A&A, 635, L13
Anthony-Twarog, B. J., Deliyannis, C. P., Harmer, D., et al. 2018, AJ, 156, 37
Baird, S. R., Roberts, W. J., Snow, T. P., & Wallerstein, G. 1975, PASP, 87, 385
Barbuy, B., Spite, M., Hill, V., et al. 2011, A&A, 534, A60
Barrado, D. 2016, Stellar Clusters: Benchmarks of Stellar Physics and Galactic Evolution - EES2015 80,; Les Ulis: EDP Science) 115
Barrado, D., Bouy, H., Bouvier, J., et al. 2016, A&A, 596, A113
Basri, G. 2000, ARA&A, 38, 485
Bazot, M., Creevey, O., Christensen-Dalsgaard, J., & Meléndez, J. 2018, A&A, 619, A172
Beck, P. G., do Nascimento Jr., J.-D., Duarte, T., et al. 2017, A&A, 602, A63
Boesgaard, A., & Tripicco, M. J. 1986, ApJ, 302, L49
Boesgaard, A. M., Lum, M. G., Deliyannis, C. P., et al. 2016, ApJ, 830, 49
Boesgaard, A. M., Deliyannis, C. P., Lum, M. G., & Chontos, A. 2022a, ApJ, 941, 21
Boesgaard, A. M., Lum, M. G., Chontos, A., & Deliyannis, C. P. 2022b, ApJ, 927, 118
Boffin, H. M. J., Abia, C., Isern, J., & Rebolo, R. 1993, A&AS, 102, 361
Bouvier, J., Barrado, D., Moraux, E., et al. 2018, A&A, 613, A63
Bretz, M. C. 1966, Abundance Determinations in Stellar Spectra IAUS 26 eds. Hubenet H. (London: Academic Press) 304
Brown, J. A., Sneden, C., Lambert, D. L., & Dutchover, E., Jr. 1989, ApJS, 71, 293
Butler, R. P., Cohen, R. D., Duncan, D. K., & Marcy, G. W. 1987, ApJ, 319, L19
Cameron, A. G. W., & Fowler, W. A. 1971, ApJ, 164, 111
Cannon, A. J. 1917, PA, 25, 521
Cantat-Gaudin, T., Anders, F., Castro-Ginard, A., et al. 2020, A&A, 640, A1
Carlos, M., Meléndez, J., Spina, L., et al. 2019, MNRAS, 485, 4052
Carlos, M., Meléndez, J., do Nascimento, J.-D. Jr., & Castro, M. 2020, MNRAS, 492, 245
Castro, M., Duarte, T., Pace, G., & do Nascimento, J.-D. 2016, A&A, 590, A94
Cayrel, R., Cayrel de Strobel, G., Campbell, B., & Dappen, W. 1984, ApJ, 283, 205
Cayrel de Strobel, G., Knowles, N., Hernandez, G., & Bentolila, C. 1981, A&A, 94, 1
Cummings, J. D., Deliyannis, C. P., Maderak, R. M., & Steinhauer, A. 2017, AJ, 153, 128
Curtis, J. L., Agüeros, M. A., Matt, S. P., et al. 2020, ApJ, 904, 140
Dantas, M. L. L., Guiglion, G., Smiljanic, R., et al. 2022, A&A, 668, L7
Delgado Mena, E., Israelian, G., González Hernández, J. I., et al. 2014, A&A, 562, A92
Domínguez, I., Abia, C., Straniero, O., Cristallo, S., & Pavlenko, Ya V. 2004, A&A, 422, 1045
dos Santos, L. A., Meléndez, J., do Nascimento, J.-D., et al. 2016, A&A, 592, A156
Duncan, D. K., & Jones, B. F. 1983, ApJ, 271, 663
Forestini, M., & Charbonnel, C. 1997, A&AS, 123, 241
García López, R. J., Rebolo, R., & Martín, E. L. 1994, A&A, 282, 518
Gao, X. 2018, ApJ, 869, 9
Gratton, R. G., & D'Antona, F. 1988, A&A, 215, 66
Gruner, B., & Barnes, S. A. 2020, A&A, 644, A16
Hänni, L. 1984, SvAL, 10, 51
Hänni, L., & Nomoto, K. 1988, Atmospheric Diagnostics of Stellar Evolution Lecture Notes in Physics (Berlin: Springer) 305, 15

Hartoog, M. R. 1978, PASP, 90, 167
Hatzidimitriou, D., Morgan, D. H., Cannon, R. D., & Croke, B. F. W. 2003, MNRAS, 341, 1290
Herbig, G. H. 1965, ApJ, 141, 588
Hinkel, N. R., Timmes, F. X., Young, P. A., et al. 2014, AJ, 148, 54
Hirai, M. 1969, PASJ, 21, 91
Howes, L. M., Lindegren, L., Feltzing, S., Church, R. P., & Bensby, T. 2019, A&A, 622, A27
Iben, I. 1967, ARA&A, 5, 571
Jeffries, R. D., Jackson, R. J., Wright, N. J., et al. 2023, MNRAS, 523, 802
Jones, B. F., Fischer, D., Shetrone, M., & Soderblom, D. R. 1997, AJ, 114, 352
Kirby, E. N. 2015, HiA, 16, 278
Lambert, D. L., & Sawyer, S. R. 1984, ApJ, 283, 192
Langer, N., & Kudritzki, R. P. 2014, A&A, 564, A52
Lattanzio, J., Forestini, M., Le Bertre, T., et al. 1999, Asymptotic Giant Branch Stars IAU Symp. 191 (San Francisco, CA: ASP) 31
Liu, F., Asplund, M., Yong, D., et al. 2016, MNRAS, 463, 696
Lorenzo-Oliveira, D., Freitas, F. C., Meléndez, J., et al. 2018, A&A, 619, A73
Luck, R. E. 1982, PASP, 94, 811
Luck, R. E., & Sneden, C. 1986, PASP, 98, 310
Mahdi, D., Soubiran, C., Blanco-Cuaresma, S., & Chemin, L. 2016, A&A, 587, A131
Martín, E. L., Pavlenko, Y., & Rebolo, R. 1997, A&A, 326, 731
Martos, G., Meléndez, J., Rathsam, A., & Carvalho, S. G. 2023, MNRAS, 522, 3217
McKellar, A. 1940, PASP, 52, 407
Meibom, S., Mathieu, R. D., & Stassun, K. G. 2009, ApJ, 695, 679
Menzies, J. W., Whitelock, P. A., & Feast, M. W. 2015, MNRAS, 452, 910
Michaud, G. 1986, ApJ, 302, 650
Michaud, G., et al. 2015, Atomic Diffusion in Stars, Astron. and Astrophys. Library (Berlin: Springer)
Nepal, S., Guiglion, G., de Jong, R. S., et al. 2023, A&A, 671, A61
Newton, E. R., Rampalli, R., Kraus, A. L., et al. 2022, AJ, 164, 115
Olivares, J., Lodieu, N., Béjar, V. J. S., et al. 2023, A&A, 675, A28
Ortega, V. G., Jilinski, E., de La Reza, R., & Bazzanella, B. 2007, MNRAS, 377, 441
Palacios, A., Charbonnel, C., & Forestini, M. 2001, A&A, 375, L9
Palmerini, S., Cristallo, S., Busso, M., et al. 2011, ApJ, 741, 26
Pasquini, L., Liu, Q., & Pallavicini, R. 1994, A&A, 287, 191
Phan-Bao, N., Bessell, M. S., Nguyen-Thanh, D., et al. 2017, A&A, 600, A19
Pilachowski, C. A., & Hobbs, L. M. 1988, PASP, 100, 336
Plez, B., Hill, V., Cayrel, R., et al. 2004, A&A, 428, L9
Randich, S., Sestito, P., & Pallavicini, R. 2003, A&A, 399, 133
Randich, S., Pace, G., Pastori, L., & Bragaglia, A. 2009, A&A, 496, 441
Randich, S., Pasquini, L., Franciosini, E., et al. 2020, A&A, 640, L1
Randich, S., & Magrini, L. 2021, FrASS, 8, 6
Reeves, H. 1991, A&A, 244, 294
Robles, J. A., Lineweaver, C. H., Grether, D., et al. 2008, ApJ, 684, 691
Sackmann, I. J., & Boothroyd, A. I. 1992, ApJ, 392, L71
Sandage, A., & Eggen, O. J. 1969, ApJ, 158, 685
Scalo, J. M., & Ulrich, R. K. 1973, ApJ, 183, 151

Sengupta, S., Izzard, R. G., & Lau, H. H. B. 2013, A&A, 559, A66
Sestito, P., Randich, S., & Pallavicini, R. 2004, A&A, 426, 809
Skumanich, A. 1972, ApJ, 171, 565
Smith, V. V., & Lambert, D. L. 1990, ApJ, 361, L69
Sneden, C. 1983, PASP, 95, 745
Sneden, C., Afşar, M., Bozkurt, Z., et al. 2022, ApJ, 940, 12
Soderblom, D. R., Jones, B. F., Balachandran, S., et al. 1993, AJ, 106, 1059
Soderblom, D. R. 2010, ARA&A, 48, 581
Somers, G., & Pinsonneault, M. H. 2015, MNRAS, 449, 4131
Somers, G., & Stassun, K. G. 2017, AJ, 153, 101
Stanford-Moore, S., Adam, N., Eric, L., De Rosa, R. J., Macintosh, B., & Czekala, I. 2020, ApJ, 898, 27
Steinhauer, A., & Deliyannis, C. P. 2004, ApJ, 614, L65
Wallerstein, G., & Sneden, C. 1982, ApJ, 255, 577
Wheeler, A. J., Hogg, D. W., & Ness, M. 2021, ApJ, 908, 247
Yi, Sukyoung, Demarque, Pierre, Kim, Yong-Cheol, et al. 2001, ApJS, 136, 417

Chapter 6

Lithium in Halo Stars and Chemical Evolution

Stars belonging to the halo population of the Milky Way are called Population II, to distinguish them from the Population I, which belong to the disk of the Galaxy.

Population II stars are older, more metal-poor, and kinematically different from Population I stars. They formed in the first generations of stars when the interstellar material had a chemical composition not very contaminated by many star-formation cycles, characterized by a low iron abundance of $[Fe/H] < -1$.

Lithium has the most complex history of all the chemical elements in the periodic table. A primordial level of lithium coming from the origin of the universe accounts for about 30% of the Li atoms of the present-day cosmic environment. Exactly what is the primordial Li abundance is still under scrutiny because the yields from Standard Big Bang Nucleosynthesis are higher than the lithium content observed among the oldest stars in the Milky Way.

The lithium abundance started to rise some time after the formation of the primitive Galaxy. An early mechanism of Li enrichment might have been production brought about by highly accelerated cosmic rays. Interactions of the cosmic rays with the cool ISM induced spallation reactions that broke up CNO atoms and produced Li atoms. These reactions, which started early in the history of the Galaxy, have continued to make lithium, but their net enrichment may have been diminished by astration of Li-poor material.

The evolution of lithium associated with cosmic rays can be traced with ^{9}Be observations in different stellar populations. This stable isotope is thought to come only from spallation reactions in the ISM. It is possible to infer from the ^{9}Be abundance how little ^{7}Li could have been produced by spallation reactions at the time of the formation of the halo stars.

Later in the Galactic history low-mass stars with their long evolutionary timescales and large numbers started to dominate the chemical evolution. These stars produce flares and nova explosions.

doi:10.1088/2514-3433/acd8bbch6

Lithium overabundances above the cosmic level have been observed in low-mass X-ray binaries with compact relativistic objects (neutron stars and black holes) and red giants.

It has been argued that the dominant source of lithium enrichment in the disk of the Milky Way is nova explosions. Further investigation of this possibility and comparative studies with other lithium enrichment sources in the Galaxy are warranted.

6.1 The Primordial Lithium Abundance and the Cosmological Tension

It was realized over four decades ago that the observational abundance of lithium in Population II stars could be used as a constraint on Big Bang nucleosynthesis (Wagoner 1967; Boesgaard 1976), particularly among old stars of very metal-poor composition.

Spite & Spite (1982) carried out the first study of lithium abundances in a sample of 13 Population II stars using high-resolution spectra obtained at the coudé focus of the CFHT telescope at the MKO. A representative set of their spectra in the lithium spectral region is shown in Figure 6.1.

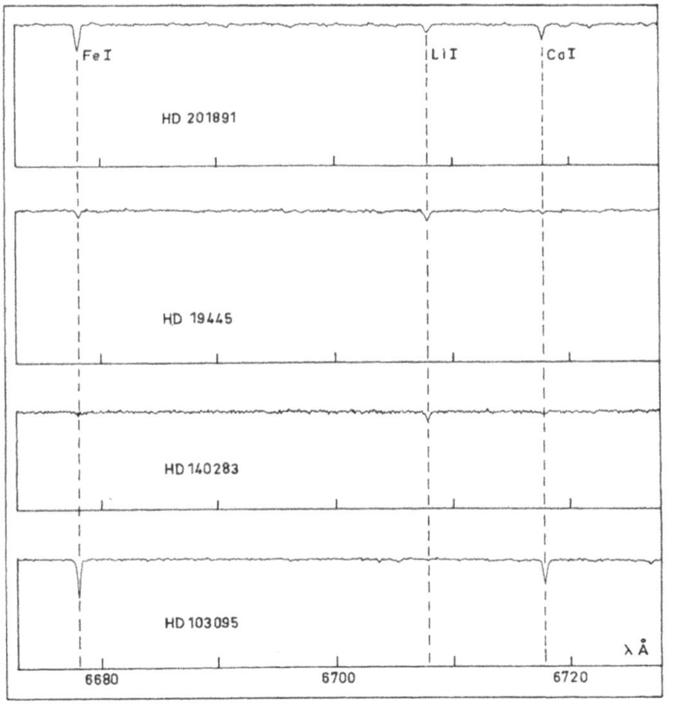

Figure 6.1. CFHT spectra of Population II stars. The main atomic lines are indicated. Note that in very metal-poor stars the Ca I and Fe I lines are very weak, and the Li I resonance doublet becomes the strongest feature in this spectral region. Figure credit: Spite, F., & Spite, M. 1982 "Abundances of lithium in unevolved halo stars and old disk stars: Interpretation and consequences" A&A 115 357, reproduced with permission © ESO.

The lithium abundances for the stars were determined with CoG analysis under the assumption of LTE using the atmospheric models of Peytremann (1974). The errors in Li abundance arising from stellar parameter determination and model imperfections were estimated.

The main conclusion of the Spite and Spite study is that the lithium abundance is constant over a wide range of spectral type (F7–G4) or effective temperature (6252–5495 K) among metal-deficient stars belonging to the halo of the Milky Way (Population II).

This remarkable invariance of lithium is frequently dubbed the Spite plateau and it has been interpreted as evidence for a primordial threshold coming directly from the origin of the universe via Big Bang nucleosynthesis (BBN).

From the primordial value coming from BBN, lithium has increased to the present cosmic value in the Galaxy. Rebolo et al. (1988) studied for the first time the relation between lithium and iron abundance in Galactic stars, including Population II stars.

The original value of the lithium abundance in the Spite plateau was log N (Li) = 2.0, but it has been slightly raised by many studies of larger samples (Spite & Spite 1986; Bonifacio & Molaro 1997; Rebolo et al. 1987; Gutiérrez et al. 1999; Ryan et al. 2000; Bonifacio et al. 2007, 2018; Shi et al. 2007; Aoki et al. 2009; Hosford et al. 2009; Monaco et al. 2010; Sbordone et al. 2010; Schaeuble & King 2012; Molaro et al. 2020; Kielty et al. 2021).

Nowadays it is accepted that the level of the Spite plateau is located around log N (Li) = 2.2, and does not exceed log N(Li) = 2.4 (Meléndez & Ramírez 2004), and it has become clear that the plateau is not really flat, there being a wide spread in Li abundance.

The upper envelope of the lithium abundance in metal-poor stars is about a factor of three lower than the prediction for the primordial lithium abundance from SBBN constrained by the baryonic density derived from the D/H ratio in QSOs intervening galaxies and cosmic microwave background anisotropies measured with the Wilkinson Microwave Anisotropy Probe (WMAP) space mission led by NASA and confirmed by the European Planck mission (Cyburt et al. 2016; Coc & Vangioni 2017).

This discrepancy has been referred to as the lithium cosmological problem. A recent graphic representation of the tension between the SBBN prediction and the observed lithium abundances is illustrated in Figure 6.2.

Possibilities for bridging the gap between the SBBN predictions and the lithium abundances of Population II stars include the following:

1. That the SBBN scenario needs to be revised. For a review of the SBBN the reader is referred to Iocco et al. (2009). Non-standard primordial lithium production has been addressed by Kusakabe et al. (2008), and Jedamzik & Pospelov (2009).
2. That there has been some lithium depletion between SBBN and the formation of the Population II stars. The depletion could have taken place when the primordial gas was processed in the first generation of stars (Population III). It has been estimated that this process could account for about a factor of two depletion, although with large uncertainties due to the

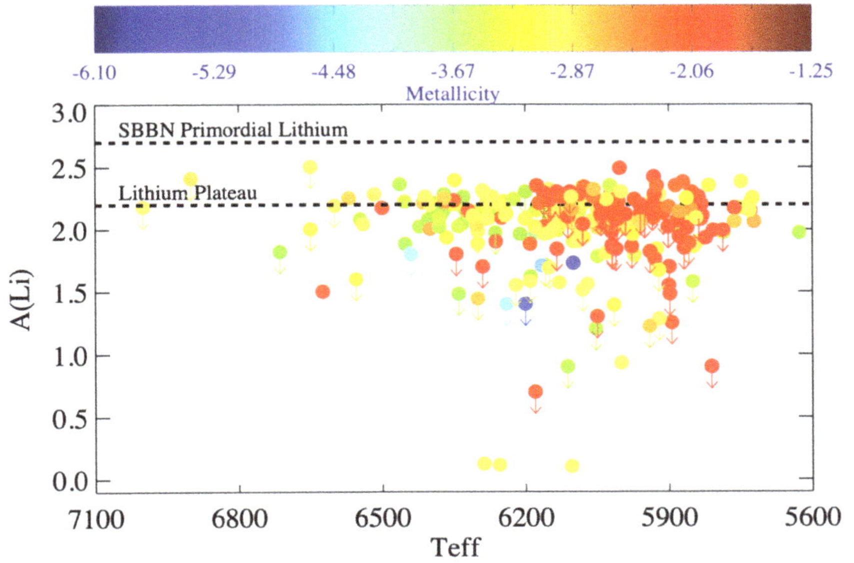

Figure 6.2. Lithium abundance of Population II main-sequence (defined as log $g > 3.7$) stars versus effective temperature. The color scale provides information on the metallicity. The data includes stars in the halo of the Milky Way, globular clusters, and accreted galaxies. The flat level of the Spite plateau is marked with the lower horizontal dashed straight line. It appears to be universal, although there is a significant spread around it. The prediction for the primordial lithium abundance from Standard Big Bang Nucleosynthesis (SBBN) is marked by the upper horizontal dashed straight line. No Population II stars reach the SBBN Li abundance level. Figure credit: Matteucci, F. et al. 2021 "The evolution of lithium: Implications of a universal Spite Plateau" MNRAS 505 200.

unknown properties of the first stars (Piau et al. 2006). Massive population III stars may have also seeded the ISM with some trace amount of the rarer ^{6}Li isotope (Reeves 2005).

The main problem of this mechanism is that it requires that of order of 50% of all the ISM material to have been recycled through Population III stars, an implausible process that would have left other chemical imprints that are not observed.

3. That the Population II stars have experienced some surface lithium depletion caused by the mixing of the material at the bottom of the atmospheric convection zone. A combination of microscopic diffusion and turbulence could do the trick (Richard et al. 2005; Korn et al. 2007; Piau 2008; Lind et al. 2009; Meléndez et al. 2010; Fields & Olive 2022), but it requires fine tuning that is difficult to make consistent with the abundances of other chemical elements (Romano et al. 2003; Bonifacio et al. 2007).

It has been noted that the Li abundances of main-sequence stars in the old (5 Gyr) metal-poor (factor of two below solar) open cluster Berkely 32 display a pattern that does not depend much on $T_{\rm eff}$, and suggests the presence of an unidentified Li depletion mechanism that is quite insensitive to stellar mass and metallicity (Randich et al. 2009).

The degree of flatness of the Spite plateau has been examined carefully using high-quality spectra and state-of-the-art abundance analysis including NLTE and 3D effects. It has been found that there are statistically significant correlations of lithium abundances both with effective temperature and with metallicity among halo stars (Meléndez et al. 2010; Hansen et al. 2014). The relations between lithium and $T_{\rm eff}$ for different types of metal-poor stars are shown in Figure 6.3, while the relation with metallicity can be seen in Figure 6.4.

Deviations from invariance of the Spite plateau could be interpreted as hints at mechanisms of lithium depletion acting during the lifetime of Population II stars. The relationship between Li abundance and $T_{\rm eff}$ can be interpreted as mostly standard Li depletion. The relationship between Li and Fe abundances for extremely metal-poor stars seems to be a more complex phenomenon that is not well understood (Aguado et al. 2019, 2022; González Hernández et al. 2020).

4. Last, but not least, there is the possibility that the lithium abundances from the Li I resonance doublet in metal-poor stars are systematically affected by poorly understood NLTE effects. Photoionization of the ground level of the neutral lithium atom may be subject to strong NLTE conditions. For example, these effects are well studied in the aluminum (Al) atom, which

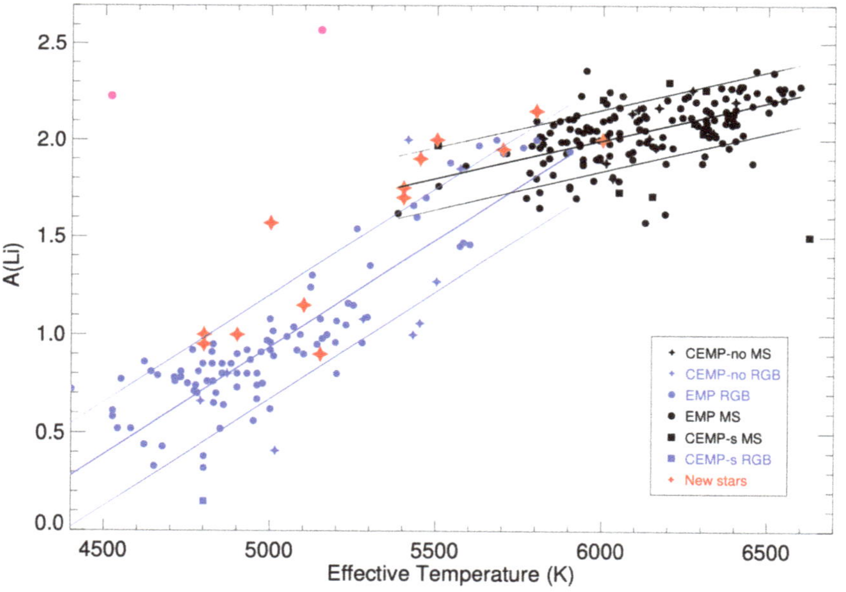

Figure 6.3. Distribution of lithium abundances versus effective temperature for different families of metal-poor stars. The key to the inset acronyms is the following: MS means main-sequence, RGB means red giant branch, EMP means extremely metal-poor ([Fe/H] < −3), CEMP means carbon-enhanced metal-poor star. The blue straight line denotes the best-fit correlation between lithium and $T_{\rm eff}$ for giants (bracketed by the 1σ limits as thinner blue lines). The black straight line denotes the best-fit correlation for dwarfs and the 1σ limits. Figure credit: Bandyopadhyay, A. et al. 2022 "Li distribution, kinematics and detailed abundance analysis among very metal-poor stars in the Galactic halo from the HESP-GOMPA survey" ApJ 937 52.

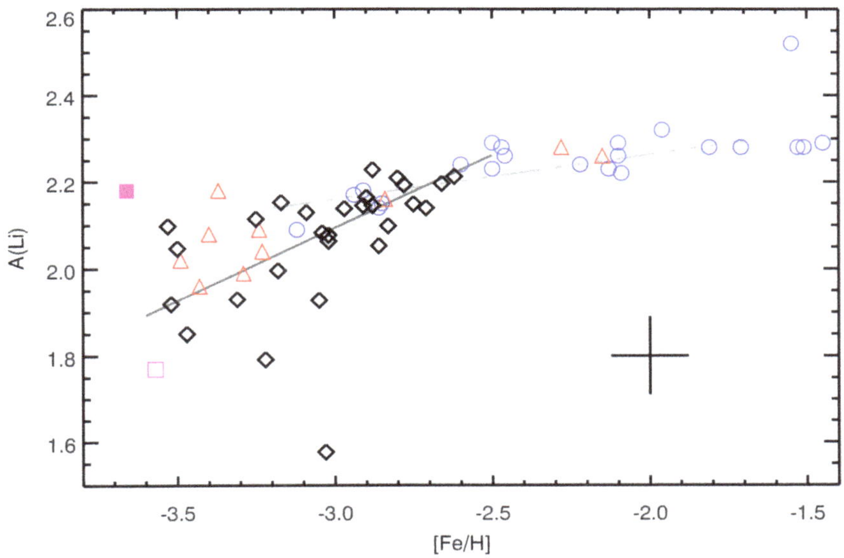

Figure 6.4. Distribution of lithium abundances for extremely metal-poor main-sequence stars. The data sources are the following: blue circles (Asplund et al. 2006), red triangles (Aoki et al. 2009), magenta squares (González Hernández et al. 2008), and black diamonds (Sbordone et al. 2010). The linear regression best-fit to the data changes in slope at an iron abundance of Fe/H = −2.5 as indicated by the solid and dot-dashed gray lines. Figure credit: Sbordone, L. et al. 2010 "The metal-poor end of the Spite plateau. I. Stellar parameters, metallicities and lithium abundances" A&A 522 A26, reproduced with permission © ESO.

has a low first ionization potential (5.99 eV), but still slightly higher than that of lithium (5.39 eV). Both Al^+ and Li^+ are the dominant ionization states of these two atoms in metal-poor atmospheres.

Spectra of metal-poor stars obtained with the Space Telescope Imaging Spectrograph (STIS) allowed Roederer & Lawler (2021) to detect the Al II line at 2669.155 Å. They determined the Al abundances with spectral synthesis under LTE conditions using both the Al II line and the resonance lines from Al I at 3944 Å and 3961 Å.

Large systematic discrepancies were found as illustrated in Figure 6.5, in the sense that the Al abundances obtained from the Al II line are higher than those coming from the resonance Al I lines. In a study of Ca II lines versus Ca I lines in metal-poor stars, Mashonkina et al. (2007) also found significant positive abundance corrections due to NLTE effects when using the Ca I line at 4226 Å.

Could such strong and systematic NLTE effects also occur in the determination of lithium abundances from the Li I lines? Would they be enough to bridge the SBBN/Spite gap and account for the trend to apparently lower lithium abundances among the most metal-poor stars known?

Extensive NLTE calculations of the formation of lithium lines have been presented by several teams (Steenbock & Holweger 1984; Magazzù et al. 1992; Carlsson et al. 1994; Martín et al. 1994; Pavlenko & Magazzù 1996; Barklem et al.

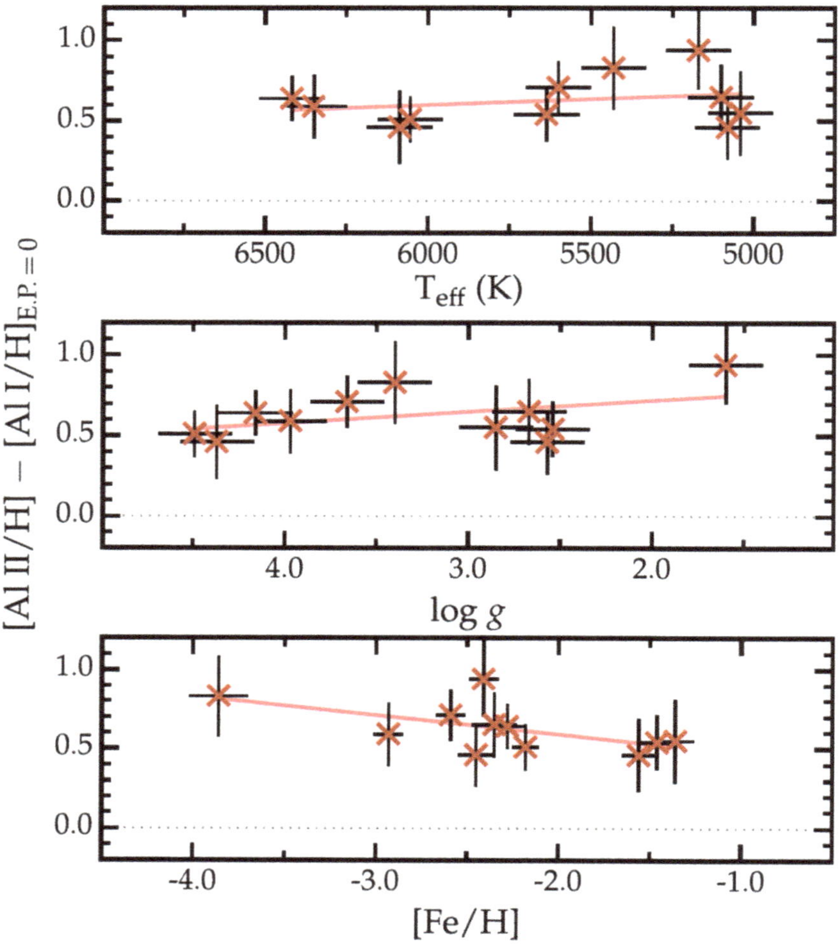

Figure 6.5. Discrepancy in aluminum abundances in metal-poor stars derived from different lines as a function of effective temperature, surface gravity and iron abundance. The dotted line indicates the level of zero discrepancy. Best-fit linear regression to the data are represented with solid straight lines. Figure credit: Roederer, I. U., & Lawler, J. E. 2021 "Detection of Al II in the ultraviolet spectra of metal-poor stars: An empirical LTE test of NLTE aluminum abundance calculations" ApJ 912 119.

2003; Takeda & Kawanomoto 2005; Lind et al. 2009; Osorio et al. 2011), but they lack the empirical test of observations of ionized lithium lines or lines from highly excited levels of the neutral lithium atom.

So far, the spectral diagnostics of lithium are based on only three lines (6104 Å, 6708 Å, and 8126 Å), all of them coming from the ground level or the first excited level of the neutral Li atom and not known to experience strong NLTE effects in metal-poor atmospheres.

An active area of theoretical effort is the computation of 3D hydrodynamic effects (Kiselman 1997; Asplund et al. 2003; Barklem et al. 2003; Lind et al. 2013; Klevas et al. 2016; Harutyunyan et al. 2018; Mott et al. 2020).

The computational cost of NLTE 3D calculations is high, and hence the range of parameter space they can cover is limited. A 3D NLTE grid spanning a very wide range of stellar parameters has only recently become available (Wang et al. 2021). An example of their results is shown in Figure 6.6. The 3D NLTE corrections to the lithium abundances obtained using 1D LTE approximation tend to be negative for most of the parameter space explored in these simulations, but they become mildly positive among the most metal-poor high-gravity hot models. Still, these predictions are far from what would be required to bridge the SBBN/Spite gap. For the most common cases, the abundance corrections due to 3D NLTE effects are quite small (<0.05 dex).

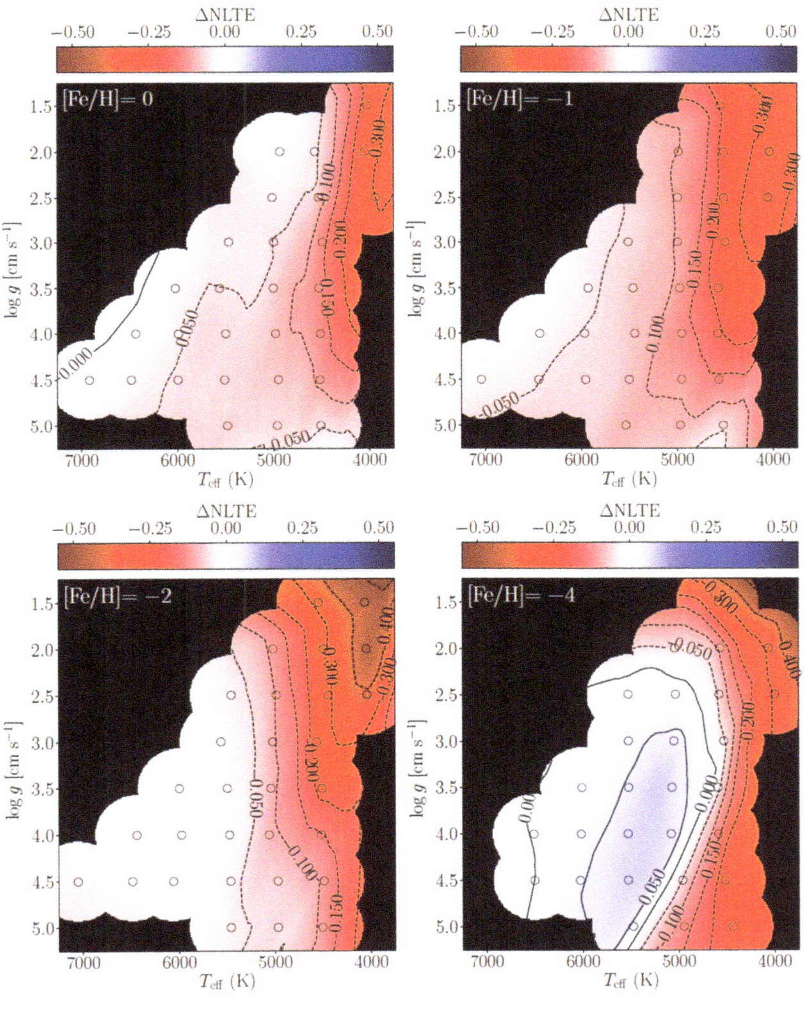

Figure 6.6. Corrections to 1D LTE lithium abundance obtained using 3D NLTE models spanning a wide range of stellar parameters and a single abundance of log N(Li) = 2.0, close to the level of the Spite plateau. Figure credit: Wang, E. X. et al. 2021 "3D NLTE spectral line formation of lithium in late-type stars" MNRAS 500 2159.

6.2 Evolution of Lithium on Cosmic Scale

Both the upper envelope of lithium abundance in Population II stars and the SBBN prediction for the primordial lithium content in the universe are below the cosmic abundance of lithium in the present Galaxy; hence sources of enrichment must be more efficient than sources of destruction during the evolution of the universe. The first recognized source of production of Li and other fragile elements such as beryllium (Be) and boron (B) on galactic scales comes from calculations of the interaction of cosmic rays with the ISM (Reeves et al. 1970; Meneguzzi et al. 1971).

Be is produced by spallation reactions only, and hence constraints on its abundance place limits on the amount of Li that is made in the ISM (Molaro & Beckman 1984).

Cosmic rays have a ratio of Be to Li that is much higher than what is observed in the solar system (Figure 6.7) and thus they cannot be the main source of Li in the Galaxy.

Current understanding of cosmic ray induced nucleosynthesis indicates that this mechanism contributes only < 20% of the lithium present in the ISM and that stellar sources of lithium enrichment are required. An example of the relative contributions of different lithium production sources with respect to iron is provided in Figure 6.8.

Around 70% of the lithium atoms on Earth, the solar system, and the ISM around us could come from relatively low-mass stars (1–4 $M_{\odot}$) via red giant evolution or nova explosions, but the yields predicted from those sources are too low to enrich the whole disk of the Milky Way (Romano et al. 2001; Travaglio et al. 2001; Karakas 2010).

Observations of ^{7}Be in nova explosions suggests that their actual lithium yields may be much higher than theoretically expected, but the direct detection of lithium in those systems is still elusive (Molaro et al. 2016, 2023). If the real yields of ^{7}Li in nova explosions are as high as those suggested by the ^{7}Be observations of nova

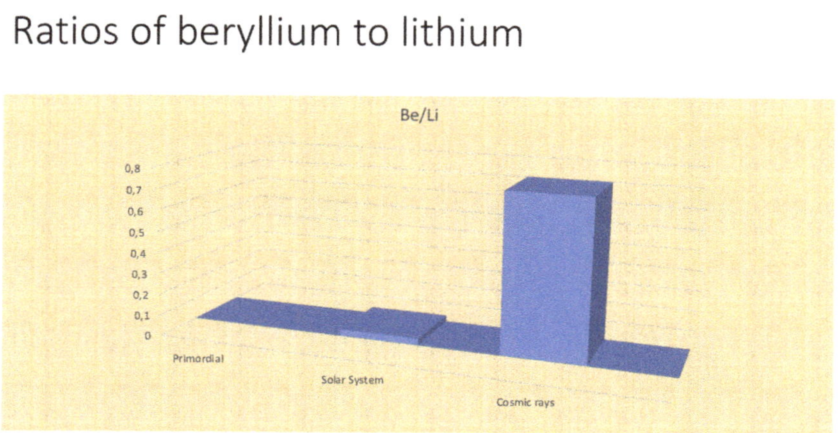

Figure 6.7. Schematic representation of Be/Li ratios predicted by SBBN (no Be at all), observed in the solar system (Be much less abundant than Li) and observed in Galactic cosmic rays (Be almost as abundant as Li). Figure credit: the author.

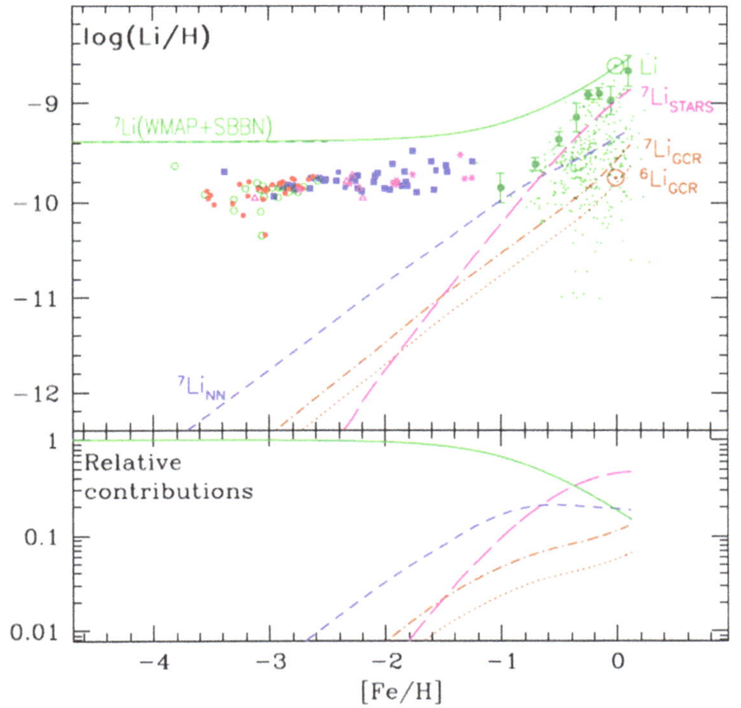

Figure 6.8. Lithium versus iron abundances in the Milky Way. The selection of stars is biased toward making the Spite plateau appear very clearly as the locus of the stars with iron abundance lower than a tenth of the solar level ([Fe/H] < −1). The contributions of different sources of lithium enrichment are represented by different colors and line types: Big Bang (solid green), Galactic cosmic rays (dot-dashed brown for ^{7}Li and dotted brown for ^{6}Li), core collapse supernovae (dashed blue) and evolved stars (AGBs, red giants and/or novae; long dashed pink). Figure credit: Prantzos, N. 2012 "Production and evolution of Li, Be and B isotopes in the Galaxy" A&A 542 A67, reproduced with permission © ESO.

ejecta, these events could be the dominant site of Li enrichment in the Galaxy (Cescutti & Molaro 2019; Romano et al. 2021; Kemp et al. 2022).

Very young stars and low-mass active orange and red dwarfs could contribute lithium via fast particle-induced nucleosynthesis in flares (Deliyannis & Malaney 1995; Tatischeff & Thibaud 2007). They represent the dominant class (by number) of stars in the Galaxy. Kelly et al. (2020) have estimated that these stars may have contributed up to 5% of the present Galactic lithium abundance, and are therefore probably not the main source of Li enrichment.

Sources of lithium associated with long-lived low-mass stars that cause an enrichment of the ISM delayed with respect to iron are most promising (Casuso & Beckman 2000; Romano et al. 2005). A contender in this regard are low-mass X-ray binaries (LMXBs). Galactic chemical evolutionary models including a population of LMXBs are needed to quantify their role in the lithium enrichment of the Milky Way.

The models of lithium evolution in the Milky Way predict that the Li abundance in the ISM should increase from the time the Sun was born to the present, and also toward the Galactic center.

The first prediction has been confirmed by machine-learning studies of large spectroscopic samples (Nepal et al. 2023).

The second predictions remain unconfirmed by observational data. Observational studies do not find any trend of lithium abundance with respect to kinematic or dynamical properties (Boesgaard et al. 2005). In fact, the theoretical predictions of closed-box galactic chemical evolutionary models are likely not realistic. Guiglion et al. (2019) and Zhang et al. (2023) have discussed the importance of including inside-out disk formation and radial migration in Galactic chemical evolution models.

The distribution of the lithium abundances among solar-type stars in the Milky Way may be a tracer of Galactic scale events that affected the gas accretion history or the star-formation rate. A review of the abundances of B, Be, and Li in different types of stars, including correlations between them, can be found in Boesgaard (2023).

Using a galactic chemical evolution model, Roca-Fàbrega et al. (2021) have searched for the imprint that major events in the star-formation history of the Milky Way may have left on the pattern of lithium abundances of solar-type stars. They construct the following general sequence of events:

(1) > 7 Gyr ago: The Galaxy contained an ISM with low iron abundance and low star-formation rate. Stars were forming at a low rate with a constant lithium abundance at the level of the Spite plateau.
(2) 7 to 6 Gyr ago: Massive pristine gas flow and mixing with the slightly polluted ISM material already present in the Galaxy. New stellar population with low iron abundance and lithium abundance similar to that produced by SBBN.
(3) 6 to 3 Gyr ago: Very strong episode of star formation induced by strong interaction of the Milky Way with a satellite Galaxy (Mor et al. 2019). New stellar population with enhanced lithium and iron abundances.
(4) From 3 Gyr ago to the present: Two more interactions with satellite galaxies have kept star formation going (Ruiz-Lara et al. 2020). Lithium-rich and iron-rich stellar populations keep forming.

The lack of evidence for a net lithium enrichment of the local ISM in nearby star formation regions could indicate that the Sun may be going through a region of the Milky Way which is somewhat mixed with pristine material, and may not reflect the maximum lithium abundance of the Galactic disk.

Young dwarf stars (age < 3 Gyr) that are high in α elements content (α elements are those that can be created by α particle capture such as Mg, Si, S and Ca) with respect to iron display lithium abundances that are about a factor of 2 higher than the proto-solar value (Borisov et al. 2022), suggesting that some star-forming regions in the Galaxy have higher lithium content than those in the solar vicinity.

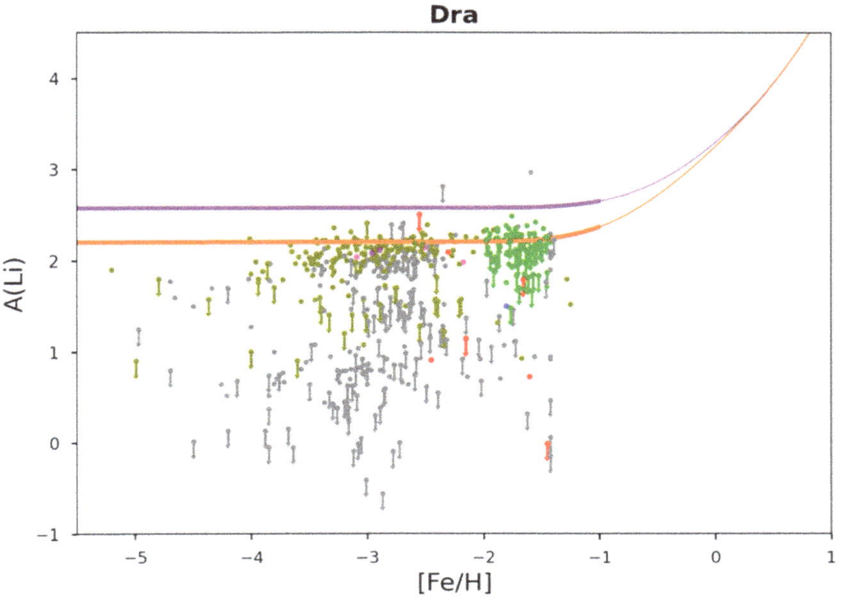

Figure 6.9. The predicted relationship between the abundance of lithium versus that of iron for the Draco spheroidal dwarf Galaxy, a satellite of the Milky Way. The purple line refers to a primordial lithium abundance from SBBN using the cosmological parameters from WMAP and Planck. The orange line refers to a primordial lithium abundance equal to the Spite plateau. Gray dots are data from the halo of the Milky Way. Green dots are data points from the Omega Centauri globular cluster. Red dots represent data from the S2 stream (probably a disrupted dwarf Galaxy). Magenta dots come from the Gaia-Enceladus stream. The data were taken from Molaro et al. (2020) and Aguado et al. (2021). Figure credit: Matteucci, F. et al. 2021 "The evolution of lithium: Implications of a universal Spite plateau" MNRAS 505 200.

The galactic evolution of lithium has been calculated for dwarf galaxies that are satellites of or that have been accreted by the Milky Way (Molaro et al. 2020; Matteucci et al. 2021). For example, the evolution of lithium with respect to iron for the Draco dwarf spheroidal Galaxy is illustrated in Figure 6.9. This Galaxy is located at a distance of 80 kpc and has a metallicity of $[Fe/H] = -1.8 \pm 0.2$. It is dominated by an old stellar population that was formed about 10 Gyr ago (Aparicio et al. 2001).

The galactic evolution of Be has also been studied and suggests that the conditions for chemical enrichment in the early Galaxy were quite inhomogeneous (Smiljanic et al. 2021).

Galactic evolutionary models for dwarf galaxies are able to fit the rather flat relation between lithium and iron, and predict a strong increase in lithium for metallicities close to solar and higher. This expectation appears to have been confirmed by lithium abundance studies of large samples of hot main-sequence stars (Nepal et al. 2023).

A schematic illustration of the sequence of events that are thought to have happened in the Galaxy is provided in Figure 6.10. This time sequence may be

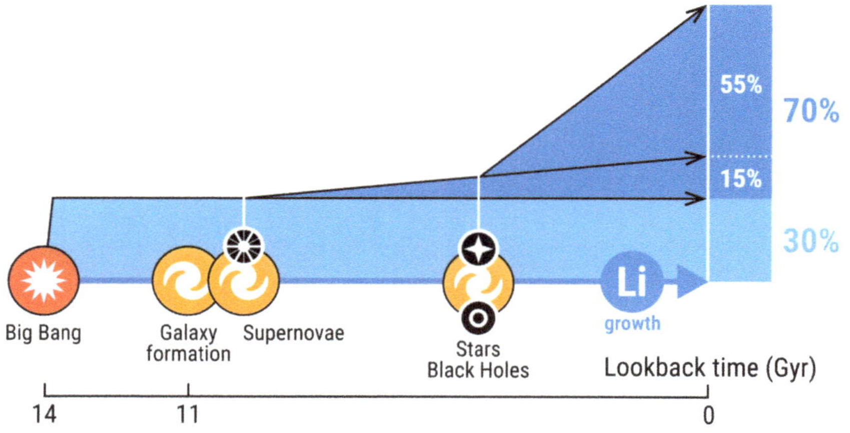

Figure 6.10. The schematic history of lithium enrichment in the Galaxy. Figure credit: the author with assistance by Gaby Pérez.

universal for all galaxies with a chemical evolution similar to that of the Milky Way, but it may be different for galaxies with different chemical evolution such as, for example, the Small Magellanic Cloud, which is more metal-poor and has been found to have a lower ISM lithium abundance than the Galaxy (Howk et al. 2012).

Bibliography

Aguado, D. S., González Hernández, J. I., Allende Prieto, C., & Rebolo, R. 2019, ApJ, 874, L21
Aguado, D. S., Myeong, G. C., Belokurov, V., et al. 2021, MNRAS, 500, 889
Aguado, D. S., Molaro, P., Caffau, E., et al. 2022, A&A, 668, A86
Aoki, W., Barklem, P. S., Beers, T. C., et al. 2009, ApJ, 698, 1803
Aparicio, A., Carrera, R., & Martínez-Delgado, D. 2001, AJ, 122, 2524
Asplund, M., Carlsson, M., & Botnen, A. V. 2003, A&A, 399, L31
Asplund, M., Lambert, D. L., Nissen, P. E., Primas, F., & Smith, V. V. 2006, ApJ, 644, 229
Bandyopadhyay, A., Sivarani, T., Beers, T. C., et al. 2022, ApJ, 937, 52
Barklem, P. S., Belyaev, A. K., & Asplund, M. 2003, A&A, 409, L1
Boesgaard, A. 1976, PASP, 88, 353
Boesgaard, A. M., Stephens, A., & Deliyannis, C. P. 2005, ApJ, 633, 398
Boesgaard, A. 2023, ApJ, 943, 40
Bonifacio, P., & Molaro, P. 1997, MNRAS, 285, 847
Bonifacio, P., Molaro, P., Sivarani, T., et al. 2007, A&A, 462, 851
Bonifacio, P., Caffau, E., Spite, M., et al. 2018, A&A, 612, A65
Borisov, S., Prantzos, N., & Charbonnel, C. 2022, A&A, 668, A181
Carlsson, M., Rutten, R. J., Bruls, J. H. M. J., & Shchukina, N. G. 1994, A&A, 288, 860
Casuso, E., & Beckman, J. E. 2000, PASP, 112, 942
Cescutti, G., & Molaro, P. 2019, MNRAS, 482, 4372
Coc, A., & Vangioni, E. 2017, IJMPE, 26, 1741002
Cyburt, Richard H., Fields, Brian D., Olive, Keith A., & Yeh, Tsung-Han 2016, RvMP, 88, 5004
Deliyannis, C. P., & Malaney, R. A. 1995, ApJ, 453, 810

Fields, B. D., & Olive, K. A. 2022, JCAP, 10, 78
González Hernández, J. I., Bonifacio, P., Ludwig, H.-G., et al. 2008, A&A, 480, 233
González Hernández, J. I., Aguado, D. S., Allende Prieto, C., Burgasser, A. J., & Rebolo, R. 2020, ApJ, 889, L13
Guiglion, G., Chiappini, C., Romano, D., et al. 2019, A&A, 623, A99
Gutiérrez, C. M., García López, R. J., Rebolo, R., Martín, E. L., & François, P. 1999, A&AS, 137, 93
Hansen, T., Hansen, C. J., Christlieb, N., et al. 2014, ApJ, 787, 162
Harutyunyan, G., Steffen, M., Mott, A., et al. 2018, A&A, 618, A16
Hosford, A., Ryan, S. G., García Pérez, A. E., Norris, J. E., & Olive, K. A. 2009, A&A, 493, 601
Howk, J. Ch., Lehner, N., Fields, B. D., & Mathews, G. J. 2012, Natur, 489, 121
Iocco, F., Mangano, G., Miele, G., Pisanti, O., & Serpico, P. D. 2009, PhR, 472, 1
Jedamzik, K., & Pospelov, M. 2009, NJPh, 11, 105028
Karakas, A. I. 2010, MNRAS, 403, 1413
Kelly, D. E., Christian, D. J., Mathioudakis, M., & Jevremović, D. 2020, RAA, 20, 104
Kemp, A. J., Karakas, A. I., Casey, A. R., et al. 2022, ApJ, 933, L30
Kielty, C. L., Venn, K. A., Sestito, F., et al. 2021, MNRAS, 506, 1438
Kiselman, D. 1997, ApJ, 489, L107
Klevas, J., Kučinskas, A., Steffen, M., Caffau, E., & Ludwig, H.-G. 2016, A&A, 586, A156
Korn, A. J., Grundahl, F., Richard, O., et al. 2007, ApJ, 671, 402
Kusakabe, M., Kajino, T., Boyd, R. N., Yoshida, T., & Mathews, G. J. 2008, ApJ, 680, 846
Lind, K., Asplund, M., & Barklem, P. S. 2009, A&A, 503, 541
Lind, K., Melendez, J., Asplund, M., Collet, R., & Magic, Z. 2013, A&A, 554, A96
Magazzù, A., Rebolo, R., & Pavlenko, Y. V. 1992, ApJ, 392, 159
Martín, E. L., Rebolo, R., Magazzù, A., & Pavlenko, Ya. V. 1994, A&A, 282, 503
Mashonkina, L., Korn, A. J., & Przybilla, N. 2007, A&A, 461, 261
Matteucci, F., Molero, M., Aguado, D. S., & Romano, D. 2021, MNRAS, 505, 200
Meléndez, J., & Ramírez, I. 2004, ApJ, 615, L33
Meléndez, J., Casagrande, L., Ramírez, I., Asplund, M., & Schuster, W. J. 2010, A&A, 515, L3
Meneguzzi, M., Audouze, J., & Reeves, H. 1971, A&A, 15, 337
Molaro, P., & Beckman, J. 1984, A&A, 139, 394
Molaro, P., Izzo, L., Mason, E., Bonifacio, P., & Della Valle, M. 2016, MNRAS, 463, L117
Molaro, P., Izzo, L., Bonifacio, P., Hernanz, M., Selvelli, P., & della Valle, M. 2020, MNRAS, 496, 2902
Molaro, P., Izzo, L., Selvelli, P., et al. 2023, MNRAS, 518, 2614
Monaco, L., Bonifacio, P., Sbordone, L., Villanova, S., & Pancino, E. 2010, A&A, 519, L3
Mor, R., Robin, A. C., Figueras, F., Roca-Fàbrega, S., & Luri, X. 2019, A&A, 624, L1
Mott, A., Steffen, M., Caffau, E., & Strassmeier, K. G. 2020, A&A, 638, A58
Nepal, S., Guiglion, G., de Jong, R. S., et al. 2023, A&A, 671, A61
Osorio, Y., Barklem, P. S., Lind, K., & Asplund, M. 2011, A&A, 529, A31
Pavlenko, Y. V., & Magazzù, A. 1996, A&A, 311, 961
Peytremann, E. 1974, A&AS, 18, 81
Piau, L., Beers, T. C., Balsara, D. S., et al. 2006, ApJ, 653, 300
Piau, L. 2008, ApJ, 689, 1279
Prantzos, N. 2012, A&A, 542, A67
Randich, S., Pace, G., Pastori, L., & Bragaglia, A. 2009, A&A, 496, 441

Rebolo, R., Beckman, J., & Molaro, P. 1987, A&A, 172, L17
Rebolo, R., Molaro, P., & Beckman, J. E. 1988, A&A, 192, 192
Reeves, H., Fowler, W. A., & Hoyle, F. 1970, Natur, 226, 727
Reeves, H. 2005, ESA Publication Series, (Les Ulis: EDP Science) 17, 15
Richard, O., Michaud, G., & Richer, J. 2005, ApJ, 619, 538
Roca-Fàbrega, S., Llorente de Andrés, F., Chavero, C., Cifuentes, C., & de la Reza, R. 2021, A&A, 656, A64
Roederer, I. U., & Lawler, J. E. 2021, ApJ, 912, 119
Romano, D., Matteucci, F., Ventura, P., & D'Antona, F. 2001, A&A, 374, 646
Romano, D., Tosi, M., Matteucci, F., & Chiappini, C. 2003, MNRAS, 346, 295
Romano, D., Chiappini, C., Matteucci, F., & Tosi, M. 2005, A&A, 430, 491
Romano, D., Magrini, L., Randich, S., et al. 2021, A&A, 653, A72
Ruiz-Lara, T., Gallart, C., Bernard, E. J., & Cassisi, S. 2020, NatAs, 4, 965
Ryan, S. G., Beers, T. C., Olive, K. A., Fields, B. D., & Norris, J. E. 2000, ApJ, 530, L57
Sbordone, L., Bonifacio, P., Caffau, E., et al. 2010, A&A, 522, A26
Schaeuble, M., & King, J. R. 2012, PASP, 124, 164
Shi, J. R., Gehren, T., Zhang, H. W., Zeng, J. L., & Zhao, G. 2007, A&A, 465, 587
Smiljanic, R., Zych, M. G., & Pasquini, L. 2021, A&A, 646, A70
Spite, F., & Spite, M. 1982, A&A, 115, 357
Spite, F., & Spite, M. 1986, A&A, 163, 140
Steenbock, W., & Holweger, H. 1984, A&A, 130, 319
Takeda, Y., & Kawanomoto, S. 2005, PASJ, 157, 45
Tatischeff, V., & Thibaud, J. P. 2007, A&A, 469, 265
Travaglio, C., Randich, S., Galli, D., et al. 2001, ApJ, 559, 909
Wagoner, R. V. 1967, Sci., 155, 1369
Wang, E. X., Nordlander, T., Asplund, M., et al. 2021, MNRAS, 500, 2159
Zhang, H., Chen, Y., Zhao, G., Bi, S., Zhang, X., & Xue, X. 2023, MNRAS, 520, 4815

Chapter 7

Lithium in Failed Stars (Brown Dwarfs)

This chapter deals with lithium in objects that never settle on the stellar main sequence and keep cooling forever. These failed stars are also called brown dwarfs or substellar-mass objects. They are formed in the same molecular clouds as stars but do not reach enough mass to shine as stars do.

After plunging into complete destruction owing to very efficient convective mixing in red dwarfs, lithium makes a spectacular comeback at the ultracool temperatures typical of the brown dwarfs because of their low central temperatures. Such behavior is very well documented in the benchmark open clusters that are nearest to us (Hyades and Pleiades).

The development of degenerate cores during the gravitational contraction of brown dwarfs becomes the dominant effect in stabilizing their structures, as opposed to nuclear burning in stars. The theoretical minimum mass for stable H fusion is located around 0.075 $M_{\odot}$ for solar metallicity, which is the boundary between stars and brown dwarfs, and the theoretical minimum mass for lithium burning is located around 0.055 $M_{\odot}$, which is the predicted boundary between Li-burning and Li-preserving brown dwarfs. The theoretical limit for thermonuclear burning of deuterium is located at lower mass and has not yet been observed. It has been proposed as a possible boundary between brown dwarfs and planets.

The observational lithium test has been useful in identifying many genuine lithium-preserving brown dwarfs. Lithium in brown dwarfs has also been used as a precise nuclear chronometer for open clusters with ages in the range 100–1000 Myr, which includes the Hyades and the Pleiades.

Lithium in brown dwarf binaries with measured dynamical masses from orbital motions have been used to determine the minimum mass for thermonuclear lithium burning (MMLB), which is a very sensitive nuclear test of the validity of evolutionary models.

The presence of lithium in the atmospheres of brown dwarf candidates is a test of their substellar mass. For solar chemical abundance, the mass boundary between

doi:10.1088/2514-3433/acd8bbch7

lithium-burning objects and fully lithium-preserving objects has been determined at 0.049 $M_{\odot}$ (51.5 M_J) using observations of brown dwarf binaries, a value lower than model predictions.

Low-metallicity brown dwarfs are expected to have preserved the primordial lithium abundance. Determination of lithium abundance in those objects is very challenging but is certainly worthwhile because it could bridge the gap in our knowledge between Big Bang nucleosynthesis and the Spite plateau, and hence solve the cosmological lithium problem.

7.1 Discovery of Lithium in UX Tau C

March 1990: a half-hour inter-island flight from Tenerife to the lush volcanic island of La Palma, locally known as the Isla Bonita (beautiful island), carried two young astronomers and one graduate student among its passengers.

The purpose of that trip was to carry out a three-night run at the INT, a venerable and highly reliable telescope that had seen first light back in 1965. The INT had been first located near the gorgeous 15th-century Hertsmonceux Castle in Sussex, to the south of London, where it spent over a decade observing in mostly cloudy and hazy weather conditions.

Pushed by people like Fred Hoyle and Martin Ryle, the Royal Greenwich Observatory realized that, not too far away, there was a much better site from which to watch the skies and shipped the INT to the world-class mountain site 2400 m high ORM—this episode is commented on in the book by JE Beckman (2021).

The first night was blessed with a pitch-dark and crystal-clear sky. The three observers were busily going through the list of targets for the night that included a mix of T Tauri and halo stars, the youngest and oldest known stars in the Milky Way.

Finally, after the customary focusing of the telescope, which always seemed to eat up too much time for them in their hunger for starlight, the telescope operator announced that everything was ready to go to the first target, which was located in the Taurus—Auriga region, well-known for hosting one of the stellar cradles nearest to us.

As the targets were quickly setting, the observers had to move fast to catch some T Tauri stars. The normal procedure was to get spectra of one star at a time, carefully placing it on the slit of the Intermediate Dispersion Spectrograph (IDS). However, visual binaries with the right projected separation allowed them to get two objects for the price of one by placing both components on the slit.

The IDS rotated to align the slit with the axis joining the two T Tauri stars in the UX Tau binary system, and the operator took an additional moment to do a fine refocusing when the motors settled to the tracking mode. Routinely, the astronomers requested to increase the integration time on each target to check for possible companions, and suddenly—there was one …

The operator was staring at the acquisition TV camera and softly, almost as if he was talking to himself, said something in Spanish … "I think there is another, fainter object, very close to the bright star". The three observers jumped as a single body to check on the TV monitor what he was pointing at.

They had stumbled upon UX Tau C, the third component of the UX Tau multiple system (see Figure 7.1), hitherto unknown to them, but not to George Howard Herbig, who had spotted it during a night of good visibility at the 3 m Shane Telescope at Lick Observatory (California). Katherine Robbins Bell had included it as a footnote in the Third Catalog of Emission-Line Stars of the Orion Population (Herbig & Bell 1988), which was like a bible of T Tauri stars at that time.

UX Tau C was also known to astronomers Andrea Moneti and Hans Zinnecker, who had observed the system in January 1988 with an infrared camera on the 1.5 m telescope at Cerro Tololo Interamerican Observatory in Chile, but they did not publish their results until three years later (Moneti & Zinnecker 1991).

The three observers of this tale abandoned part of the observing program that they had carefully planned before catching sight of UX Tau C; they focused instead on getting spectra of it with the IDS.

At the end to those three magnificent nights they had detected the Li I resonance feature for the very first time in a very young and cool low-mass object of spectral

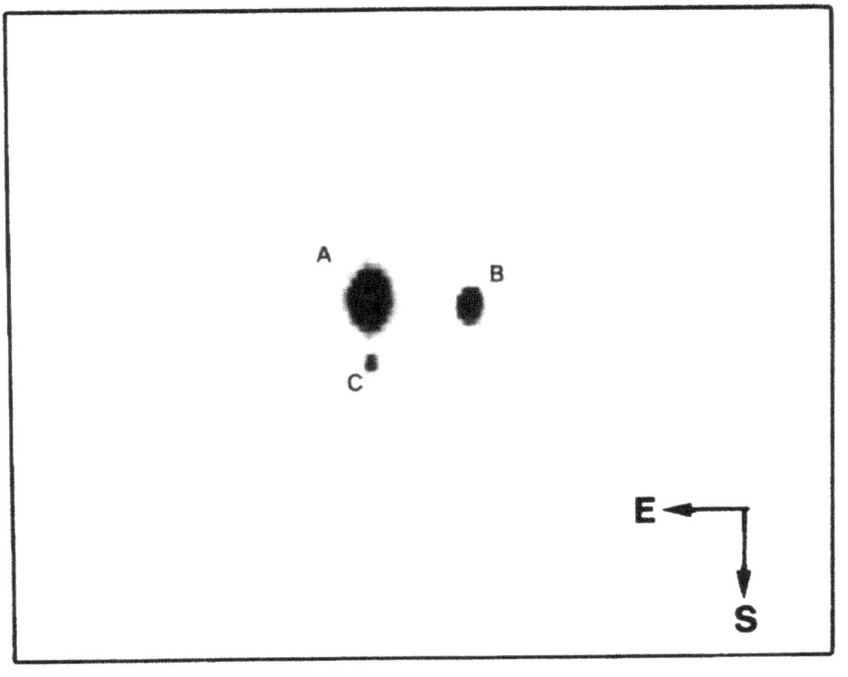

Figure 7.1. Image of the UX Tau triple system taken with the TV camera of the INT on a clear night of March 1990 with good visibility. The brown dwarf candidate UX Tau C is the faintest object of the three, located at just 2.7 arcsec from the bright primary UX Tau A and about 4 magnitudes fainter. Figure credit: Magazzù, A., Martín, E. L., & Rebolo, R. 1991 "Lithium in the pre-main sequence triple system UX Tauri" A&A 249 149, reproduced with permission © ESO.

type later than M5 (see Figure 7.2), close to the coolest stars that were known at the time, and they started thinking that it might be possible to find lithium in even cooler objects, the elusive brown dwarfs.

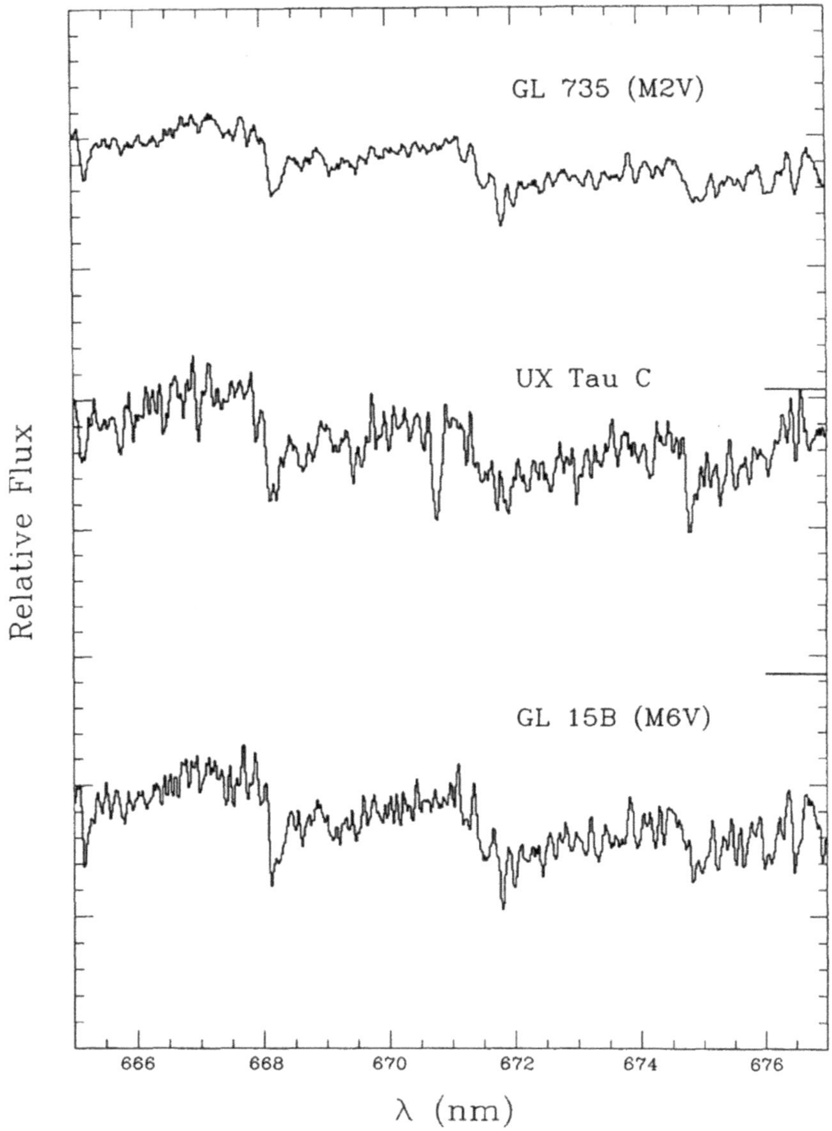

Figure 7.2. Spectrum of UX Tau C obtained with the IDS at the INT compared with two old M-type stars from the solar neighborhood. Note the presence of broad TiO molecular bands that dominate the shape of the spectra in this region. The Li I resonance doublet at 670.8 nm is clearly seen only in UX Tau C but not at all in the other two M dwarfs. Figure credit: Magazzù, A., Martín, E. L., & Rebolo, R. 1991 "Lithium in the pre-main sequence triple system UX Tauri" A&A 249 149, reproduced with permission © ESO

7.2 The Lithium Test for Brown Dwarf Candidates

The discovery of UX Tau C fired the starting pistol in a race to find and confirm the first brown dwarfs with the lithium test. Inspired by the lithium detection in UX Tau C, Magazzù et al. (1991) and Rebolo et al. (1992) proposed that the detection of the Li I resonance feature at 670.8 nm could be used to verify the nature of brown dwarf candidates and test evolutionary models of brown dwarfs.

Magazzù et al. (1993) carried out the first spectroscopic search for lithium among brown dwarf candidates in binaries. None of the five binaries in their study showed a detectable Li I resonance feature. Figure 7.3 shows the spectra of two of those brown dwarf binary candidates compared with the spectrum of UX Tau C.

The lithium test did not support the substellar dynamical masses of the components of GL 473 (alternative name Wolf 424) estimated by Heintz (1989) from astrometric measurements, but it was consistent with the mass estimates made from speckle imaging observations (Henry et al. 1992). High-precision astrometry carried out with the Fine Guidance Sensors on the HST demonstrated that the component masses of GL 473 are clearly stellar (Torres et al. 1999), in agreement with their failure to pass the lithium test.

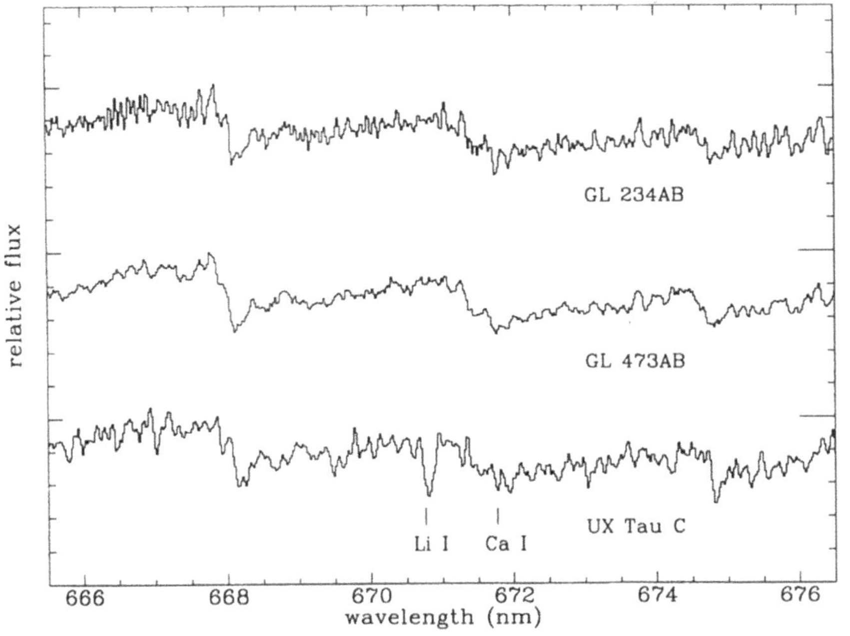

Figure 7.3. Observational spectra of two brown dwarf binary candidates compared with the spectrum of UX Tau C. Note the presence of the strong Li I resonance doublet in UX Tau C but its absence in the other two objects. These two binaries failed to pass the lithium test and hence their masses were estimated to be larger than 0.06 $M_{\odot}$. Figure credit: Magazzù, A., Martín E. L., & Rebolo, R. 1993 "A spectroscopic test for substellar objects" ApJ 404 L17.

The lithium test was extended to 10 single brown dwarf candidates in the solar vicinity, four brown dwarf candidates in open clusters (Hyades and Pleiades) and two brown dwarf candidates in binaries with thus far negative results (Marcy et al. 1994; Martín et al. 1994). The term "**lithium test**" was coined (Nelson et al. 1993) and all brown dwarf candidates to which it had been applied between 1993 and 1994 failed to pass it.

7.3 The First Lithium Brown Dwarfs

Nearby young open clusters such as α Persei and the Pleiades are prime hunting grounds for brown dwarfs. Deep imaging surveys looking for them were carried out by several teams (Jameson & Skillen 1989; Stauffer et al. 1989; Hambly & Jameson 1991; Hambly et al. 1991; Rebolo et al. 1992), fostering theoretical progress in modeling young brown dwarfs (Stringfellow 1991; Laughlin & Bodenheimer 1993).

Low-resolution reconnaissance spectroscopy proved to be good enough for obtaining the fundamental parameters of brown dwarf candidates, such as effective temperature and luminosity, but not for applying the lithium test (Rebolo et al. 1992; Hamilton & Stauffer 1993).

Back in those days, there was only one telescope in the world that could provide the combination of large aperture and high spectral resolution that was needed to apply the lithium test to the faintest brown dwarf candidates known in open clusters, the mighty 10 m Keck I telescope of the MKO on the summit of the 4207 m high Maunakea volcano, a sacred mountain for the native Hawaiian people.

Two brown dwarfs in the Pleiades were the smoking gun that finally confirmed beyond any reasonable doubt that brown dwarfs form in open clusters and that they preserve their initial lithium content in stark contrast with very low-mass stars.

The two lithium brown dwarfs were named Teide 1 and Calar 3 because they had been found in deep imaging surveys carried out at the Teide and Calar Alto observatories, respectively (Rebolo et al. 1995; Martín et al. 1996; Zapatero Osorio et al. 1997).

They were observed in November 1995 and February 1996 with the Low Resolution Imaging Spectrograph (LRIS) at the Keck II telescope. The final spectra obtained are shown in Figure 7.4. Strong Li I resonance features are clearly present in both of them (Rebolo et al. 1996). The subordinate Li I line at 812.6 nm was also detected in Teide 1. These lithium detections were a confirmation of their membership in the Pleiades and they are consistent with their masses estimated at 55 M_J using evolutionary models.

A brighter brown dwarf candidate named PPl 15 from the survey carried out at Palomar observatory by Stauffer et al. (1989) had been previously observed with HIRES at the Keck I telescope and showed a weaker Li I feature, suggesting partial lithium depletion and a mass closer to the substellar limit (Basri et al. 1996), which prompted Zapatero Osorio et al. (1997) to call it a transition object and suggest that it could be a binary owing to its position in the CMD of the Pleiades cluster.

PPl 15 was confirmed as a double-lined spectroscopic binary by analysis of HIRES spectra in Basri & Martín (1999) with an orbital period of 5.8 days, a mass

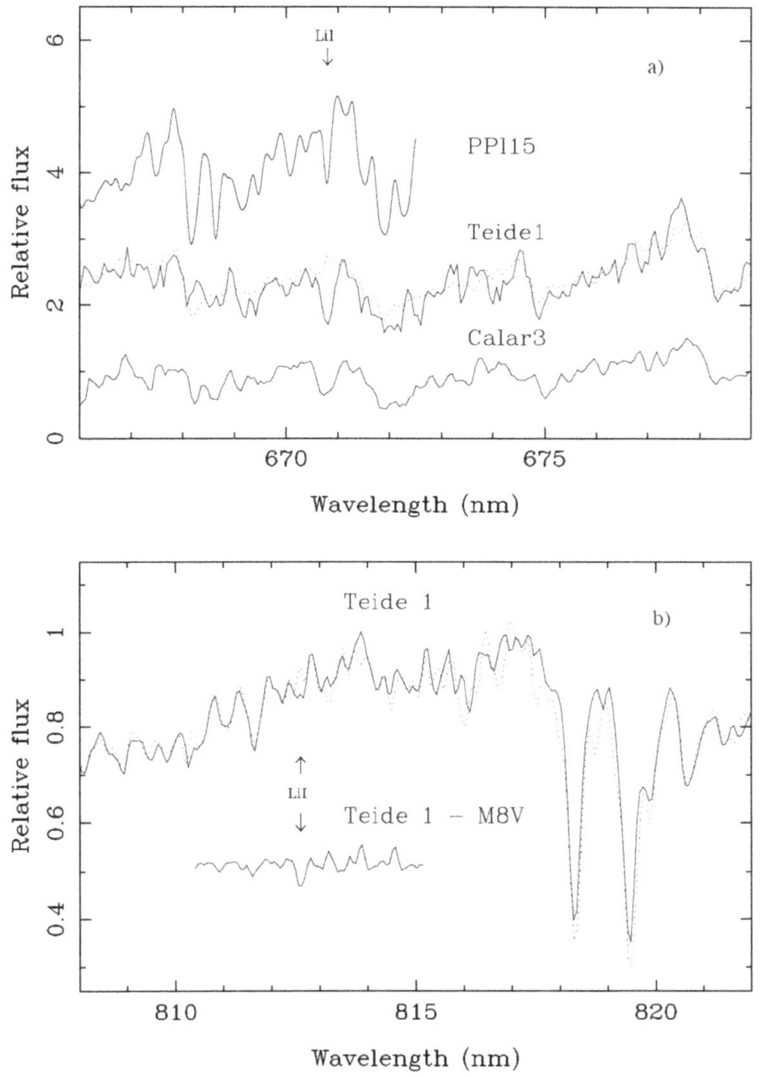

Figure 7.4. (a) Upper panel: the spectral region around the Li I resonance doublet at 670.8 nm (position indicated with an arrow) observed with LRIS on the 10 m Keck II telescope. Continuous lines are used to draw the spectra of the three Pleiades objects PPl15, Calar 3, and Teide 1. A dotted line traces the spectrum of the field dwarf vB 10, which has a spectral type (M8) and an effective temperature (2600 K) similar to Teide 1. Note that lithium is clearly detected in the Pleiades objects, but not in vB10. (b) lower panel: the spectral region around the Li I subordinate line at 812.6 nm (position indicated with an arrow and labeled) observed with the Keck telescope. Continuous lines are used to trace the spectrum of the Pleiades object Teide 1. A dotted line is used to trace the average spectrum of the field dwarfs LP412-31 and vB 10, which are of spectral type (M8) and have an effective temperature of 2600 K similar to Teide 1. Note that lithium is clearly detected in the Pleiades object, but not in the M8 field dwarfs. Even though the subordinate Li I line is located in a region where brown dwarfs are brighter, the fact that it is much weaker than the resonance line makes it less favorable for lithium detection. Figure credit: Rebolo, R. et al. 1996 "Brown dwarfs in the Pleiades cluster confirmed by the lithium test" ApJ 469 L53.

ratio of 0.4, and a total mass in the range 0.06–0.07 $M_{\odot}$ according to theoretical models. The double-lined nature of the spectra may have diluted the Li I resonance feature and the binary may possibly have undepleted lithium content, a possibility still not fully clarified until now.

Soon after lithium was discovered among brown dwarfs in the Pleiades, news arrived concerning similar discoveries in brown dwarfs unattached to any known association or cluster. Thackrah et al. (1997) announced the detection of lithium in a very red and faint object, which they named 296A, found in photographic plates. They used the RGO spectrograph at the AAT in December 1995 to record the spectra that showed the Li I resonance feature at 670.8 nm. They estimated a spectral type of M6, an effective temperature of 2800 K and a mass and age similar to PPl15 in the Pleiades. This object became the first young field brown dwarf to pass the lithium test.

A search for high proper motion objects using the ESO Schmidt camera at La Silla Observatory produced the discovery of Kelu 1 (Ruiz et al. 1997), named after the word for red in the Mapuche language, which displayed a strong Li I resonance feature and weak TiO bands in spectra obtained in March 1997 with EFOSC1 at the 3.6 m ESO telescope. Analysis of the spectral energy distribution (SED) of Kelu 1 indicated an effective temperature of 1900 K and suggested the condensation of dust grains in the photosphere. In fact, Kelu 1 was one of the first young L-type lithium brown dwarfs ever discovered, and it was later found to be a binary resolved by the Keck sodium laser guide star adaptive optics system (Liu & Leggett 2005).

Almost coincident with the announcement of Kelu 1 was that of three even cooler and fainter field brown dwarfs from the DENIS survey (Delfosse et al. 1997). One of them, named DENIS-P J1228-1547, showed a very prominent Li I resonance doublet in spectra obtained in 1997 June at the AAT (Tinney et al. 1997) and at the Keck I telescope (Martín et al. 1997). With a pseudo-equivalent width of 3.5 Å, the Li I resonance doublet of this object was the largest ever measured in any brown dwarf. Another DENIS object, named DENIS-P J1058.7-1548, had no obvious Li I resonance doublet. The resonance lines of all the alkali elements present in the optical spectra (Cs, Li, Na, K, and Rb) were very strong and broad due to van der Waals forces in high-pressure and low-temperature conditions (see Figure 7.5).

The very broad alkali lines, together with the weakening of TiO bands, prompted the suggestion of a new spectral class cooler than M that was dubbed L (Martín et al. 1997). The L classification became established a few years later with discoveries of more brown dwarfs in the field and in the Pleiades cluster (Kirkpatrick et al. 1999, 2000; Martín et al. 1998, 1999).

The lithium test showed that some very late-M type dwarfs in the solar vicinity that were known for a long time actually were young brown dwarfs. A remarkable case is that of LP 944-20, which had long been cataloged as a high proper motion star for a long time (Luyten & Kowal 1975; Irwin et al. 1991) and was demonstrated to be a lithium-bearing brown dwarf by spectroscopic observations obtained with CASPEC on the 3.6 m ESO telescope at La Silla in 1994 February (Tinney 1998). The lithium detection constrained its age to lie in the range 475–650 Myr and its mass to be in the range 0.057–0.063 $M_{\odot}$.

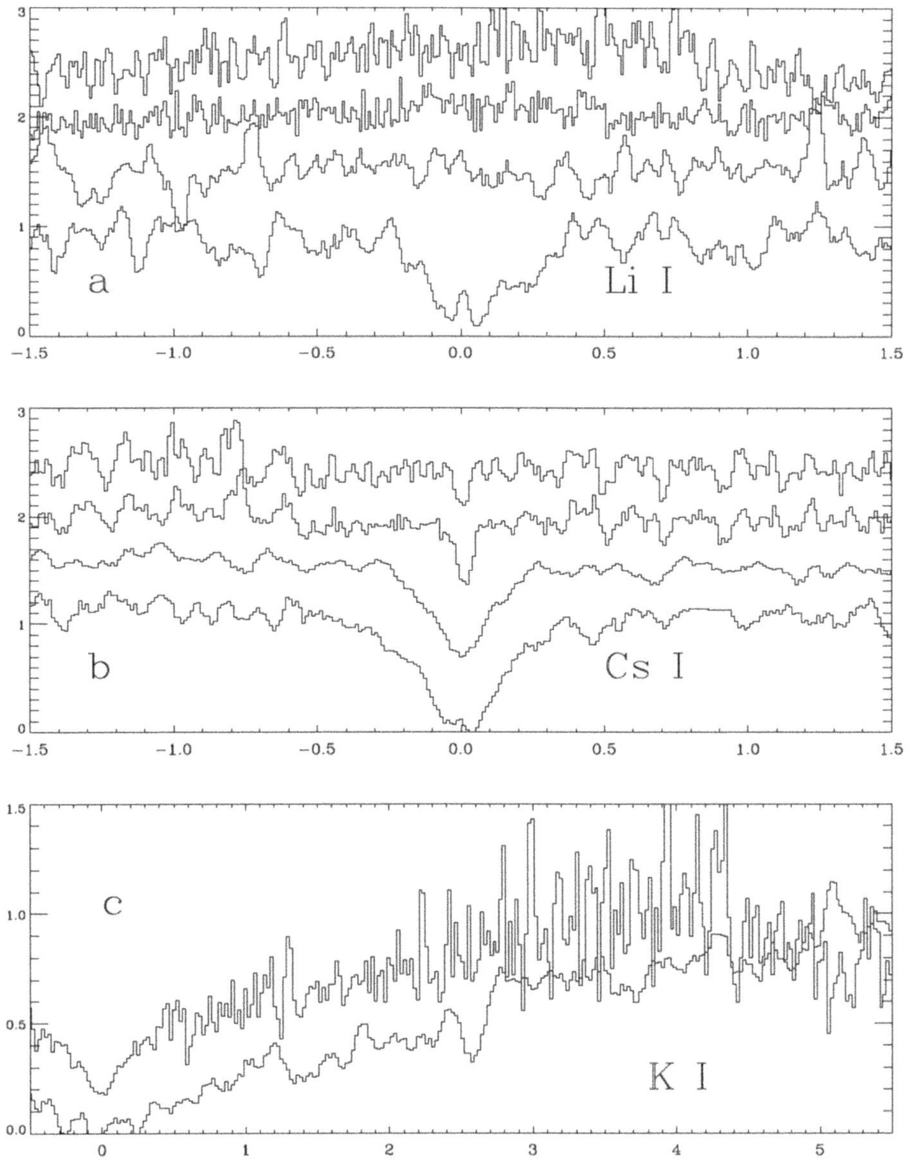

Figure 7.5. High resolution (R = 31,000) spectra of four ultracool dwarfs obtained with HIRES at the Keck I telescope in 1997 June. From top to bottom the four objects shown with offsets of 0.5 are Gliese 406 (M6), LHS 2924 (M9), DENIS-P J1058.7-1548 (L3), and DENIS-P J1228-1547 (L5); they represent a sequence of decreasing effective temperature from about 2600 K to about 1600 K. The spectra of the L dwarfs are smoothed with a 5-point boxcar. The wavelength scale is given in nm from the center of each alkali resonance line, namely Li I at 670.8 nm in panel a; Cs I at 852.1 nm in panel b; and K I at 769.9 nm in panel c. Note that Li I is detected only in DENIS-P J1228-1547, and that all three resonance lines in this object have extremely broad wings owing to high-pressure effects caused by van der Waals forces on the spectral line formation (e.g. Kuhn 1937) in the brown dwarf atmosphere. Figure credit: Martín, E. L. et al. 1997 "Keck HIRES spectra of the brown dwarf DENIS-P J1228.2-1547" A&A 327 L2, reproduced with permission © ESO.

Another outstanding case is PC 002507.3+044714, which was identified in the Palomar CCD grism survey as an unusual emission-line object. Low-resolution follow-up spectroscopy of this object obtained in August 1990 showed a very late-M and peculiar spectrum with very strong emission lines of the hydrogen Balmer series (Schneider et al. 1991). Spectroscopic monitoring of PC 002507.3+044714 carried out with LRIS on the Keck II telescope from November 1995 to December 1998 showed that the emission lines are variable and persistent and that the object is a lithium brown dwarf (Martín et al. 1999). The cause of the variability remains unknown.

7.4 Lithium Abundances in Brown Dwarfs

The determination of lithium abundances from the spectra of M-type and early L-type brown dwarfs is tricky because the Li I resonance feature is formed in a region dominated by TiO molecular bands. The effect of switching the TiO absorption on and off for an effective temperature of 2000 K has dramatic consequences on the appearance of the Li I feature. Therefore, only pseudo-equivalent widths (pEWs) can be measured in brown dwarfs, meaning that EWs of atomic lines are measured with respect to a local pseudo-continuum, which is strongly affected by the presence of molecular band absorption.

The determination of absolute lithium abundance in brown dwarfs depends critically on the knowledge of molecular band formation in ultracool atmospheres. Based on progress in TiO band data, it has been estimated that the nearby brown dwarf LP944-20 has log N(Li) = 3.25, consistent with undepleted cosmic abundance (Pavlenko et al. 2007). However, these authors note that it would be desirable to confirm this Li abundance using the subordinate Li line at 812.6 nm, which was detected in Teide Pleiades 1. There has not yet been any such confirmation.

It is reassuring that M-type stars and brown dwarfs in the very young σ Orionis cluster (age estimated at 2–4 Myr) show cosmic lithium abundance in agreement with theoretical expectations (Zapatero Osorio et al. 2002). Their lithium abundances have been derived from comparison of theoretical pEW curves of growth with pEW measurements from mid-resolution spectra (see Figure 7.6).

The large scatter in measured pEWs of the Li I feature for the coeval population of M-type members of the σ Orionis cluster is intriguing. A wide scatter is also seen among late M-type members in the Beta Pictoris association (age = 22 Myr, Shkolnik et al. 2017), and older open clusters such as IC2391 (age = 50 Myr, Barrado y Navascués et al. 2004, see Figure 7.7) and the Pleiades (age = 120 Myr, Rebolo et al. 1996; Stauffer et al. 1998). Since the scatter is seen in associations and clusters of very different ages, from just a few Myr to over 100 Myr, it seems appealing to invoke the effects of age on PMS Li depletion, but the question of the effects of varying opacities in the TiO bands that dominate the spectral region where the Li I resonance doublet is located cannot be neglected.

A search for variability in and away from TiO bands using a tunable filter technique with the AAT gave a positive detection in the late-M lithium brown dwarf LP944-20 (Tinney & Tolley 1999) indicative of rotationally driven temperature

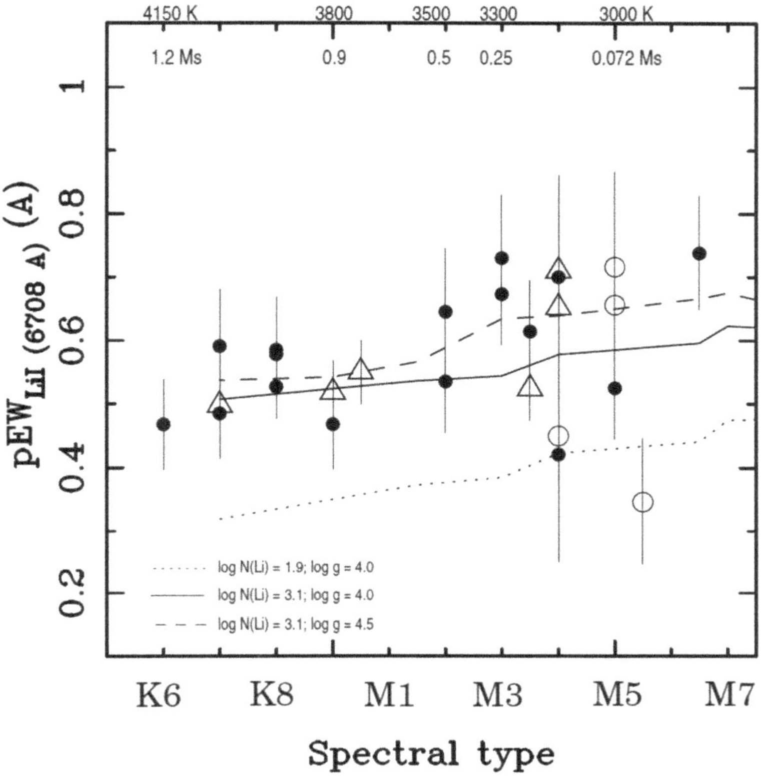

Figure 7.6. Li I resonance doublet pEWs measured in stellar and brown dwarf members in the very young σ Orionis open cluster as a function of spectral type. The substellar mass limit at 0.075 $M_{\odot}$ corresponds to spectral class M5. Filled circles denote objects with weak Hα in emission, open circles denote objects with strong Hα in emission and open triangles denote objects with forbidden emission lines indicative of mass outflows, which are likely to have their pEWs diminished by non-photospheric veiling. Theoretical curves of growth of the Li I feature as a function of $T_{\rm eff}$ are shown with a dashed line for high surface gravity and cosmic Li abundance, a continuous line for lower surface gravity and cosmic Li abundance, and a dotted line for lower surface gravity and depleted Li abundance. The correspondence between spectral type, effective temperature, and mass estimated for this cluster are labeled. Figure credit: Zapatero Osorio, M. R. et al. 2002 "Lithium and Hα in stars and brown dwarfs of sigma Orionis" A&A 384 937, reproduced with permission © ESO.

fluctuations of just 20 K. However, the variability of the same object was not confirmed in low-resolution spectroscopic monitoring at the ESO 3.6 m telescope with EFOSC2, although the sensitivity of those observations was estimated at only 50 K in terms of the variability of the effective temperature (Guenther et al. 2009). The influence of TiO variability in the spread in the observed pEWs of the Li I resonance doublet among the M-type members in young open clusters and the field remains to be investigated in more detail.

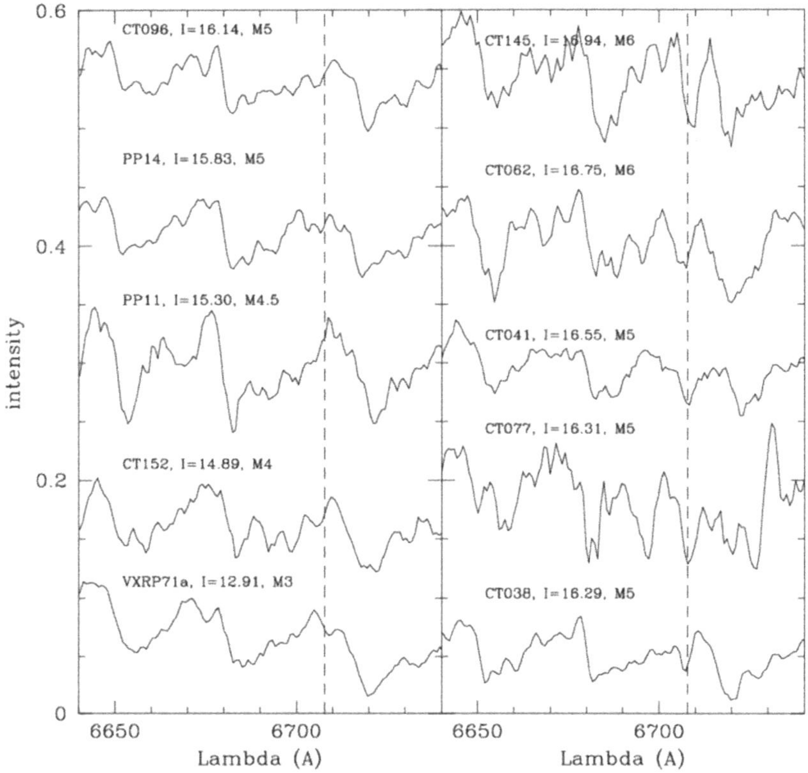

Figure 7.7. Spectra of M-type members in the young open cluster IC2391 (50 Myr) obtained in 1999 with the Hydra II spectrograph at the CTIO 4 m telescope. The position of the Li I resonance doublet is indicated with a dashed vertical line. Note the wide range of Li I feature strengths among cluster members of similar spectral subclass (between M5 and M6). Figure credit: Barrado y Navascués, D. et al. 2004 "Spectroscopy of very low mass stars and brown dwarfs in IC 2391: Lithium depletion and Hα emission" ApJ 614 386.

L-type objects are thought to have a significant contribution of dust opacity in their optical spectrum (Allard et al. 2001), which imposes a challenge in deriving absolute lithium abundances from the Li I resonance doublet. A comparative approach to deriving lithium abundances in L-type brown dwarfs was introduced by Martín et al. (2018), who proposed that the upper envelope of Li I resonance doublet pEW values in brown dwarfs could be taken as a reference for the undepleted cosmic Li abundance. Values of measured pEWs below the upper envelope could be interpreted as evidence of Li depletion, the amount of which could be inferred using synthetic spectra.

This method has been applied to L dwarfs in the Hyades cluster. The caveat to this comparative approach is that it assumes that all brown dwarfs have similar metallicities and surface gravities, an assumption that may hold approximately for the majority of objects in the solar vicinity, but not for all of them because approximately 10% of field brown dwarfs have been shown to have ages < 100 Myr and probably have lower surface gravity (Cruz et al. 2009; Martín et al. 2010).

Very young L-type brown dwarfs are likely to have lower surface gravity because they have smaller masses and larger radii than their older counterparts of the same spectral type. The effect of low surface gravity may be a decrease in the pEW of the Li I resonance line, although the comparative analysis of late-M and L dwarfs with low and high surface gravity spectroscopic indicators has not been conclusive (Kirkpatrick et al. 2008). Special care must be taken when trying to infer lithium abundances of very young brown dwarfs, although it is likely that they have preserved their initial cosmic composition because they have lower masses than the older counterparts of the same spectral type.

The strength of the Li I resonance feature in brown dwarfs depends strongly on the spectral type, which is a proxy for the effective temperature. The upper envelope of pEW values increases with later spectral type (cooler temperature) up to spectral class between L5 and L6, and then decreases toward the latest L dwarfs and the early T dwarfs. This behavior can be interpreted as a result of the increasing abundance of neutral Li because of its low ionization potential, and the decrease in TiO opacity down to the temperatures of the L5–L6 dwarfs (around 1500 K), and then the decrease in neutral Li for cooler temperatures due to the formation of Li bearing molecules such as LiH, LiF, LiBr, LiCl, and LiOH (Lodders 1999); and also the increase in the wings of the very strong Na I and K I resonance doublets, which may provide a significant source of opacity in the region of the Li I resonance doublet (Allard et al. 2007).

Weck et al. (2004) performed detailed calculations of the impact of LiCl in the infrared spectra of brown dwarfs with temperatures in the range between 1500 K and 900 K and surface gravities between $\log g = 5.0$ and $\log g = 3.0$. They concluded that the strongest effects (at the level of 20%) were found at $T_{\rm eff} = 1200$ K and $\log g = 3.0$ in the spectral region around the LiCl fundamental vibrational band at 15.8 μm.

The coolest brown dwarfs for which lithium detections have been secured are the following three T dwarfs:

1) WISE J104915.57-531906.1 B (also known as Luhman 16B), the angularly resolved T0.5 secondary in the nearest brown dwarf binary system which was observed with the Magellan Echellette spectrograph on the 6.5 m Clay Magellan telescope by Faherty et al. (2014) and with the X-Shooter on the VLT by Lodieu et al. (2015). The pEW (Li I) values measured independently by those authors are 3.8 ± 0.4 Å and 8.4 ± 1.5 Å, respectively. Lodieu et al. (2015) checked whether or not these different values could be produced by variability by directly comparing the Magellan data with the VLT data, and concluded that there were no significant differences. They attributed the discrepancy in the pEW values to different methods of measurement. The alkali resonance features in L dwarfs have extremely broad wings, making it difficult to decide where to place the integration limits and the pseudo-continuum level to obtain the pEW measurement. Lodieu et al. also

attempted to measure the lithium isotope ratio in the X-Shooter spectrum of Luhman 16A, but failed to detect the line asymmetry that could indicate the presence of ^{6}Li.

2) SIMP J013656.5+093347.3, a T2/T2.5 single dwarf discovered by Artigau et al. (2006) with a near-infrared proper motion survey carried out with the Observatoire du Mont Mégantic wide field NIR camera installed on the CTIO 1.5 m telescope. Optical spectra were obtained for it with LRIS at the Keck I and DEIMOS at the Keck II telescopes, respectively (Pineda et al. 2016). A clear detection of the Li I resonance doublet was reported and an average pEW of 7.2 ± 0.7 Å was measured by direct integration of the absorption line region.
3) DENIS J225210.7-173013 AB, an ultracool dwarf identified by Kendall et al. (2004) using CGS4 at the 3.9 m United Kingdom Infrared Telescope in the MKO, and resolved as a tight binary (angular separation 0.13 arcsec) using the Near-Infrared Camera and Multi-Object Spectrometer (NICMOS) onboard the HST (Reid et al. 2006). Hereafter this binary is referred to as Reid 1AB, following the convention of the Washington Double Star catalog of naming objects after the discoverer of the binary nature. A spectral type of T3.5 and a dynamical mass of 41 M_J were obtained for Reid 1B in the study of Dupuy & Liu (2017). A detection of the Li I resonance feature in the combined light of Reid 1 AB was reported in optical spectra obtained with the long-slit mode on the OSIRIS spectrograph at the GTC (Martín et al. 2022). These authors attributed the presence of lithium to the contribution of the secondary, and derived a pEW = 10.2 ± 2.5 Å for the Li I feature in Reid 1B.

These three T lithium brown dwarfs, as well as some late-L lithium brown dwarfs, are excellent candidates for investigating the presence of lithium-bearing molecules in their atmospheres. Theoretical calculations of the opacities of those molecules are gradually becoming available (Liang et al. 2020; Gharib-Nezhad et al. 2021).

7.5 The Substellar Lithium Depletion Boundary

Determination of the lithium abundances in the coeval populations of several open clusters has provided a nuclear chronometer that is orthogonal to other dating methods such as the upper main-sequence turnoff point. It was first realized by Basri et al. (1996) that the Li abundances among brown dwarf candidates in the Pleiades cluster implied an age in excess of 100 Myr, significantly older than the canonical age of 70 Myr that was commonly adopted until then.

The method of locating the boundary between Li depletion and Li preservation among fully convective members in open clusters is the LDB, and for the age of the Pleiades and older this boundary is located in the brown dwarf domain.

As time progresses, lithium depletion carves out an increasingly wide canyon in the Li abundance versus effective temperature plane (see Figure 7.8) among the populations of open clusters.

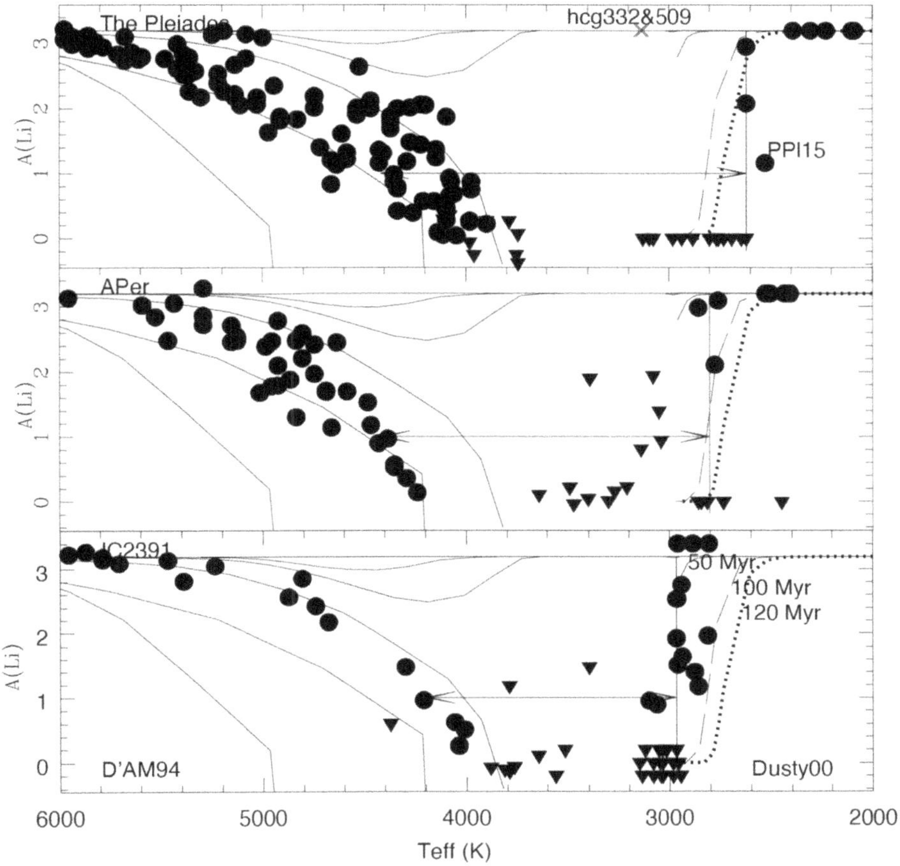

Figure 7.8. The lithium abundance versus effective temperature plane in three open clusters of decreasing age, namely, from top to bottom, the Pleiades (age = 130 ± 20 Myr), α Persei (age 85 ± 10 Myr), and IC 2391 (age 50 ± 5 Myr). These ages were estimated by Barrado y Navascués Stauffer and Jayawardhana (2004). Filled circles denote lithium abundances estimated for individual cluster members from spectroscopic measurements of the strength of the Li I resonance doublet. Down-pointing triangles denote upper limits on the lithium abundances obtained from non-detections of the Li I resonance doublet. Theoretical curves of lithium depletion for stars (left side, D'Antona & Mazzitelli 1994 for ages 1, 3, 5, 10, 20, and 100 Myr) and for brown dwarfs (right side, Chabrier et al. 2000 for ages 50, 100 and 120 Myr) are shown. The width of the lithium depletion canyon is indicated with arrows and it increases for older clusters. Figure credit: Barrado y Navascués, D. et al. 2004 "Spectroscopy of very low mass stars and brown dwarfs in IC 2391: Lithium depletion and Hα emission" ApJ 614 386.

The most recent LDB study of the Pleiades age using mid-resolution spectra obtained at the 10 m Keck I telescope for a sample of objects covering the narrow temperature range from 2800 K to 2650 K (spectral types between M6 and M7) has given 112 Myr with a formal error bar of only 4.4% (Dahm 2015).

By the age of about 200 Myr, lithium depletion comes to a halt because the interiors are totally degenerate, which means that further gravitational contraction not only does not increase the central temperature, but actually decreases it. It was thought for a long time that at this point the LDB method could not be applied for older ages (e.g. Burke et al. 2004). However, even though the lithium abundances may remain unchanged, the canyon keeps opening up because the cool side of it is made of brown dwarfs that keep cooling down.

This effect has made it possible to obtain the LDB age for the nearest open cluster, namely the Hyades, at 650 ± 70 Myr (Martín et al. 2018; Lodieu et al. 2018, see Figure 7.9), and a lower limit of 550 Myr for the second-closest one, namely Coma Berenices (Martín et al. 2020).

Since the LDB is located at just about the hydrogen-burning limiting mass (0.075 $M_{\odot}$) in the Pleiades cluster, and at just about the lithium-burning limiting mass (0.06 $M_{\odot}$) in the Hyades cluster, it is safe to state that the range of validity of the LDB method in brown dwarfs has been proved to range from about 100 Myr to about

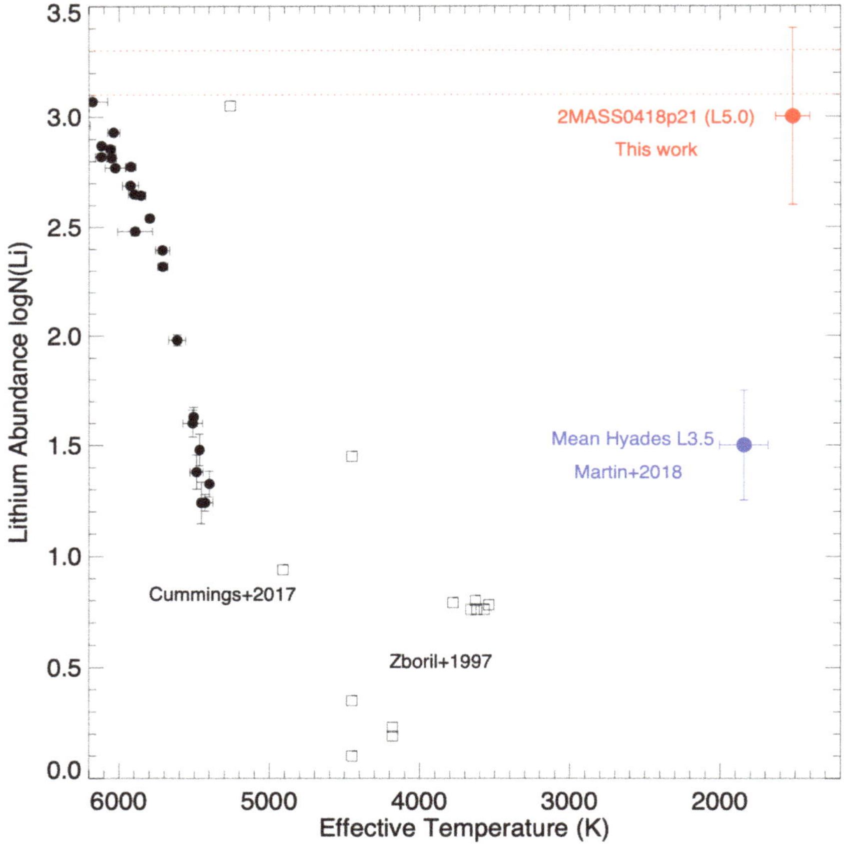

Figure 7.9. The lithium canyon for the Hyades cluster (650 ± 70 Myr). The stellar side is represented by black dots for lithium abundances in solar-type members (Cummings et al. 2017) and empty squares for upper limits in K-type and M-type members (Zboril et al. 1997). The substellar realm is represented by the mean value of the lithium abundance in two L3.5—L4 members (blue point; Martín et al. 2018) and the single value of one L5 member (red point; Lodieu et al. 2018). Horizontal red dotted lines indicate the initial lithium abundance of cluster members. Figure credit: Lodieu, N., Rebolo, R., & Pérez Garrido, A. 2018 "Lithium in the Hyades L5 brown dwarf 2MASS J04183483+2131275" A&A 615 L12, reproduced with permission © ESO.

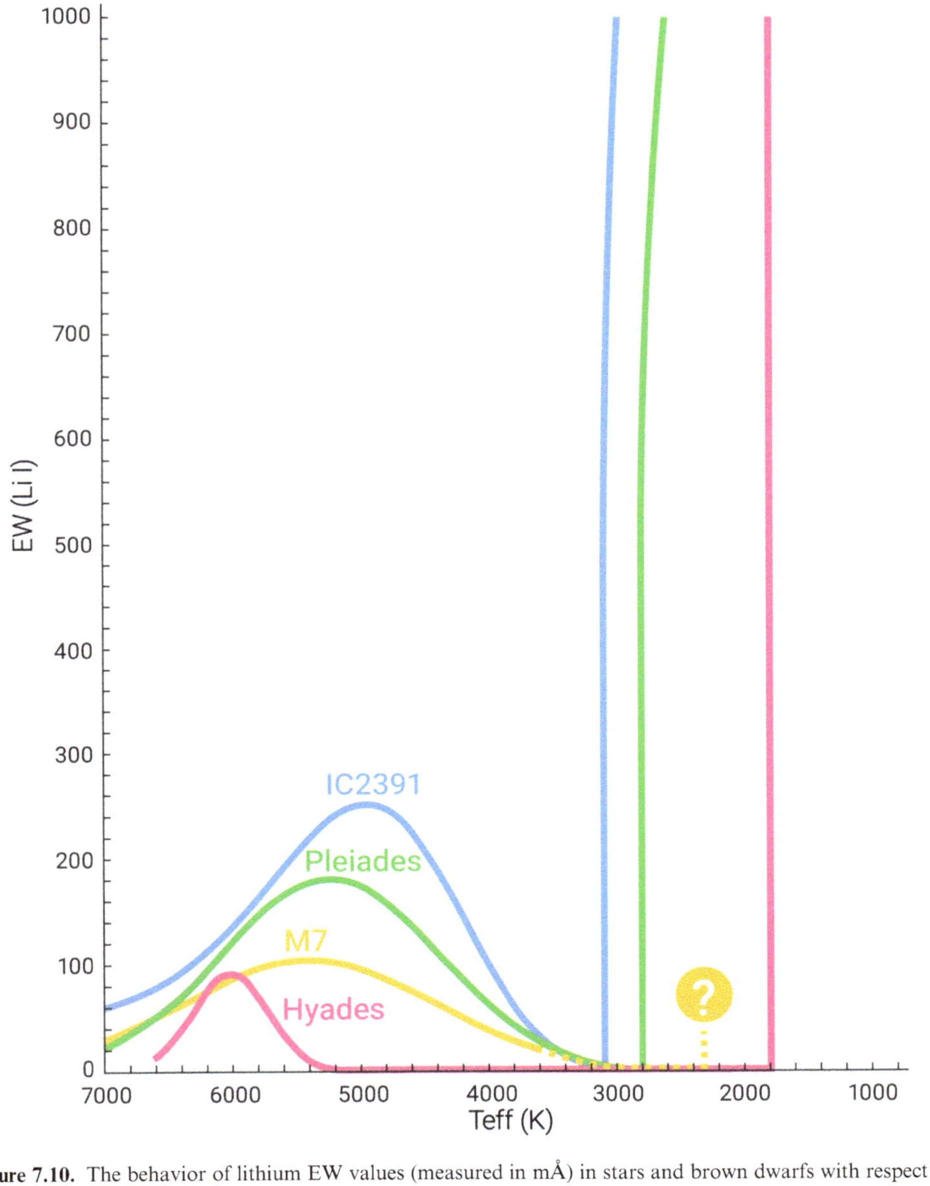

Figure 7.10. The behavior of lithium EW values (measured in mÅ) in stars and brown dwarfs with respect to effective temperature in four nearby open clusters representative of increasing age from IC 2391 (50 Myr) to the Pleiades (110 Myr), M7 (300 Myr) and the Hyades (650 Myr). The effects of lithium depletion as a function of age and mass are clearly seen. The resurgence of lithium in M7 has not been observed but it is expected at a temperature of around 2500 K. Figure credit: the author with assistance from Gaby Pérez.

700 Myr. For younger populations the LDB lies in the stellar domain, and for older populations the validity of the LDB has not been explored yet.

Figure 7.10 represents a schematic view of the current state of affairs of the behavior of lithium as a function of effective temperature across the stellar and substellar domains in four benchmark open clusters with ages ranging from 50 to 650 Myr.

7.6 Lithium in Brown Dwarf Binaries

Multiple systems are like mini-associations to which the study of lithium depletion can also be applied to constrain the ages and masses of the components. The first brown dwarf multiple system for which dynamical masses and lithium depletion have been determined is Gliese 569B, a companion of a nearby red dwarf (Henry & Kirkpatrick 1990; Martín et al. 2000; Lane et al. 2001; Zapatero Osorio et al. 2004). Furthermore, a kinematic age estimate of 300 Myr was reported because of membership in the Ursa Major moving group (Kenworthy et al. 2001).

The Gliese 569B brown dwarf system is a distant (49 AU separation) companion to the M2.5V nearby (d = 9.8 pc) star Gliese 569 (Forrest et al. 1988). Angularly resolved near-infrared spectra of each component were obtained with NIRSPEC using the adaptive optics system of the Keck II telescope and provided radial velocity data that enabled the determination of the individual masses.

Simon et al. (2006) proposed that Gliese 569 is a quadruple system with a total mass of 0.48 ± 0.03 $M_{\odot}$ consisting of a primary red dwarf with a mass of 0.35 ± 0.02 $M_{\odot}$, and three brown dwarfs with a total mass of 0.125 ± 0.007 $M_{\odot}$. The three brown dwarfs are separated by 5 arcsec (50 AU) from the primary and their orbital period is thought to be around 500 yr. Two of the brown dwarfs are angularly resolved with an angular separation of 90 milliarcsec (0.89 AU) and an orbital period of 864 days. The brighter of these two objects might itself be an unresolved brown dwarf binary with a total mass of 0.105 ± 0.011 $M_{\odot}$. Simon et al. considered that the most likely scenario is that the three brown dwarfs have dynamical masses in the range 0.04–0.05 $M_{\odot}$; they noted that the lithium test would be useful for confirming those estimates.

Magazzù et al. (1993) and Zapatero Osorio et al. (2005) have reported there is no lithium detected in the combined light of Gliese 569, and that all the brown dwarfs in the system therefore fail to pass the lithium test because it is thought that they contribute similar proportions of flux to the total light. This makes the triple system scenario highly unlikely.

According to theoretical models, after about 200 Myr, the maximum mass that a lithium brown dwarf binary could have been reached when the system consists of two components of identical mass and each mass is just at the limit of lithium burning. If the lithium-burning minimum mass is located at 0.06 $M_{\odot}$ (Pozio 1991), the maximum lithium brown dwarf binary would then be 0.120 $M_{\odot}$, and that of a lithium brown dwarf triple system would be 0.180 $M_{\odot}$.

The dynamical masses found for Gliese 569B are below those values, but they do not present the Li I resonance doublet in the spectra. The total mass in the Gliese 569B has been revised upwards to 0.132 ± 0.007 $M_{\odot}$ by Dupuy & Liu (2017), which helps to explain the lack of lithium, but it is still not consistent with a triple system scenario.

One of the first two field L-type lithium brown dwarfs, namely DENIS-P J1228.5-1547, became the first brown dwarf binary to be angularly resolved into its two components with the HST (Martín et al. 1999). It is composed of two nearly-identical brown dwarfs of spectral types L4.5 (Martín et al. 2006). The orbital period is estimated at about 50 years and the total dynamical mass is 0.101 ± 0.018 $M_{\odot}$ (Brandner et al. 2004; Dupuy & Liu 2017).

This object is located below the theoretical maximum mass for a lithium brown dwarf binary and is thus consistent with the detection of a strong Li I resonance doublet (Martín et al. 1997; Tinney et al. 1997).

Lithium burning in brown dwarfs was used to constrain the age of a very low-mass binary by Burgasser & Blake (2009) for the system 2MASS J03202839-0446358 AB (Cruz et al. 2003) which is composed of an M8.5 primary and a T5 secondary. They considered that the lack of lithium detection in spectra reported by Cruz et al. (2003) ruled out an age < 300 Myr and a mass < 0.065 $M_{\odot}$ for the late-M primary.

These model-dependent lithium depletion constraints are shown in Figure 7.11, together with other constraints coming from the luminosities and radial velocities of this system, which still lacks dynamical mass determinations.

At just 3.62 pc from us, Epsilon Indi consists of a triple system composed of an orange main-sequence star (K4.5 spectral type) and a brown dwarf binary at 1500 AU (Scholz et al. 2003; McCaughrean et al. 2004). The total mass in the brown dwarf binary system has been estimated at 0.138 ± 0.0010 $M_{\odot}$, and the mass ratio is $q = M_2/M_1 = 0.93$ from astrometric monitoring by the Carnegie Astrometric Planet Search and the Cerro Tololo Observatory Parallax Investigation (Dieterich et al. 2018).

Both components are T dwarfs and they do not present lithium in mid-resolution optical spectra ($R = 1000$) obtained with FORS2 at the VLT (King et al. 2010). These authors estimated that lithium is depleted by a factor of 1000 in the T1 dwarf Epsilon Indi Ba (see Figure 7.12). They suggested, however, that the absence of the Li I resonance line in the spectrum of the T6 dwarf Epsilon Indi Bb could be due to the effect of lithium being locked into molecules. This is now ruled out owing to the high dynamical masses of both components (Dieterich et al. 2018) that place them in a regime where brown dwarfs are expected to deplete lithium very efficiently.

Dupuy & Liu (2017) obtained dynamical masses for the components in 31 brown dwarf binary systems from a decade-long monitoring using the HST, the Keck telescopes, and the CFHT.

Only two of their systems with well-constrained ages from the dynamical masses and evolutionary models had lithium detections from combined light spectra available in the literature, namely Gliese 417BC (total dynamical mass 0.095 ± 0.003 $M_{\odot}$; spectral types of the components: L4.5 and L6, pEW (Li I) = 11.5 Å, Kirkpatrick et al. 2001) and SDSS J0423-04 (total dynamical mass 0.079 ± 0.003 $M_{\odot}$; spectral types of the components: L6.5 and T2; pEW (Li I) = 11 Å, Kirkpatrick et al. 2008).

They are shown as upward pointing blue triangles in Figure 7.13. Both of these systems have total masses well below the theoretical maximum lithium brown dwarf

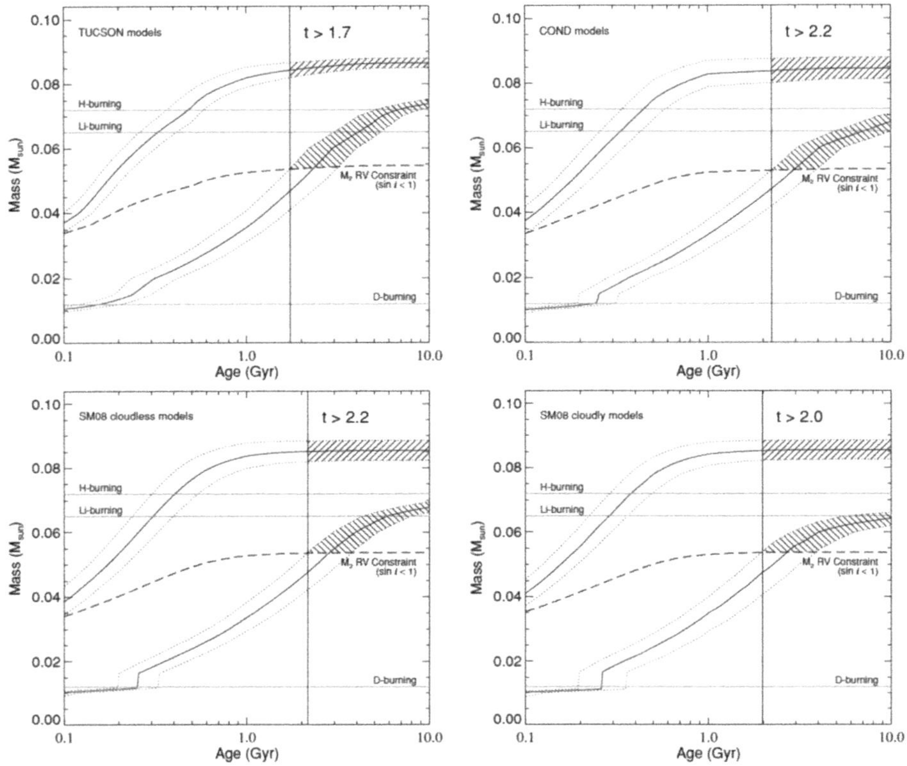

Figure 7.11. Model-dependent minimum age constraints for the binary system 2MASS J03202839-0446358 AB range from 1.7 Myr (TUCSON models; Burrows et al. 2001) to 2.2 Myr (COND models; Baraffe et al. 2003 and SM08 cloudless models from Saumon & Marley 2008). Each panel corresponds to a different set of models. Hydrogen, lithium, and deuterium burning mass limits are considered as model-independent constraints and are denoted with horizontal gray lines for solar-metallicity. Figure credit: Burgasser, A. J., & Blake, C. H. 2009 "An age constraint for the very-low-mass stellar/brown dwarf binary 2MASS J03202839-0446358AB" AJ 137 4621.

binary mass between 0.120 and 0.110 $M_{\odot}$, and hence they are consistent with total preservation of their initial lithium content.

Dupuy & Liu (2017) found six binary systems with well-constrained ages and masses and no lithium detections in their combined light. Their total dynamical masses range from 0.158 ± 0.007 $M_{\odot}$ (LP 349-25 AB) to 0.096 ± 0.007 $M_{\odot}$ (DENIS-P J2252); the spectral types of the components are L6 and T2; and no lithium was noted in an 8 m Gemini South spectrum obtained with the R400_G5325 disperser obtained by Reid et al. (2008).

The presence of lithium in Gliese 417BC but not in DENIS-P J2252 suggests that the maximum lithium brown dwarf binary mass is located at a total binary mass of just below 0.100 $M_{\odot}$, which is somewhat below the theoretical nuclear limits between 0.120 and 0.110 $M_{\odot}$. This discrepancy may be significant given the

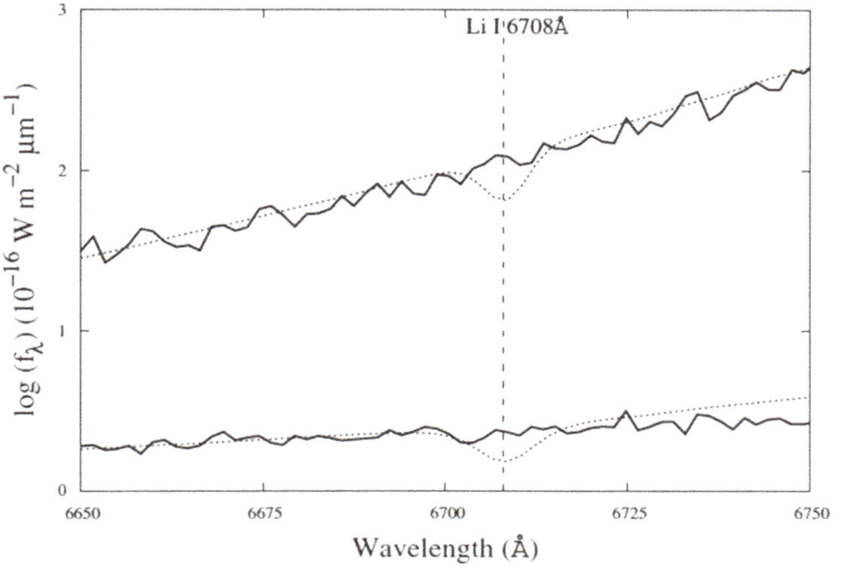

Figure 7.12. FORS2 spectra around the Li I resonance doublet of Epsilon Indi Ba (T1.5, upper thick line) and Epsilon Indi Bb (T6, lower thick line) compared with synthetic spectra with the following parameters: upper dotted line: $T_{\mathrm{eff}} = 1300$ K, $\log g = 5.50$, metallicity $= -0.2$, Li abundance 1000 times below cosmic; lower dotted line: $T_{\mathrm{eff}} = 920$ K, $\log g = 5.25$, metallicity $= -0.2$, cosmic Li abundance. Figure credit: King, R. R. et al. 2010 "Epsilon Indi Ba, Bb: A detailed study of the nearest known brown dwarfs" A&A 510 A99, reproduced with permission © ESO.

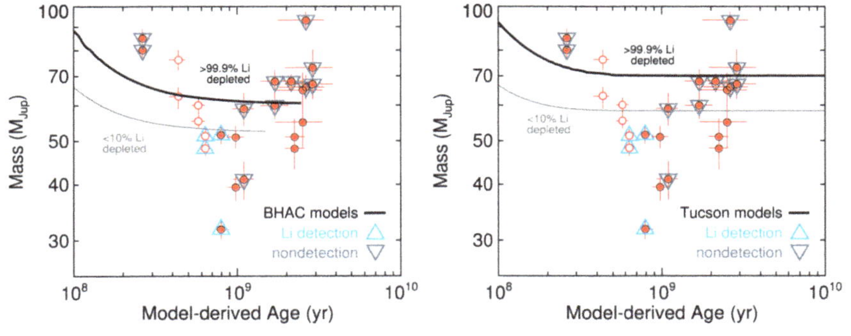

Figure 7.13. Age and mass estimates for the components in brown dwarf binaries. Isolithium depletion curves are shown as black curves (total depletion) and gray curves (little depletion) for two different sets of models (Lyon and Tucson groups). Objects with lithium observations are indicated with triangles as labeled. Figure credit: Dupuy, T. J. & Liu, M. C. 2017 "Individual dynamical masses of ultracool dwarfs" ApJS 231 15.

relatively small error bars on the observational dynamical masses and the theoretical nuclear masses.

Dupuy & Liu (2017) noted that the Tucson models (Burrows et al. 2001) predict that some brown dwarf binaries should have preserved lithium, but the Lyon models

(Baraffe et al. 2015) predict substantial lithium depletion for the same systems. This difference may be attributed to the Lyon models having more opacity in the atmospheres of brown dwarfs, resulting in higher central temperatures and lower masses for lithium thermonuclear fusion than the Tucson models.

A very recent example of the comparison between dynamical and nuclear masses is the case of DENIS J063001.4-184014 AB (abridged here to Sahlmann 2), which is an ultracool dwarf identified in the DENIS database as a low galactic latitude brown dwarf candidate with optical spectral type M8.5 obtained from low-resolution spectra collected at the 2.3 m telescope at Siding Spring Observatory (Phan-Bao et al. 2008).

It was found to be a binary system with an orbital period of 3.07 yr by astrometric monitoring using the FORS2 instrument at the VLT (Sahlmann et al. 2015). The total mass of Sahlmann 2AB is estimated at 0.104 ± 0.013 $M_{\odot}$, which is again below the maximum total mass of a lithium brown dwarf binary at 0.120 $M_{\odot}$, and thus it is expected that this system should have preserved all of its initial lithium abundance.

The individual masses of the components in Sahlmann 2AB have been estimated at 0.052 $M_{\odot}$ for each of them using the astrometric data plus radial velocity information from the primary obtained with NIRSPEC on the 10 m Keck II telescope and resolved imaging with NIRC2 and laser guide adaptive optics on the same telescope (Sahlmann et al. 2021).

The masses of both components in the Sahlmann 2AB system lie below the minimum mass for lithium burning according to the models of Nelson et al. (1993). When the orbital error bars on the masses are considered, the primary has to be below 0.061 $M_{\odot}$ and the secondary must be below 0.057 $M_{\odot}$, so they should not have completely destroyed their initial lithium content, and it is consequently expected that at least some residual amount of lithium could be observed in the photospheres.

Sahlmann et al. (2021) reported a non-detection of lithium based on a high-resolution UVES spectrum of Sahlmann 2AB obtained with the VLT, but placed an upper limit of only 1 Å on the pEW of the Li I resonance doublet. For comparison, the M9.5 brown dwarf LP944-20 which keeps its initial cosmic abundance according to Pavlenko et al. (2007) has a Li I pEW = 0.65 ± 0.05 Å, so that an upper limit of 1 Å does not necessarily imply any Li depletion in Sahlmann 2AB.

In order to carry out a more sensitive search for lithium in Sahlmann 2AB, spectroscopic observations were secured in service time at the GTC with the OSIRIS spectrograph in long-slit mode (1″ slit width) and grating R1000R (Martín et al. 2022). These authors did not detect lithium and they improved the upper limit of the Li I pEW by almost a factor of six.

Taking into account the measured flux ratio at near-infrared wavelengths and the likely spectral types of each component from binary fitting of the low-resolution spectrum obtained with SpeX on the 3 m NASA Infrared Telescope Facility in Hawaii, these authors derived upper limits of pEW Li I < 0.18 Å for the primary and pEW Li I < 2.65 Å for the secondary, implying that both have depleted their initial lithium content by at least a factor of 10 because they clearly lie well below the upper envelope of the pEW Li I for their spectral types (M8.5 and L4.5, respectively).

Martín et al. (2022) used their results on lithium depletion in the individual components of Sahlmann 2AB, together with their analysis of Li in the components of Reid 1AB, and data from the literature, to constrain the observational MMLB and compare with the theoretical calculations. Their comparison between the data and two sets of evolutionary models is illustrated in Figure 7.14, and it led to the determination that the MMLB is located at 0.049 $M_{\odot}$ (51.5 M_J), which is a somewhat lower mass than predicted by the models, including the recent ones by Phillips et al. (2020).

More work is needed to follow up on lithium in brown dwarf binaries. The dynamical masses inferred from the astrometric and radial velocity studies of the orbital parameters are crucial. It is also crucial to improve on the observations of lithium, which are poor or even missing in some cases. Brown dwarf binaries with

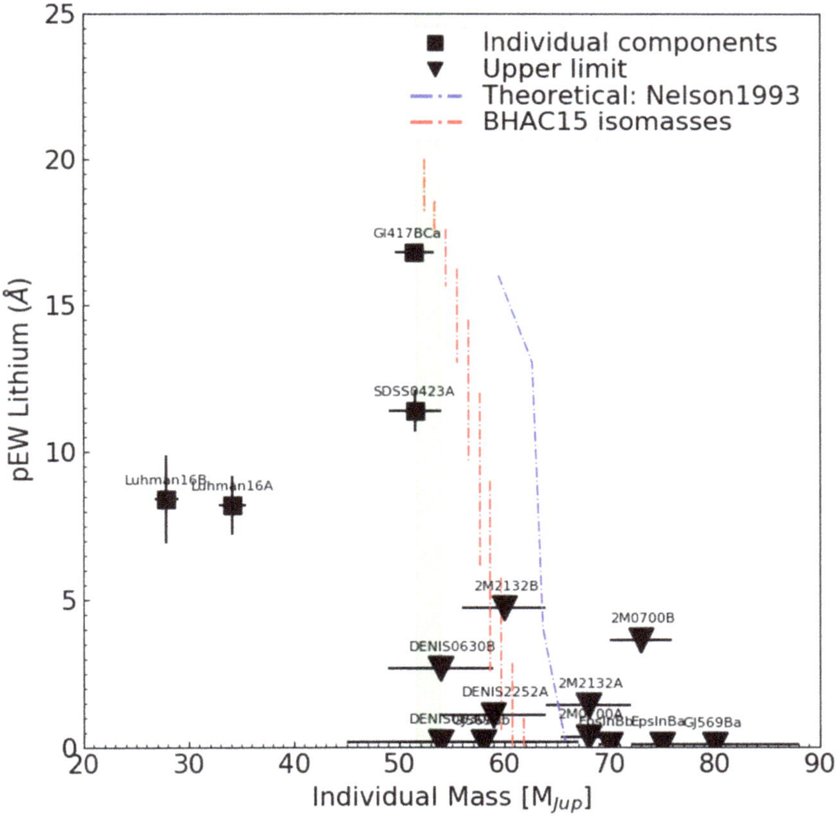

Figure 7.14. The pEW of the Li I resonance doublet versus dynamical mass for the components in brown dwarf binaries. Two sets of models are shown for comparison (Nelson et al. 1993 and Baraffe et al. 2015, alias BHAC15). It appears that the observed MMLB (green band) is sharper and slightly shifted toward lower mass compared to the theoretical predictions. Note that the dynamical masses are given in Jovian units (1 $M_{\odot}$ = 1047 M_J). Figure credit: Adapted from Figure 5 in Martín, E. L. et al. 2022 "New constraints on the minimum mass for thermonuclear lithium burning in brown dwarfs" MNRAS 510 2841.

dynamical masses and lithium data are cornerstones for validating evolutionary models.

The number of brown dwarf binary systems with both dynamical and nuclear constraints on the total mass is still scarce, but it seems already quite suggestive that the data implies more lithium depletion than predicted by the models, suggesting that the central temperatures required for Li burning in the cores of brown dwarfs may be sustained for longer than previously thought. A larger sample of brown dwarf binaries could enable the possibility of testing the role of additional parameters such as metallicity, rotation, and interactions with dark matter particles if they exist.

7.7 Lithium in Metal-Poor Brown Dwarfs

In the previous sections, it has been assumed that the chemical composition of the brown dwarfs is similar to that of the Sun, i.e., that they are of solar composition, which is a reasonable assumption for most of the objects that populate the solar neighborhood. The effects of low metallicity on Li burning and the first attempts to detect them are briefly discussed here.

As a limiting case Burke et al. (2004) calculated the effects of zero-metallicity opacity on the LDB of the Pleiades and found a decrease in the time it takes to destroy lithium with respect to the solar-metallicity opacities. The central temperature reached in very low-mass stars with metallicity 100 times lower than the Sun, however, is estimated to be lower than for their solar-metallicity counterparts (Chabrier & Baraffe 1997, 2000), and thus it is reasonable to expect that the lithium-burning limiting mass increases with decreasing metallicity. Nevertheless, detailed calculations of lithium depletion as a function of metallicity are not yet available.

On the observational front, a search for lithium in two metal-poor L-type subdwarfs using FORS2 at the VLT and OSIRIS at the GTC was reported by Lodieu et al. (2015), who placed an upper limit on the lithium abundance of an L3.5 subdwarf ($T_{\rm eff} = 2000$ K) at a level of a million times below the cosmic value, and they estimated from it a limit on the amount of Li-rich circumstellar or insterstellar material that could have been accreted onto the surface of the object during its lifetime. These authors discuss the possibility of detecting lithium in metal-poor T dwarfs, which might have retained the primordial lithium abundance if they have a mass below the minimum lithium burning mass for low metallicity.

There is a long-standing interest in deriving primordial lithium abundances in metal-poor substellar objects (Rebolo et al. 1995), which could lead to a solution of the discrepancy between the SBBN lithium abundance and the upper envelope of lithium in Population II stars.

Zhang et al. (2018) estimated that the lithium-burning minimum mass for metallicity 10 times lower than solar could be located at 0.065 $M_{\odot}$, which could correspond to a brown dwarf of $T_{\rm eff} = 900$ K (spectral class T5) at an age of 10 Gyr, which might be appropriate for an object belonging to the halo of the Milky Way. The formation of lithium lines, monatomic or molecular, in such low-metallicity, low-temperature, and high-gravity conditions has not yet been calculated.

The first optical spectrum of a metal-poor brown dwarf candidate was obtained with OSIRIS at the GTC (Lodieu et al. 2022). These authors report the detection of the Cs I resonance line at 894.3 nm, but the S/N ratio of their spectrum was not enough to place any significant constraint on the presence of the Li I resonance doublet at 670.8 nm.

Bibliography

Allard, F., Hauschildt, P. H., Alexander, D. R., Tamanai, A., & Schweitzer, A. 2001, ApJ, 556, 357

Allard, F., Allard, N. F., Homeier, D., et al. 2007, A&A, 474, L21

Artigau, É., Doyon, R., Lafrenière, D., et al. 2006, ApJ, 651, L57

Baraffe, I., Chabrier, G., Barman, T. S., Allard, F., & Hauschildt, P. H. 2003, A&A, 402, 701

Baraffe, I., Homeier, D., Allard, F., & Chabrier, G. 2015, A&A, 577, A42

Barrado y Navascués, D., Stauffer, J. R., & Jayawardhana, R. 2004, ApJ, 614, 386

Basri, G., Marcy, G. W., & Graham, J. R. 1996, ApJ, 458, 600

Basri, G., & Martín, E. L. 1999, AJ, 118, 2460

Beckman, J. E. 2021, Multimessnger Astronomy (Berlin: Springer)

Brandner, W., Martín, E. L., Bouy, H., et al. 2004, A&A, 428, 205

Burgasser, A. J., & Blake, C. H. 2009, AJ, 137, 4621

Burke, C. J., Pinsonneault, M. H., & Sills, A. 2004, ApJ, 604, 272

Burrows, A., Hubbard, W. B., Lunine, J. I., & Liebert, J. 2001, RvMP, 73, 719

Chabrier, G., & Baraffe, I. 1997, A&A, 327, 1039

Chabrier, G., & Baraffe, I. 2000, ARA&A, 38, 337

Chabrier, G., Baraffe, I., Allard, F., & Hauschildt, P. 2000, ApJ, 542, 464

Cruz, K. L., Reid, I. N., Liebert, J., Kirkpatrick, J. D., & Lowrance, P. J. 2003, AJ, 126, 2421

Cruz, K. L., Kirkpatrick, J. D., & Burgasser, A. J. 2009, AJ, 137, 3345

Cummings, J. D., Deliyannis, C. P., Maderak, R. M., & Steinhauer, A. 2017, AJ, 153, 128

Dahm, S. 2015, ApJ, 813, 108

D'Antona, F., & Mazzitelli, I. 1994, ApJS, 90, 467

Delfosse, X., Tinney, C. G., Forveille, T., et al. 1997, A&A, 327, L25

Dieterich, S. B., Weinberger, A. J., Boss, A. P., et al. 2018, ApJ, 865, 28

Dupuy, T. J., & Liu, M. C. 2017, ApJS, 231, 15

Faherty, J. K., Beletsky, Y., Burgasser, A. J., et al. 2014, ApJ, 790, 90

Forrest, W. J., Skrutskie, M. F., & Shure, M. 1988, ApJ, 330, L119

Gharib-Nezhad, E., Marley, M. S., Batalha, N. E., et al. 2021, ApJ, 919, 21

Guenther, E. W., Zapatero Osorio, M. R., Mehner, A., & Martín, E. L. 2009, A&A, 498, 281

Hambly, N. C., Hawkins, M. R. S., & Jameson, R. F. 1991, MNRAS, 253, 1

Hambly, N. C., & Jameson, R. F. 1991, MNRAS, 249, 137

Hamilton, D., & Stauffer, J. 1993, AJ, 105, 1855

Heintz, W. D. 1989, A&A, 217, 145

Henry, T. J., & Kirkpatrick, J. D. 1990, ApJ, 354, L29

Henry, T. J., Johnson, D. S., McCarthy, D. W., & Kirkpatrick, J. D. 1992, A&A, 254, 116

Herbig, G. H., & Bell, K. R. 1988, Third Catalog of Emission-line Stars of the Orion Population Lick Bulletin (Santa Cruz, CA: Lick Observatory)1111,

Irwin, M., McMahon, R. G., & Reid, N. 1991, MNRAS, 252, 61

Jameson, R. F., & Skillen, I. 1989, MNRAS, 239, 247

Kendall, T. R., Delfosse, X., Martín, E. L., & Forveille, T. 2004, A&A, 416, L17
Kenworthy, M., Hofmann, K.-H., Close, L., et al. 2001, ApJ, 554, L67
King, R. R., McCaughrean, M. J., Homeier, D., et al. 2010, A&A, 510, A99
Kirkpatrick, J. D., Reid, I. N., Liebert, J., et al. 1999, ApJ, 519, 802
Kirkpatrick, J. D., Reid, I. N., Liebert, J., et al. 2000, AJ, 120, 447
Kirkpatrick, J. D., Dahn, C. C., Monet, D. G., et al. 2001, AJ, 121, 3235
Kirkpatrick, J. D., Cruz, K. L., Barman, T. S., et al. 2008, ApJ, 689, 1295
Lane, B. F., Zapatero Osorio, M. R., Britton, M. C., Martín, E. L., & Kulkarni, S. R. 2001, ApJ, 560, 390
Laughlin, G., & Bodenheimer, P. 1993, ApJ, 403, 303
Liang, G. Y., et al. 2020, ChPhL, 37, 123101
Liu, M. C., & Leggett, S. K. 2005, ApJ, 634, 616
Lodders, K. 1999, ApJ, 519, 793
Lodieu, N., Burgasser, A. J., Pavlenko, Y., & Rebolo, R. 2015, A&A, 581, A73
Lodieu, N., Zapatero Osorio, M. R., Rebolo, R., et al. 2015, A&A, 579, A58
Lodieu, N., Rebolo, R., & Pérez-Garrido, A. 2018, A&A, 615, L12
Lodieu, N., Zapatero Osorio, M. R., Martín, E. L., Rebolo López, R., & Gauza, B. 2022, A&A, 663, A84
Luyten, W. J., & Kowal, C. T. 1975, Proper Motion Survey with the Forty-Eight Inch Schmidt Telescope. XLIII. One Hundred and Six Faint Stars with Large Proper Motions Separate Print (Minneapolis, MN: Univ. Minnesota Press)
Magazzu, A., Martin, E. L., & Rebolo, R. 1991, A&A, 249, 149
Magazzù, A., Martín, E. L., & Rebolo, R. 1993, ApJ, 404, L17
Marcy, G., et al. 1994, ApJ, 428, L57
Martín, E. L., Rebolo, R., & Magazzù, A. 1994, ApJ, 436, 262
Martín, E. L., Rebolo, R., & Zapatero Osorio, M. R. 1996, ApJ, 469, 706
Martín, E. L., Basri, G., Delfosse, X., & Forveille, T. 1997, A&A, 327, L29
Martín, E. L., Basri, G., Zapatero Osorio, M. R., Rebolo, R., García López, R. J., et al. 1998, ApJ, 507, L41
Martín, E. L., Basri, G., & Zapatero Osorio, M. R. 1999, AJ, 118, 2466
Martín, E. L., Delfosse, X., Basri, G., et al. 1999, AJ, 118, 1005
Martín, E. L., Brandner, W., & Basri, G. 1999, Sci., 283, 1718
Martín, E. L., Koresko, C. D., Kulkarni, S. R., Lane, B. F., & Wizinowich, P. L. 2000, ApJ, 529, L37
Martín, E. L., Brandner, W., Bouy, H., et al. 2006, A&A, 456, 253
Martín, E. L., Phan-Bao, N., Bessell, M., et al. 2010, A&A, 517, A53
Martín, E. L., Lodieu, N., Pavlenko, Y., & Béjar, V. J. S. 2018, ApJ, 856, 40
Martín, E. L., Lodieu, N., & Béjar, V. J. S. 2020, A&A, 640, A9
Martín, E. L., Lodieu, N., & del Burgo, C. 2022, MNRAS, 510, 2841
McCaughrean, M. J., Close, L. M., Scholz, R.-D., et al. 2004, A&A, 413, 1029
Moneti, A., & Zinnecker, H. 1991, A&A, 242, 428
Nelson, L. A., Rappaport, S., & Chiang, E. 1993, ApJ, 413, 364
Pavlenko, Ya V, Jones, H. R. A., Martín, E. L., et al. 2007, MNRAS, 380, 1285
Phan-Bao, N., Bessell, M. S., Martín, E. L., et al. 2008, MNRAS, 383, 831
Pineda, J. S., Hallinan, G., Kirkpatrick, J. D., et al. 2016, ApJ, 826, 73
Phillips, M. W., Tremblin, P., Baraffe, I., et al. 2020, A&A, 637, A38

Pozio, F. 1991, MmSAI, 62, 171
Rebolo, R., Zapatero Osorio, M. R., & Martín, E. L. 1995, Natur, 377, 129
Rebolo, R., Martin, E. L., Basri, G., Marcy, G. W., & Zapatero Osorio, M. R. 1996, ApJ, 66, 375
Rebolo, R., Martín, E. L., & Magazzù, A. 1992, ApJ, 389, L83
Reid, I. N., Lewitus, E., Burgasser, A. J., & Cruz, K. L. 1996, ApJ, 469, L53
Reid, I. N., Lewitus, E., Burgasser, A. J., & Cruz, K. L. 2006, ApJ, 639, 1114
Reid, I. N., Cruz, K. L., Burgasser, A. J., & Liu, M. C. 2008, AJ, 135, 580
Ruiz, M. T., Leggett, S. K., & Allard, F. 1997, ApJ, 491, L107
Sahlmann, J., Lazorenko, P. F., Ségransan, D., et al. 2015, A&A, 577, A15
Sahlmann, J., Dupuy, T. J., Burgasser, A. J., et al. 2021, MNRAS, 500, 5453
Saumon, D., & Marley, M. S. 2008, ApJ, 689, 1327
Schneider, D. P., Greenstein, J. L., Schmidt, M., & Gunn, J. E. 1991, AJ, 102, 1180
Scholz, R. -D., McCaughrean, M. J., Lodieu, N., & Kuhlbrodt, B. 2003, A&A, 398, L29
Simon, M., Bender, C., & Prato, L. 2006, ApJ, 644, 1183
Shkolnik, E. L., Allers, K. N., Kraus, A. L., Liu, M. C., & Flagg, L. 2017, AJ, 154, 69
Stauffer, J., Hamilton, D., Probst, R., Rieke, G., & Mateo, M. 1989, ApJ, 344, L21
Stauffer, J. R., Schultz, G., & Kirkpatrick, J. D. 1998, ApJ, 499, L199
Stringfellow, G. S. 1991, ApJ, 375, L21
Thackrah, A., Jones, H., & Hawkins, M. 1997, MNRAS, 284, 507
Tinney, C. G., Delfosse, X., & Forveille, T. 1997, ApJ, 490, L95
Tinney, C. G. 1998, MNRAS, 296, L42
Tinney, C. G., & Tolley, A. J. 1999, MNRAS, 304, 119
Torres, G., Henry, T. J., Franz, O. G., & Wasserman, L. H. 1999, AJ, 117, 562
Weck, P. F., Schweitzer, A., Kirby, K., Hauschildt, P. H., & Stancil, P. C. 2004, ApJ, 613, 567
Zapatero Osorio, M. R., Rebolo, R., Martín, E. L., et al. 1997, ApJ, 491, L81
Zapatero Osorio, M. R., Béjar, V. J. S., Pavlenko, Ya., et al. 2002, A&A, 384, 937
Zapatero Osorio, M. R., Lane, B. F., Pavlenko, Ya., et al. 2004, ApJ, 615, 958
Zapatero Osorio, M. R., Martín, E. L., Lane, B. F., et al. 2005, AN, 326, 948
Zboril, M., Byrne, P. B., & Rolleston, W. R. J. R. 1997, MNRAS, 284, 685
Zhang, Z. H., Pinfield, D. J., Gálvez-Ortiz, M. C., et al. 2018, MNRAS, 479, 1383

Chapter 8

Lithium in Exoplanets and Their Stellar Hosts

The study of planets around stars other than the Sun (exoplanets) has rapidly expanded in the last three decades, becoming a field in itself with multidisciplinary connections. The interest of the study of lithium in exoplanet science is manifold, but it can be divided into two main categories:

1. Lithium abundances in stars hosting different types of exoplanets provide clues on the age, rotational history, and possibly planetary migration that may result in the engulfment of the planet by the parent star. The stars can be on the main sequence part of their evolution or on the post-main sequence (giants). Because of the expansion of the stellar size when the stars evolve into giants, the possibility of accreting surrounding planets or brown dwarfs is more likely than during the main sequence phase.
2. Lithium in the atmospheres of the exoplanets themselves gives information about their atmospheres and the mechanisms of planet formation.

Taking advantage of the large body of high-resolution spectra obtained for high-precision radial velocity exoplanet searches, it has been shown that solar-type stars with detectable exoplanets tend to experience larger lithium depletion than stars without them. In this respect, the Sun may not be exceptional among its peers, which is an important long-standing question.

The mechanism that links lithium depletion with the presence of exoplanets remains unclear. A connection between angular momentum evolution, lithium depletion and planet formation does not seem to be plausible because of the very different timescales of these phenomena. A promising avenue of future research may be the exploration of exoplanet-induced tidal effects on the internal chemical mixing of solar-type stars.

Engulfment of exoplanets or brown dwarfs during stellar evolution may leave imprints in the chemical composition of stars. These events are more likely to

doi:10.1088/2514-3433/acd8bbch8

happen during the post-main sequence than on the main-sequence, but the telltale signs of them are easier to detect in main-sequence stars because they have shallower convective regions than red giants.

Chemical abundance determination in the atmospheres of exoplanets is still in its infancy, but rapidly expanding thanks to transit spectroscopy. The first detections of lithium in the atmospheres of hot Jupiters are puzzling.

8.1 Lithium in Exoplanet Hosts

The first studies of lithium abundance among stars hosting exoplanets discovered via the radial velocity method led to controversial results. Ryan (2000) concluded that 17 stars known to host planets, including the Sun, were indistinguishable in their pattern of lithium abundance from lone stars without known planets and otherwise similar in age, chemical composition, effective temperature, and mass, and that this conclusion was the opposite of that reached in the works of King et al. (1997) and Gonzalez & Laws (2000).

Israelian et al. (2004) carried out high-resolution spectroscopic observations of 79 solar-type stars with planets detected in the CORALIE radial velocity survey, and 38 comparable stars with no planets detected in the same survey. They used the following combinations of instruments and telescopes: Utrecht Echelle Spectrograph on the WHT; SARG at the TNG; the coudé spectrograph at the ESO 1.5 m telescope; and CORALIE on the 1.2 m Swiss telescope.

The resolving power of the data was between 55,000 and 57,000. They also compared their data with the studies of Chen et al. (2001) and Santos et al. (2001) to increase the numbers of stars under scrutiny. A statistical comparison of the distribution of Li abundance in samples of stars with and without known exoplanets is shown in Figure 8.1.

There seems to be a difference between the two populations of stars, which is clearer in the temperature range between 5600 K and 5850 K, but not in the range from 5850 K to 6350 K. This difference hints at higher lithium depletion in stars with known exoplanets. No correlation between lithium and other chemical elements, such as iron and oxygen, was found, in agreement with previous studies of solar-type stars in the disk of the Milky Way (Pasquini et al. 1994; Pompéia et al. 2002).

A difference in lithium abundance between stars with/without exoplanets has been confirmed with additional samples and/or independent analysis by Chen & Zhao (2006), Delgado Mena et al. 2014, Figueira et al. (2014), Gonzalez (2015), and Mishenina et al. (2016, 2020), but it has been questioned by Luck & Heiter (2006), Baumann et al. (2010), Ghezzi et al. (2010), Ramírez et al. (2012), Carlos et al. 2016, Bordé et al. (2020), and Mishenina et al. (2021).

A tricky issue is what age to assign to the stars because it is notoriously difficult to obtain accurate ages for solar-type stars that do not belong to associations or clusters. The age–lithium connection (Charbonnel & Talon 2005) needs to be properly removed from the search for an exoplanet–lithium relation. One popular

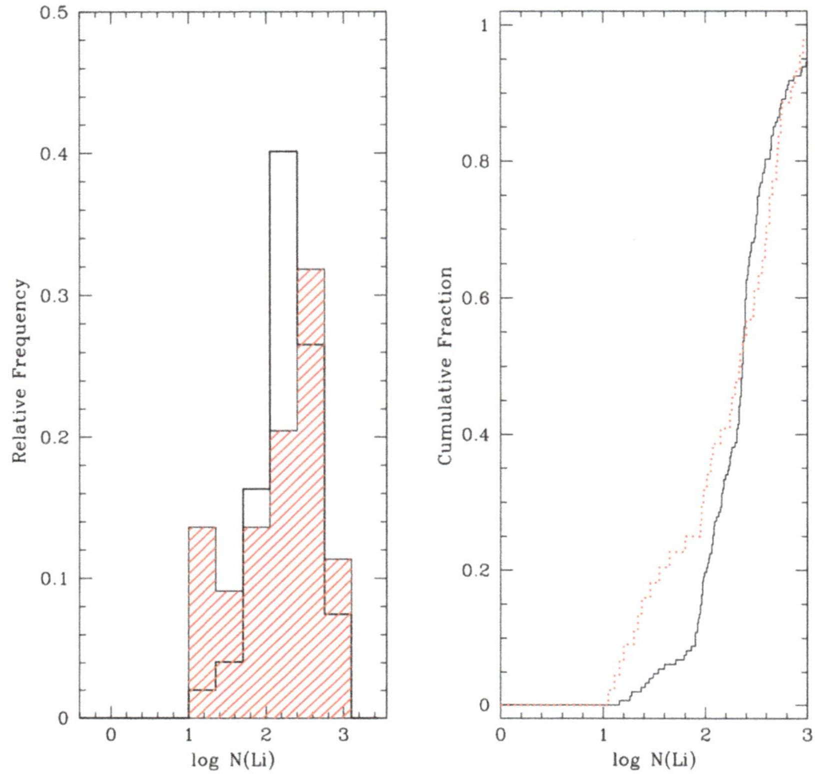

Figure 8.1. Distributions of lithium abundance in stars with known exoplanets (hatched region, red lines) and without known exoplanets (empty region, black lines). A Kolmogorov–Smirnov test indicates a probability of 20% that the distributions of these two samples are drawn from the same parent distribution of stars. Figure credit: Israelian, G. et al. 2004 "Lithium in stars with exoplanets" A&A 414 601, reproduced with permission © ESO.

method of dating Sun-like stars is gyrochronology, but it has certain limitations because stellar spin down is a complex phenomenon.

If there is a real effect of enhanced lithium depletion among solar-type stars that harbor larger planets and closer-in planets, which are more readily detectable by radial velocity, it could be interpreted as evidence for a primordial cosmogonical connection. Large gas giant planets located at small orbital distances could come from large long-lived protoplanetary disks. Such massive disks may affect the rotational evolution of the parent stars (Bouvier 2008) and could also induce enhanced accretion events (Baraffe & Chabrier 2010; Théado & Vauclair 2012).

For the Sun and other stars of similar mass, the timescale for coupling the rotation in the core with the rotation in the envelope has been calculated to be shorter than about 200 Myr (MacGregor & Brenner 1991; Krishnamurthi et al. 1997; Denissenkov et al. 2010; Gallet & Bouvier 2015; Spada & Lanzafame 2020).

The timescale for main-sequence lithium depletion in Sun-like stars is much longer than that of rotational coupling, and hence it remains unclear how rotational effects may explain the behavior of lithium at ages comparable to the age of the Sun.

Detailed models of the evolution of low-mass stars considering a variety of mixing processes (gravity waves, meridional circulation, penetrative convection, rotation-induced turbulence, turbulent shear) indicate that there is not much difference in the lithium abundance predicted for initially slow or fast rotators at ages than about 1 Gyr. For example, models that try to predict the correct amount of lithium depletion observed in open clusters are shown in Figure 8.2. They fail to reproduce the large spread in lithium abundance observed among the Sun-like members in the old M67 open cluster (age = 3.65 Gyr). The predictions of lithium

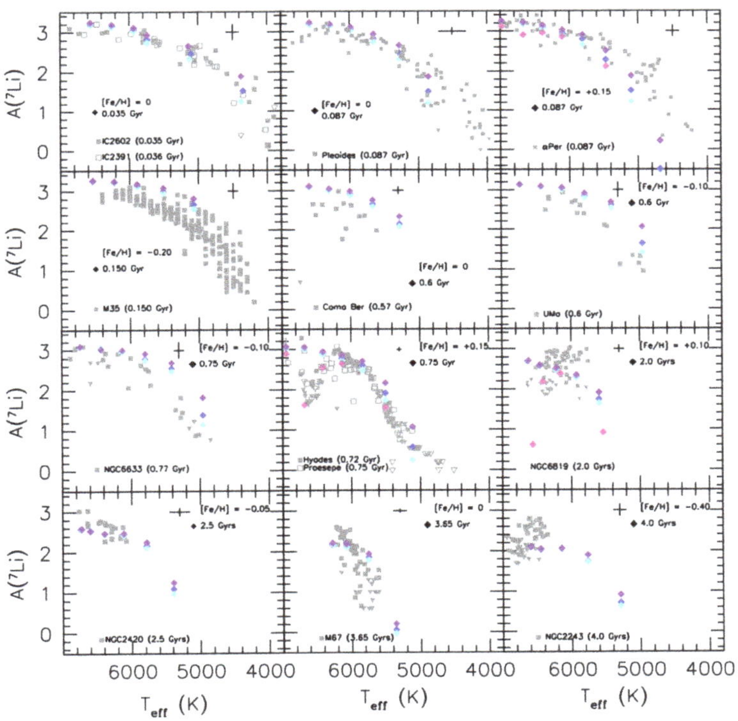

Figure 8.2. Patterns of lithium abundances with respect to effective temperature in open clusters of different ages and iron abundances. The cross in each panel indicates the typical observational error bar for each cluster. Colored diamonds indicate the predictions of the evolutionary models for different initial rotational velocities. Note that in the lower central panel, corresponding to the old M67 cluster, the range of lithium depletion predicted by models with different initial rotation is much smaller than the dispersion seen in the observational points. Figure credit: Dumont, T. et al. 2021 "Lithium depletion and angular momentum transport in F-type and G-type stars in Galactic Open Clusters" A&A 654 A46, reproduced with permission © ESO.

depletion after this length of time are insensitive to the choice of the initial rotational velocity.

A factor that plays a significant role in predicting the lithium depletion of mature Sun-like stars is the choice of the turbulent shear prescription. Whether or not stars of similar age, chemical composition, mass and rotation could experience quite different diffusion mixing is an interesting open question.

Studies of chemical abundances in wide binaries with solar-type components indicate that the depletion of lithium on the main sequence depends on additional factors other than age, mass and metallicity (Martín et al. 2002; Nissen & Gustafsson 2018; Hawkins et al. 2020).

A search for chemical signatures of atomic diffusion in the atmospheres of stars in seven binaries did not yield clear results, but a weak correlation between orbital separation and subtle abundance differences was reported (Liu et al. 2021). A much larger sample of binaries is clearly required.

Tidal interactions between giant planets and their parent stars could induce diffusive coefficients in the stellar interiors that may lead to enhanced chemical mixing and lithium depletion on long timescales. It has been discussed for over a century that the Hale 11 yr solar magnetic cycle and its long-term modulation may be influenced by tiny planetary-induced tidal forcing (Wolf 1859; Bollinger 1952; Abreu et al. 2012; Stefani et al. 2019; Albert et al. 2021) but the subject is controversial and it seems that an unidentified amplification mechanism is required (Charbonneau 2022).

The possibility of planets exciting different types of waves (Alfvén, Kelvin, Poincaré, Rossby, and gravity waves) in their host stars and their role in stellar evolution has been recognized and it has been studied by several authors (Ogilvie & Lin 2007; Mathis et al. 2013; Barker 2020, 2022; Ahuir et al. 2021; Lazovik 2021; Rao et al. 2021; Terquem 2021; Lanza 2022; Tayar et al. 2022; Wei 2022; Horstmann et al. 2023). These effects are found to be plausible but their amplitude depends on unknown parameters, such as the frictional damping efficiency. A search for a relationship between magnetic activity cycles in exoplanet hosts and the tidal influence of the exoplanets did not yield a positive result (Obridko et al. 2022), but the number of stars that are known to show cyclic activity and that have a detected exoplanet is very small.

The relevance of long-term planet–star interactions for lithium depletion in main-sequence stars including the Sun is certainly worth future investigation.

Whether or not the depletion of lithium in the Sun may be connected to the presence of the planets in the solar system is still an important open question (Gustafsson 2008; Meléndez et al. 2014) that deserves additional scrutiny.

8.2 Lithium as a Tracer of Planetary Accretion onto Main-sequence Stars

Stellar surfaces may show chemical abundance anomalies that could be due to pollution by planetary material falling onto them. The first report of a possible planet-driven enhancement of ^{6}Li in a solar-type host of several close-in giant planets was made by Israelian et al. (2001), but it was questioned by Reddy et al. (2002).

The lithium isotope ratio in this star was reassessed and a weakly significant detection of ^{6}Li was claimed by Israelian et al. (2003) using very high-resolution ($R = 105{,}000$) and high quality (S/N $= 10^3$) observations obtained with UVES at the VLT, and a careful analysis of the absorption lines in the spectral region of the Li I resonance feature.

As can be seen in Figure 8.3, even with the excellent UVES data, the presence of ^{6}Li induces just a tiny effect on the line profile of the Li I resonance doublet at 670.8 nm. Even though from the theoretical point of view the abundance of ^{6}Li is a valuable tracer of rocky material accretion onto stars (Montalbán & Rebolo 2002; Sandquist et al. 2002), from the observational side it is a very hard measurement to make.

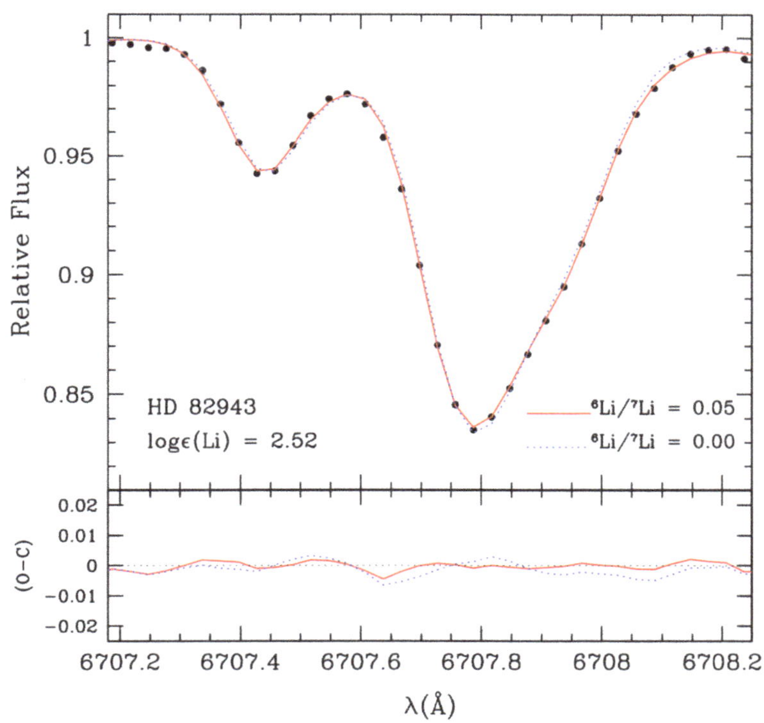

Figure 8.3. Observed line profile of the Li I resonance doublet in the solar-type exoplanet host HD 82943 obtained with UVES on the VLT (filled large dots) compared with theoretical spectra with no ^{6}Li (dotted line) and with a lithium isotope ratio $R_{\rm I} = 20$ (continuous red line). Figure credit: Israelian, G. et al. 2003 "A new measurement of the ^{6}Li/^{7}Li ratio in the extra-solar planet host star HD 82943 and the line blending in the Li I 6708 Å region" A&A 405 753, reproduced with permission © ESO.

An unsuccessful attempt to detect ^{6}Li in the atmospheres of two lithium-poor stars that are known to harbor close-in giant planets was made by Mandell et al. (2004) using high-resolution spectra (R = 115,000) obtained with the fiber coupled Echelle spectrograph at the HET. From the limits on the presence of ^{6}Li in the stellar atmospheres they were able to infer that a maximum 0.3 M_J of planetary material could have been deposited onto the star in its recent past (less than about 1 Gyr ago).

It has been proposed that close-in giant planets (hot Jupiters) may induce chemical abundance anomalies in main-sequence A-type stars that have thin convective envelopes (Jura 2015), but so far observational support for this hypothesis is scarce.

No strong chemical abundance anomalies indicating pollution of stellar photospheres by planetary material have been found even in the most favorable case of the Kelt-9 system, an A0-type star orbited by an intensely irradiated hot Jupiter planet that is observed to undergo extreme mass loss. Kama et al. (2023) reported on high-resolution (R = 67,500) spectra of this system obtained with the Gemini Remote Access to CFHT ESPaDOnS spectrograph. They reported a null detection of the Li I resonance doublet, implying no evidence for lithium enrichment due to the hot Jupiter planet. In fact, they did not find any enrichment in all the different chemical elements that were detected in the spectra.

In a comprehensive study of lithium abundance in 171 single members of the open cluster M48 (age = 420 Myr), Sun et al. (2023) found two out of five later-type A stars to have A(Li) = 3.65–3.75 dex. They proposed that the lithium enhancement above the cosmic value in these two stars could be due to accretion of planetesimals or engulfment of Li-rich and H-poor substellar object. However, they could not rule out the possibility of radiative levitation (Richer & Michaud 1993) to explain the higher than cosmic lithium abundance.

The question of whether or not accretion of planetary material onto main-sequence stars may lead to lithium enhancement remains open, and in fact some theoretical models suggest that on the contrary the accretion may induce thermohaline convection that could result in additional depletion of lithium by mixing with material from below the convective region (Théado & Vauclair 2012; Deal et al. 2015).

An extreme case scenario is the engulfment of a brown dwarf by a main-sequence star. Hydrodynamical 3D simulations for the case of a solar-mass star swallowing brown dwarfs with masses between 10 and 20 Jupiters have been carried out by Cabezón et al. (2023). They find that in the particular case of a merger between the brown dwarf and the star, a significant lithium enhancement could be produced, although it is not clear how long it may last on the stellar surface.

8.3 Lithium as a Relic of Substellar Companion Engulfment by Red Giants

The idea that there could be a connection between lithium enhancement in red giants, and the swallowing of planets around them, dates back to Alexander (1967).

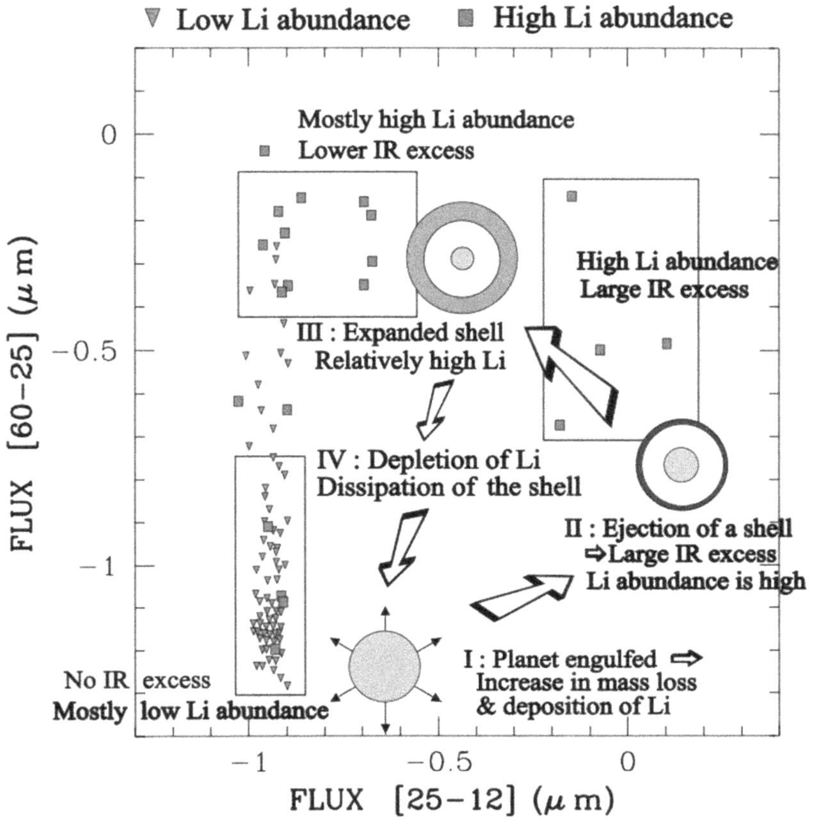

Figure 8.4. IRAS color–color diagram showing the different regions where G and K giants are located according to the model of substellar-mass object engulfment originally proposed by de la Reza et al. (1996, 1997). Normal giants start in region I and move through regions II, III, and IV and back to region I when they swallow a lithium-rich substellar companion (brown dwarf or exoplanet). Figure credit: Siess, L., & Livio, M. 1999 "The accretion of brown dwarfs and planets by giant stars—II. Solar-mass stars on the red giant branch" MNRAS 308 1133.

It was extended to giant planets and brown dwarfs by the calculations of Siess & Livio (1999), building up on previous work by Livio & Soker (1984).

These authors presented a diagram connecting the post-main sequence evolution of solar-type stars with the lithium abundance and the presence of infrared excess, which is shown in Figure 8.4. The idea for this connection came originally from the work of de la Reza et al. (1996, 1997).

The scenario represented in Figure 8.4 explains that lithium-rich G and K-type giants usually present indications of infrared excess based on colors obtained using the IRAS satellite. Three regions were identified in the IRAS color–color diagram. Region I is populated by normal K giants with neither lithium enrichment nor infrared excess.

When a substellar-mass companion (brown dwarf or giant exoplanet) with a mass below the limit of thermonuclear lithium burning is swallowed by a normal K giant, the star moves rapidly (timescale of a century) from region I to region II. Region II is populated by lithium-rich giants with large infrared excess.

These stars have atmospheres enriched in lithium by the deposition of substellar material and their infrared emission is enhanced by an increase in mass-loss rate and the ejection of a circumstellar shell.

Subsequently, the star moves from region II to region III on a timescale of about 10^4 years. Lithium depletion starts taking place in the envelope of the giant by the mixing mechanism proposed by Zahn (1992) and Charbonnel (1995), mass loss decreases, and the shell expands and cools down.

Finally, the star moves from region III through the transition region IV and back into region I on a timescale of about 80,000 years. The lithium gained in the accretion process is entirely gone from the stellar atmosphere by mixing with the interior, and the shell is also gone by mixing with the ISM.

There have been dramatic advances both in the number of giants observed (around 10^4 lithium-rich giants detected by LAMOST; Gao et al. 2019; Cai et al. 2023) and significant efforts in the theoretical understanding of the merging between substellar companions and giant stars (Soares-Furtado et al. 2021), but the picture of the evolution of lithium in post-main-sequence stars remains unclear.

The substellar engulfment scenario has been supported by some authors (Carlberg et al. 2012; Rebull et al. 2015; Aguilera-Gómez et al. 2016), but questioned by others (Fekel & Watson 1998; Martell & Shetrone 2013; Casey et al. 2016; Smiljanic et al. 2018; Martell et al. 2021).

Super-Li-rich giants with log N(Li) > 3.5 represent only a handful of the lithium-rich post-main sequence stellar population (Balachandran et al. 2000; Singh et al. 2019), and they are particularly difficult to explain with the substellar–stellar merger model. They probably need another mechanism of lithium enrichment such as hot burning nucleosynthesis (Charbonnel & Blachandran 2000; Zhou et al. 2019).

Alternative lithium production models involve tidal interactions in close binaries (Casey et al. 2019) and the merger between a helium white dwarf and a red giant (Holanda et al. 2020; Zhang et al. 2020).

An observational connection between the presence of an active chromosphere and lithium enrichment supports the idea that binary interactions play a significant role in shaping the behavior of lithium in giant stars (Sneden et al. 2022).

The serendipitous discovery of a super-lithium-rich metal-poor giant star in the halo of the Milky Way was reported by Kowkabany et al. (2022). This unusual star was also found to have a strong infrared excess using WISE colors.

Five different Li I lines were analyzed and the overall lithium abundance was determined to be log N(Li) = 5.62 ± 0.25 (including 3D NLTE effects), one of the largest ever found in any stellar atmosphere. This abundance is much larger than that expected from any substellar–stellar merger, and somewhat higher than predicted by models of lithium enrichment by mixing (Yan et al. 2018). Kowkabany et al. proposed that this star may be undergoing the process called lithium flash, which was predicted to happen when low-mass stars go through the red giant branch bump (Palacios et al. 2001).

8.4 Lithium in Exoplanetary Atmospheres

Giant exoplanets in tight orbits around their host stars are generally called hot Jupiters. They were first discovered using the high-precision radial velocity technique (Mayor & Queloz 1995; Butler et al. 1996), and offer a relatively high chance of observing transits.

The first transiting exoplanet was discovered around the solar-type star HD 209458 (Henry et al. 1999, 2000; Charbonneau et al. 2000). The combination of the light curve and the radial velocity measurements of transiting exoplanets provides information about the mass and radius (and hence the overall ρ) of the exoplanet. It was soon realized that the observed densities, masses, and radii of transiting exoplanets implied a diversity of physical conditions never seen before.

As an example of the wild diversity and strangeness of giant transiting exoplanets the distribution of radii and masses are illustrated in Figure 8.5.

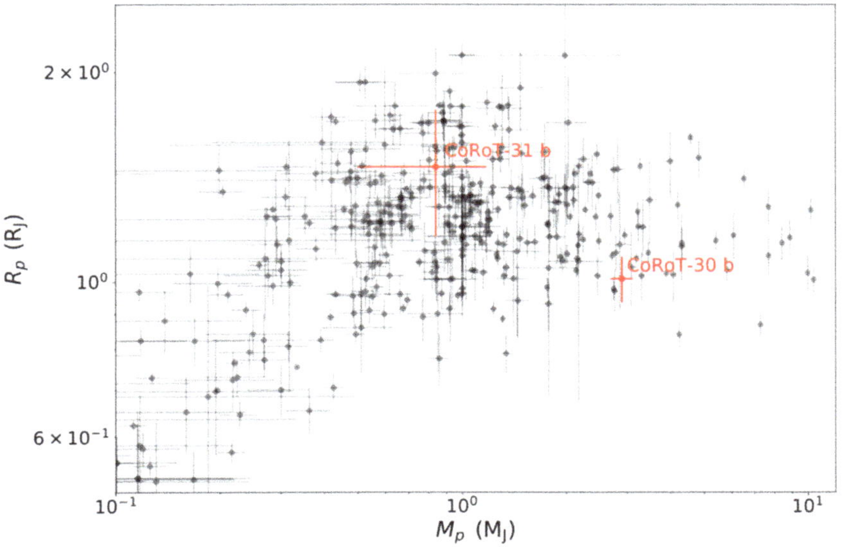

Figure 8.5. Distribution of mass and radius observed in giant exoplanets that transit across their host stars. Black dots with error bars denote data from exoplanet.org. Red dots are two exoplanets detected by CoRoT. The location of Jupiter would be in the center of the plot. Note the extremely large spread in masses and radii, much larger than the error bars. Figure credit: Bordé, P. et al. 2020 "Transiting exoplanets from the CoRoT space mission. XXIX. The Hot Jupiters CoRoT-30b and CoRoT-31b" A&A 635 A122, reproduced with permission © ESO.

During the transits, it has been predicted that absorption features, including the Li I resonance doublet, from the planet atmosphere could be detected superimposed on the stellar flux (Seager & Sasselov 2000; Brown 2001; Hubbard et al. 2001; Fortney et al. 2003; Barman et al. 2005; Charbonneau et al. 2007).

Hubble Space Telescope STIS observations of the transits of the exoplanet HD 209458b across the limb of its host star allowed Charbonneau et al. (2002) and Vidal-Madjar et al. (2003, 2004) to report the first detections of exoplanetary absorption features, namely the Na I resonance doublet, the H Lyα line, and lines of O I and C II.

Ground-based attempts to detect absorption features from transiting hot Jupiters were largely unsuccessful (Bundy & Marcy 2000; Moutou et al. 2001, 2003; Brown et al. 2002; Harrington et al. 2002; Winn et al. 2004; Deming et al. 2005; Narita et al. 2005) until the reanalysis of High Dispersion Spectrograph Subaru observations of the Na I resonance doublet in HD 209458b (Snellen et al. 2008) and observations of HD 189733b secured with the High Resolution Spectrograph at the HET (Redfield et al. 2008).

The first claim of a lithium detection in an exoplanetary atmosphere was made by Chen et al. (2018) on the basis of low-resolution transmission spectroscopy of the WASP-127 system obtained with OSIRIS at the 10.4 m Gran Telescopio Canarias in the ORM.

High-resolution follow-up spectra obtained with the Echelle Spectrograph for Rocky Exoplanets and Stable Spectroscopic Observations (ESPRESSO) at the VLT on the ESO Cerro Paranal observatory confirmed the detection of sodium but failed to recover lithium (Allart et al. 2020).

Lithium detections in the atmospheres of the ultra-hot Jupiters WASP-76b and WASP-121b have been achieved using ESPRESSO high-resolution transmission spectroscopy (Borsa et al. 2021; Tabernero et al. 2021).

WASP-76b is an inflated ($1.83 \pm 0.05\ R_J$) planet with a mass of $0.92 \pm 0.03\ M_J$ that transits across a metal-rich F7 main-sequence star ($T_{\rm eff} = 6250$ K, [Fe/H] = +0.23) with an orbital period of 1.81 days. It is a perfect target for atmospheric characterization of the exoplanet because of its large radius (almost twice that of Jupiter) and a high equilibrium temperature of 2200 K (Ehrenreich et al. 2020).

A surprising outcome of the ESPRESSO transmission spectra of WASP-76b is that the absorption due to Li I was found to be very strong (see Figure 8.6 from Tabernero et al. 2021). In fact, it was reported to be stronger than that due to K I. Potassium is much more abundant than lithium in the solar system and in the ISM. This unexpected result by Tabernero et al. calls for additional investigations regarding the implications of lithium detection in exoplanetary atmospheres.

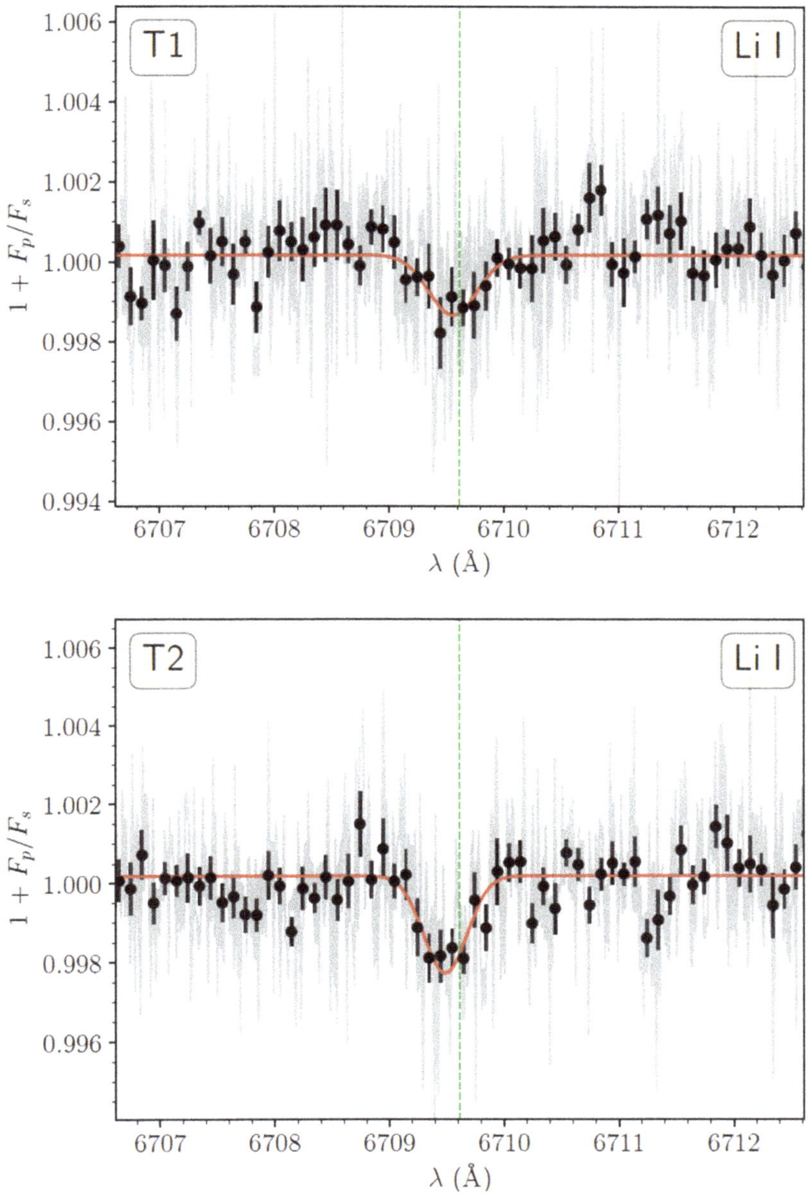

Figure 8.6. Transmission spectra of the ultra-hot Jupiter WASP-76b around the Li I resonance doublet obtained with ESPRESSO on two different transits. Each transit is shown in a different panel. Original data are denoted with thin gray lines. Binned data with error bars are represented as thick black dots. Best-fits to the data are represented with red lines. Rest-frame wavelength of the Li I feature is shown as a dashed green vertical line. Figure credit: Tabernero, H. M. et al. 2021 "ESPRESSO high-resolution transmission spectroscopy of WASP-76b" A&A 646 A158, reproduced with permission © ESO.

Bibliography

Abreu, J. A., Beer, J., Ferriz-Mas, A., McCracken, K. G., & Steinhilber, F. 2012, A&A, 548, A88
Aguilera-Gómez, C., Chanamé, J., Pinsonneault, M. H., & Carlberg, J. K. 2016, ApJ, 829, 127
Ahuir, J., Mathis, S., & Amard, L. 2021, A&A, 651, A3
Albert, C., Ferriz-Mas, A., Gaia, F., & Ulzega, S. 2021, ApJ, 916, L9
Alexander, J. B. 1967, Obs, 87, 238
Allart, R., Pino, L., Lovis, C., et al. 2020, A&A, 644, A155
Balachandran, S. C., Fekel, F. C., Henry, G. W., & Uitenbroek, H. 2000, ApJ, 542, 978
Baraffe, I., & Chabrier, G. 2010, A&A, 521, A44
Barker, A. J. 2020, MNRAS, 498, 2270
Barker, A. J. 2022, MNRAS, 927, L36
Barman, T. S., Hauschildt, P. H., & Allard, F. 2005, ApJ, 632, 1132
Baumann, P., Ramírez, I., Meléndez, J., Asplund, M., & Lind, K. 2010, A&A, 519, A87
Bollinger, C. J. 1952, Proc. Okla. Acad. Sci., 33, 307
Bordé, P., Díaz, R. F., Creevey, O., et al. 2020, A&A, 635, A122
Borsa, F., Allart, R., Casasayas-Barris, N., et al. 2021, A&A, 645, A24
Bouvier, J. 2008, A&A, 489, L53
Brown, T. M. 2001, ApJ, 553, 1006
Brown, T. M., Libbrecht, K. G., & Charbonneau, D. 2002, PASP, 114, 826
Bundy, K. A., & Marcy, G. W. 2000, PASP, 112, 1421
Butler, R. P., Marcy, G. W., Williams, E., et al. 1996, PASP, 108, 500
Cabezón, R. M., Abia, C., Domínguez, I., & García-Senz, D. 2023, A&A, 670, A155
Cai, B., Kong, X., Shi, J., et al. 2023, AJ, 165, 52
Carlberg, J. K., Cunha, K., Smith, V. V., & Majewski, S. R. 2012, ApJ, 757, 109
Carlos, M., Nissen, P. E., & Meléndez, J. 2016, A&A, 587, A100
Casey, A. R., Ruchti, G., Masseron, T., et al. 2016, MNRAS, 461, 3336
Casey, A. R., Ho, A. Y. Q., Ness, M., et al. 2019, ApJ, 880, 125
Charbonneau, D., Brown, T. M., Latham, D. W., & Mayor, M. 2000, ApJ, 529, L45
Charbonneau, D., Brown, T. M., Noyes, R. W., & Gilliland, R. L. 2002, ApJ, 568, 377
Charbonneau, D., Brown, T. M., Burrows, A., & Laughlin, G. 2007, Protostars & Planets (Tucson, AZ: Univ. Arizona Press) 951, 701
Charbonneau, P. 2022, FrASS, 9, 853676
Charbonnel, C. 1995, ApJ, 453, L41
Charbonnel, C., & Balachandran, S. C. 2000, A&A, 359, 563
Charbonnel, C., & Talon, S. 2005, Sci., 309, 2189
Chen, G., Pallé, E., Welbanks, L., et al. 2018, A&A, 616, A145
Chen, Y. Q., Nissen, P. E., Benoni, T., & Zhao, G. 2001, A&A, 371, 943
Chen, Y. Q., & Zhao, G. 2006, AJ, 131, 1816
de La Reza, R., Drake, N. A., & da Silva, L. 1996, ApJ, 456, L115
de la Reza, R., Drake, N. A., da Silva, L., Torres, C. A. O., & Martín, E. L. 1997, ApJ, 482, L77
Deal, M., Richard, O., & Vauclair, S. 2015, A&A, 584, A105
Delgado Mena, E., Israelian, G., González Hernández, J. I., et al. 2014, A&A, 562, A92
Deming, D., Brown, T. M., Charbonneau, D., Harrington, J., & Richardson, L. J. 2005, ApJ, 622, 1149
Denissenkov, P. A., Pinsonneault, M., Terndrup, D. M., & Newsham, G. 2010, ApJ, 716, 1269
Dumont, T., Charbonnel, C., Palacios, A., & Borisov, S. 2021, A&A, 654, A46

Ehrenreich, D., Lovis, C., Allart, R., et al. 2020, Natur, 580, 597
Fekel, F. C., & Watson, L. C. 1998, AJ, 116, 2466
Figueira, P., Faria, J. P., Delgado-Mena, E., et al. 2014, A&A, 570, A21
Fortney, J. J., Sudarsky, D., Hubeny, I., et al. 2003, ApJ, 589, 615
Gallet, F., & Bouvier, J. 2015, A&A, 577, A98
Gao, Q., Shi, J.-R., Yan, H.-L., et al. 2019, ApJS, 245, 33
Ghezzi, L., Cunha, K., Smith, V. V., & de la Reza, R. 2010, ApJ, 724, 154
Gonzalez, G., & Laws, C. 2000, AJ, 119, 390
Gonzalez, G. 2015, MNRAS, 446, 1020
Gustafsson, B. 2008, PhyS, 130, 014036
Hawkins, K., Lucey, M., Ting, Y.-S., et al. 2020, MNRAS, 492, 1164
Henry, G. W., Marcy, G., Butler, R. P., & Vogt, S. S. 1999, IAU Circ, 7307, 1
Henry, G. W., Marcy, G. W., Butler, R. P., & Vogt, S. S. 2000, ApJ, 529, L41
Holanda, N., Drake, N. A., & Pereira, C. B. 2020, AJ, 159, 9
Horstmann, G. M., Mamatsashvili, G., Giesecke, A., Zaqarashvili, T. V., & Stefani, F. 2023, ApJ, 944, 48
Hubbard, W. B., Fortney, J. J., Lunine, J. I., et al. 2001, ApJ, 560, 413
Israelian, G., Santos, N. C., Mayor, M., & Rebolo, R. 2001, Natur, 411, 163
Israelian, G., Santos, N. C., Mayor, M., & Rebolo, R. 2003, A&A, 405, 753
Israelian, G., Santos, N. C., Mayor, M., & Rebolo, R. 2004, A&A, 414, 601
Jura, M. 2015, AJ, 150, 166
Kama, M., Folsom, C. P., Jermyn, A. S., & Teske, J. K. 2023, MNRAS, 518, 3116
King, J. R., Deliyannis, C. P., Hiltgen, D. D., et al. 1997, AJ, 113, 1871
Kowkabany, J., Ezzeddine, R., Charbonnel, C., et al. 2022, arXiv:2209.02184
Krishnamurthi, A., Pinsonneault, M. H., Barnes, S., & Sofia, S. 1997, ApJ, 480, 303
Lanza, A. F. 2022, A&A, 665, A47
Lazovik, Y. 2021, MNRAS, 508, 3408
Liu, F., Bitsch, B., Asplund, M., et al. 2021, MNRAS, 508, 1227
Livio, M., & Soker, N. 1984, MNRAS, 208, 763
Luck, R. E., & Heiter, U. 2006, AJ, 131, 3069
MacGregor, K. B., & Brenner, M. 1991, ApJ, 376, 204
Mandell, A. M., Ge, J., & Murray, N. 2004, AJ, 127, 1147
Martell, S. L., & Shetrone, M. D. 2013, MNRAS, 430, 611
Martell, S. L., Simpson, J. D., Balasubramaniam, A. G., et al. 2021, MNRAS, 505, 5340
Martín, E. L., Basri, G., Pavlenko, Y., & Lyubchik, Y. 2002, ApJ, 579, 437
Mathis, S., Decressin, T., Eggenberger, P., & Charbonnel, C. 2013, A&A, 558, A11
Mayor, M., & Queloz, D. 1995, Natur, 378, 355
Meléndez, J., Schirbel, L., Monroe, T. R., et al. 2014, A&A, 567, L3
Mishenina, T., Basak, N., Adibekyan, V., Soubiran, C., & Kovtyukh, V. 2021, MNRAS, 504, 4252
Mishenina, T., Kovtyukh, V., Soubiran, C., & Adibekyan, V. Zh. 2016, MNRAS, 462, 1563
Mishenina, T., Soubiran, C., Charbonnel, C., Katsova, M., & Nizamov, B. 2020, MnSAI, 91, 171
Montalbán, J., & Rebolo, R. 2002, A&A, 386, 1039
Moutou, C., Coustenis, A., Schneider, J., et al. 2001, A&A, 371, 260
Moutou, C., Coustenis, A., Schneider, J., Queloz, D., & Mayor, M. 2003, A&A, 405, 341
Narita, Norio, Suto, Yasushi, Winn, Joshua N., et al. 2005, PASJ, 57, 471

Nissen, P. E., & Gustafsson, B. 2018, A&ARv, 26, 6

Obridko, V. N., Katsova, M. M., & Sokoloff, D. D. 2022, MNRAS, 516, 1251

Ogilvie, G. I., & Lin, D. N. C 2007, ApJ, 661, 1180

Palacios, A., Charbonnel, C., & Forestini, M. 2001, A&A, 375, L9

Pasquini, L., Liu, Q., & Pallavicini, R. 1994, A&A, 287, 191

Pompéia, L., Barbuy, B., Grenon, M., & Castilho, B. V. 2002, ApJ, 570, 820

Ramírez, I., Fish, J. R., Lambert, D. L., & Allende Prieto, C. 2012, ApJ, 756, 46

Rao, Suvrat, Pezzotti, Camilla, Meynet, Georges, et al. 2021, A&A, 651, A50

Rebull, L. M., Carlberg, J. K., Gibbs, J. C., et al. 2015, AJ, 150, 123

Reddy, B. E., Lambert, D. L., Hrivnak, B. J., & Bakker, E. J. 2002, AJ, 123, 1993

Redfield, S., Endl, M., Cochran, W. D., & Koesterke, L. 2008, APJ, 673, L87

Richer, J., & Michaud, G. 1993, ApJ, 416, 312

Ryan, S. G. 2000, MNRAS, 316, L35

Sandquist, E. L., Dokter, J. J., Lin, D. N. C., & Mardling, R. A. 2002, ApJ, 572, 1012

Santos, N. C., Israelian, G., & Mayor, M. 2001, A&A, 373, 1019

Seager, S., & Sasselov, D. D. 2000, ApJ, 537, 916

Siess, L., & Livio, M. 1999, MNRAS, 308, 1133

Singh, R., Reddy, B. E., Bharat Kumar, Y., & Antia, H. M. 2019, ApJ, 878, L21

Smiljanic, R., Franciosini, E., Bragaglia, A., et al. 2018, A&A, 617, A4

Sneden, C., Afşar, M., Bozkurt, Z., et al. 2022, ApJ, 940, 12

Snellen, I. A. G., Albrecht, S., de Mooij, E. J. W., & Le Poole, R. S. 2008, A&A, 487, 357

Soares-Furtado, M., Cantiello, M., MacLeod, M., & Ness, M. K. 2021, AJ, 162, 273

Spada, F., & Lanzafame, A. C. 2020, A&A, 636, A76

Stefani, F., Giesecke, A., & Weier, T. 2019, SoPh, 294, 60

Sun, Q., Deliyannis, C. P., Steinhauer, A., Anthony-Twarog, B. J., & Twarog, B. A. 2023, ApJ, 952, 71

Tabernero, H. M., Zapatero Osorio, M. R., Allart, R., et al. 2021, A&A, 646, A158

Tayar, J., Moyano, F. D., Soares-Furtado, M., et al. 2022, ApJ, 940, 23

Terquem, C. 2021, MNRAS, 503, 5789

Théado, S., & Vauclair, S. 2012, ApJ, 744, 123

Vidal-Madjar, A., Désert, J. -M., Lecavelier des Etangs, A., et al. 2004, ApJ, 604, L69

Vidal-Madjar, A., Lecavelier des Etangs, A., Désert, J. -M., et al. 2003, Natur, 422, 143

Wei, X. 2022, A&A, 664, A10

Winn, J. N., Suto, Y., Turner, E. L., et al. 2004, PASJ, 56, 655

Wolf, R. 1859, MNRAS, 19, 85

Yan, H. L., et al. 2018, NatAs, 2, 790

Zahn, J. P. 1992, A&A, 265, 115

Zhang, X., Jeffery, C. S., Li, Y., & Bi, S. 2020, ApJ, 889, 33

Zhou, Y., Yan, H., Shi, J., Blanco-Cuaresma, S., et al. 2019, ApJ, 877, 104

www.ingramcontent.com/pod-product-compliance
Ingram Content Group UK Ltd.
Pitfield, Milton Keynes, MK11 3LW, UK
UKHW061858100926
8556IPUK00008B/913

9 780750 336215